DIGITAL CMOS CIRCUIT DESIGN

THE KLUWER INTERNATIONAL SERIES IN ENGINEERING AND COMPUTER SCIENCE

VLSI, COMPUTER ARCHITECTURE AND DIGITAL SIGNAL PROCESSING

Consulting Editor

Jonathan Allen

Other books in the series:

Logic Minimization Algorithms for VLSI Synthesis. R.K. Brayton, G.D. Hachtel, C.T. McMullen, and A.L. Sangiovanni-Vincentelli. ISBN 0-89838-164-9.

Computer-Aided Design and VLSI Device Development. K.M. Cham, S.-Y. Oh, D. Chin, and J.L. Moll. ISBN 0-89838-204-1.

Adaptive Filters: Structures, Algorithms, and Applications. M.L. Honig and D.G. Messerschmitt. ISBN 0-89838-163-0.

Introduction to VLSI Silicon Devices: Physics, Technology and Characterization. B. El-Kareh and R.J. Bombard. ISBN 0-89838-210-6.

Latchup in CMOS Technology: The Problem and Its Cure. R.R. Troutman. ISBN 0-89838-215-7.

DIGITAL CMOS CIRCUIT DESIGN

by

Marco Annaratone
Carnegie-Mellon University
Pittsburgh, Pennsylvania

KLUWER ACADEMIC PUBLISHERS
Boston / Dordrecht / Lancaster

Distributors for North America:
Kluwer Academic Publishers
101 Philip Drive
Assinippi Park
Norwell, Massachusetts 02061, USA

Distributors for the UK and Ireland:
Kluwer Academic Publishers
MTP Press Limited
Falcon House, Queen Square
Lancaster LA1 1RN, UNITED KINGDOM

Distributors for all other countries:
Kluwer Academic Publishers Group
Distribution Centre
Post Office Box 322
3300 AH Dordrecht, THE NETHERLANDS

Library of Congress Cataloging-in-Publication Data

Annaratone, Marco.
Digital CMOS circuit design.

(The Kluwer international series in engineering and computer science. VLSI, computer architecture, and digital signal processing)
Includes index.
1. Metal oxide semiconductors, Complementary.
2. Digital electronics. 3. Integrated circuits—
Very large scale integration. I. Title. II. Series.
TK7871.99.M44A56 1986 621.395 86-10646
ISBN 0-89838-224-6

Printed in the United States of America

To Hanni (and Vanessa, of course)

Contents

Foreword

In light of decreasing feature size and greater sophistication of modern processing technology, CMOS has become increasingly attractive, providing low-power (at moderate frequencies), good scalability, and rail-to-rail operation. For many designers, particularly those approaching VLSI from a system viewpoint, previous experience has been mainly with ratioed NMOS design, and so there is a need to build on this experience and make a natural transition into CMOS design. Indeed, there is much that can be borrowed from NMOS experience, mainly centered around the techniques for creating N channel pulldown structures. Based on these contributions, CMOS has now grown to the point where there are several circuit styles which have evolved, and these are amply described in this book. Starting at the level of the individual MOSFET, basic building blocks are described, as well as the variety of CMOS fabrication processes in contemporary usage. Circuit style issues are then expanded to provide the user with several useful design methodologies, and much care is given to electrical performance considerations, including characteristics of interconnect, gate delay, and I/O buffering. This understanding is then applied to macro-sized components, including array multipliers, where the reader acquires a unified view of architectural performance through parallelism, and circuit performance through scrupulous attention to device sizing and control of parasitic circuit elements. In addition, layout techniques to avoid latchup, a consideration not previously encountered by NMOS designers, are given careful treatment.

Designers who are approaching CMOS from previous NMOS experience, or those who are contemplating their first designs, will find a rich treatment of major design issues centered around CMOS, in a style that is thoughtful, detailed, and broad from the system perspective. The emerging high-performance designs will partake of the benefits of increasingly exciting process and circuit innovation, forming the basis for lasting contributions to contemporary digital design.

Jonathan Allen
Consulting Editor

Preface

Complementary metal-oxide semiconductor (CMOS) technology has become the most effective fabrication process for the production of very large-scale integrated (VLSI) digital circuits. Moreover, there is wide-spread consensus that no other technology will effectively compete with CMOS for many years to come. It is therefore essential for the VLSI circuit designer to fully master the wide range of possibilities that CMOS can offer.

The purpose of this book is not simply to present and discuss the most important techniques used in the design of CMOS digital circuits. Rather, digital CMOS design is dealt with in the realm of constantly decreasing feature sizes, which poses new challenges to the designer in search of ultimate speed. A quick look at the Table of Contents exemplifies this: interconnection and off-chip communication delays are dealt with extensively in Chapters 5 and 7.

As is well-known, the intrinsic gate delay no longer dictates circuit speed. Interconnection and off-chip communication delays represent far more important limitations. This is another reason why other technologies featuring shorter gate delay than CMOS — such as Gallium Arsenide, in particular — will not necessarily prevail in the short term. Smaller feature sizes will also require lower power supply voltage. Although it is still possible to use the standard 5V power supply for micrometer technologies, sub-micron processes will necessitate lower voltages, with dramatic effects on design methodologies. Lower voltages and smaller parasitic capacitances make dynamic logic design less reliable, noise problems may call for differential interconnections, and so on. Therefore, it is crucial to realize that the approach to the design of high-performance chips must be reconsidered.

The book is intended for a reader with good background in semiconductor physics and previous experience in nMOS circuit design. Chapter 1 introduces the basic concepts and terminology of CMOS technology. Simple gates — such as NAND, NOR, and so on — are presented. Comparisons between CMOS and nMOS, ECL, and TLL are made. Chapter 2 is a review of MOS physics and presents the most important small-geometry effects with specific reference to the ones of particular interest to the logic designer. DC analysis of the CMOS inverter is carried out. Power dissipation and noise margin characteristics are studied as well. The CMOS inverter is analyzed by using both the simplified classical equations of MOS physics and a more accurate model.

Chapter 3 is devoted to CMOS fabrication processes: bulk p-well, n-well, twin-tub, and SOS are presented. Latchup is analyzed and the appropriate counter-measures to minimize its occurrence are discussed. Layout techniques for latchup avoidance are instead dealt with in Appendix 1, which discusses layout methodologies. Whenever appropriate, topics have been discussed from the logic designer's perspective. This approach has been followed throughout the book.

Chapter 4 deals with logic design. Both static and dynamic logics have been discussed and compared. Although choosing one logic style instead of another should be influenced by requirements at the chip level and by constraints at the system level, this last aspect seldom plays a role in determining the logic style. This is understandable in general purpose chips (e.g., microprocessors), where the actual "external" constraints are not clearly defined during the design stage, but it is less justifiable in dedicated circuits (e.g., floating-point units), where some constraints at the system level *do exist.* This chapter presents a survey of the logic families and stresses the above issues.

Chapter 5 is related to the previous chapter, even though at a lower level of abstraction. In fact, it deals with circuit design. Computation of parasitic parameters is presented together with models for long interconnections, and the general issue of circuit optimization — interconnection modeling, delay minimization, transistor sizing both in static and dynamic logic, etc. — is treated.

Chapter 6 presents some interesting building-blocks such as full-adders, sense amplifiers, etc. These circuits serve as good examples of some issues discussed in the previous chapters. Therefore, Chapter 6 is *not* structured as a sort of "designer's hand-book of digital CMOS circuits." Rather, few circuits are discussed, but in great detail.

Chapter 7 deals with off-chip communication, from optimum design of output pads to circuitries to protect the chip against electrostatic discharge. Noise, power dissipation, area, and speed are all parameters that play a dominant role in the design of input and output buffers, and different optimization techniques should be used on a case-by-case basis. Some of these techniques are discussed in the chapter, and the rationale for using one instead of another is presented.

Finally, layout is dealt with in Appendix 1. Layout for latchup avoidance is discussed, together with the critical issue of power and ground routing, which so significantly affects circuit behavior.

The book originated from two mini-courses taught by the author at Politecnico di Milano (Milan, Italy) and at Ing. C. Olivetti (Ivrea, Italy) in 1984, and from two lectures held at the Computer Science Department of Carnegie-Mellon in 1983 and 1984. The author would like to thank Alan Sussman and Hank Walker, Carnegie-Mellon, for their comments and Allan Fisher, Carnegie-Mellon, for his comments on an early draft of Chapter 7. The author is deeply indebted to Francesco Gregoretti, Politecnico di Torino (Turin, Italy), for the time he spent in discussing various aspects of the book and for his invaluable comments. Franco Maddalena, Politecnico di Torino, made useful suggestions that helped improve Chapter 2 and Chapter 5.

M. Annaratone

Pittsburgh, Pennsylvania

DIGITAL CMOS CIRCUIT DESIGN

Chapter 1
Introduction

CMOS (Complementary Metal Oxide Semiconductor) is the most interesting technology presently available for mass production of chips, because of its low power dissipation, good scalability and speed. After a slow start, due to difficulties in the fabrication process, CMOS became a mature technology in the early eighties and will be more and more widely used as time goes by. One of the fascinating features of CMOS is its very wide range of applications: CMOS circuits can be found both in wrist-watches and supercomputers, covering the entire range of the electronic market. The future of CMOS looks even brighter.

Although the nMOS fabrication of integrated circuits such as the Motorola 68020 (a high performance, 32-bit CMOS microprocessor) would still have been feasible — nMOS processes with feature size as small as $1\mu m$ have been reported and studied [4] — its life-span as a product would have been short because of problems in power and heat dissipation. Given the extremely high investments for a new product such as a 32-bit microprocessor, it is necessary to use a technology mature enough to give good yield[1] and, at the same time, one which is still open to scalability. This is exactly the

---

[1] One way to express the yield of a fabrication process is as the number of fully functional chips per wafer.

case of CMOS.

CMOS has been used for different products besides microprocessors: floating-point processors have been implemented in CMOS, CMOS memories with excellent size/speed ratio are available, etc.. Some of these products feature clock speed close to 20MHz, and will be even faster when CMOS is further scaled down. Indeed, CMOS is challenging both TTL and ECL technologies at their own strength: speed. Although ECL still has an edge over CMOS as far as gate delay is concerned, this is not necessarily true any more at the system (chip) level. ECL and CMOS gate arrays have been compared, with very interesting results: ECL still prevails, but CMOS will surpass ECL speed in the near future [1]. It is important to point out that for extremely fast switching speeds, CMOS power dissipation increases significantly, as we shall see in Section 2.7, and low power dissipation becomes less and less an argument in favor of CMOS when compared with other competing technologies.

As far as CMOS vs. TTL is concerned, the same minicomputer built in TTL and CMOS technologies [1] demonstrated once again that low power dissipation and higher circuit density, two of CMOS's main features, can lead to higher speed at the system level. The experiment showed that the CMOS version was running twice as fast as the TTL version, with forty-four times the system packing density and power dissipation reduced by two orders of magnitude. Last but not least, the cost of the CMOS version was thirteen times lower.

This chapter presents the important features of CMOS and compares it with nMOS, with which the reader should already be familiar. Many of these characteristics need further investigation and will be discussed in the following chapters. Comparison will be made mainly at the logic level, rather than at the electrical level; that is, the MOS transistor is considered as an ideal switching device with zero time propagation delay. The fundamental logic gates, that is, NAND, NOR, XOR, and transmission gate, will be discussed. Finally, positive logic has been used throughout the chapter: "logic 1," "high voltage level," and "V_{dd}" are interchangeable terms, likewise "logic 0," "low voltage level," and "V_{ss}."

1.1. From nMOS to CMOS

MOS technology provides us with two different kinds of devices: enhancement and depletion mode transistors. Each one can then be implemented as an n-channel or p-channel device, thus bringing the total number of MOS devices to four. The terminology "nMOS technology" — more appropriately referred to in the literature as "enhancement/depletion," E/D nMOS technology — has come to identify fabrication processes which form n-channel depletion and n-channel enhancement transistors over a p-substrate. The nMOS inverter features an n-channel enhancement transistor as a pull-down device, and an n-channel depletion transistor as a pull-up device which is always on, that is, it never switches. The logic could be implemented with a resistor as a pull-up, but for various reasons (such as area and speed) a depletion device is preferable.

While in nMOS there is only one switching device — the n-channel enhancement transistor — two switching devices are available in CMOS technology: the n-channel and p-channel enhancement transistor. The availability of two "complementary" devices affects design methodologies and layout disciplines to such an extent that a specific treatment for CMOS becomes necessary.

Conceptually, a CMOS gate is not different from an nMOS gate: both feature a pull-up and a pull-down section, like all logic gates. However, the pull-up section has a different *behavior*: in nMOS, the n-channel depletion transistor, used as a pull-up device, acts as a current source and is always on; in CMOS, the p-channel transistor is a device capable, like its n-channel counterpart, of switching on and off.

The MOS transistor is shown in Fig. 1-1 and is a four-terminal device; the terminals

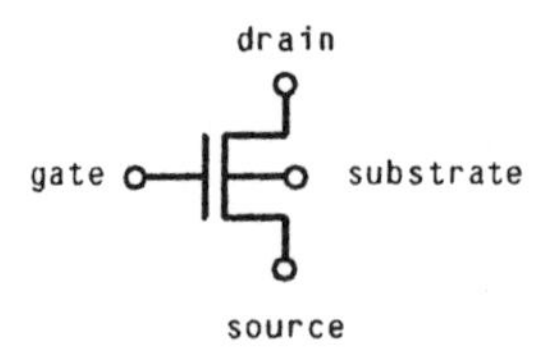

Figure 1-1: MOS transistor.

are called *gate, drain, source* and *substrate.* The substrate terminal will be explicitly shown only when required by the topic under discussion, otherwise it will be omitted. Source and drain terminals are identified by the direction of the current through the device or by their respective potentials. Usually, the source in an n-channel device is at lower potential than the drain terminal (and vice-versa for the p-channel transistor). Two different symbols for the n- and p-channel enhancement transistors can be found in the literature. These are shown in Fig. 1-2. The pair on the left will be used in this book.

Figure 1-2: Two different symbols for n-channel and p-channel enhancement transistors.

The logic behavior of the two CMOS devices is shown in Fig. 1-3. The n-channel device conducts (switch closed) when a logic "1" is applied to its gate and does not conduct (switch open) when a logic "0" is applied to its gate. The p-channel device conducts when a logic "0" is applied to its gate and is open when a logic "1" is applied. From this point of view, the two devices are indeed complementary. We shall see later that this complementarity is not perfect when the two devices are compared at the electrical, rather than at the logic level. It is worth pointing out that this different electrical behavior is indeed one of the major sources of problems for

Figure 1-3: Logic behavior of the two MOS enhancement transistors.

the CMOS designer.

A first comparison between the nMOS inverter and the CMOS inverter will help us to understand the main features that have made CMOS become such a superior technology. Fig. 1-4 shows the nMOS inverter on the left and the CMOS inverter on the right. Besides having different pull-up devices, the input signal in the CMOS inverter is connected to the gates of both devices, while the input signal goes only to the gate of the n-channel device in the nMOS inverter.

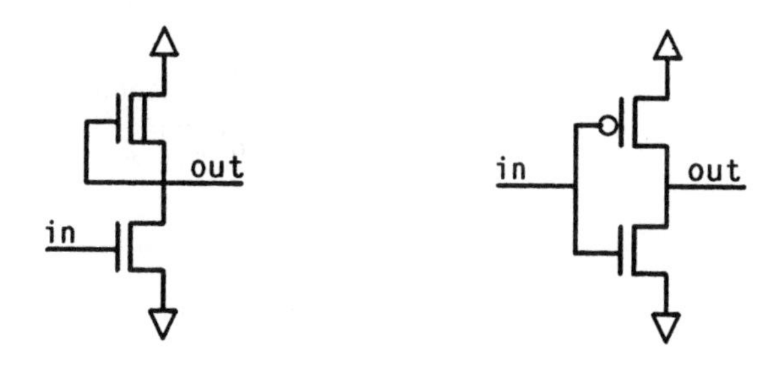

Figure 1-4: nMOS and CMOS inverters.

A comparison between the nMOS and CMOS inverter leads us to the following conclusions:

- In CMOS, rise-time and fall-time are, *theoretically*, the same; this is not true in nMOS where the dynamic behavior of the inverter is *not* symmetrical, that is, where rise-time is slower than fall-time. In fact, because nMOS is a *ratioed* logic, the rise-time and fall-time depend on the geometrical ratio of the two devices. Rise-time and fall-time would be the same if the depletion and enhancement n-channel devices had the same dimensions, something that cannot be done because of noise margin[1] and driving capability problems. Rise-time and fall-time in CMOS do not depend on the geometrical ratio of the two devices.

- In CMOS, when one transistor is conducting, the other one is not. Fig. 1-5 shows the nMOS and CMOS inverters when "1" and "0" are applied to their gates. When the input of the nMOS inverter is "1," the n-channel transistor conducts. Since the depletion device is *always conducting*, we have a static path from V_{dd} to V_{ss}, which causes *static power dissipation*. The static power dissipation of the nMOS inverter can be either zero, or, when the input is a logic "1":

$$(V_{dd} - V_{ss}) \times I_{inv},$$

where I_{inv} is the current flowing through the inverter. This power is typically less than 0.3mW for the basic inverter. As we shall see, CMOS power dissipation is almost entirely dynamic, because there is never a *static* path between V_{dd} and V_{ss}. A CMOS gate dissipates power only in the very short switching period when both devices are conducting (see Section 2.7).

- Unlike static nMOS, the output of a static CMOS gate makes a full voltage swing between V_{dd} and V_{ss} regardless of the relative size of the

[1]The noise margin is a parameter which indicates the ability of a circuit to assure correct operation under variations in the process parameters and external disturbances such as voltage spikes. See Section 2.6 for details.

two devices. In other words, CMOS is a *ratioless* logic. In the nMOS inverter, the output *only asymptotically* reaches V_{ss}. This full output swing allows us to implement logic disciplines that are unique to CMOS, as we shall see in Chapter 4.

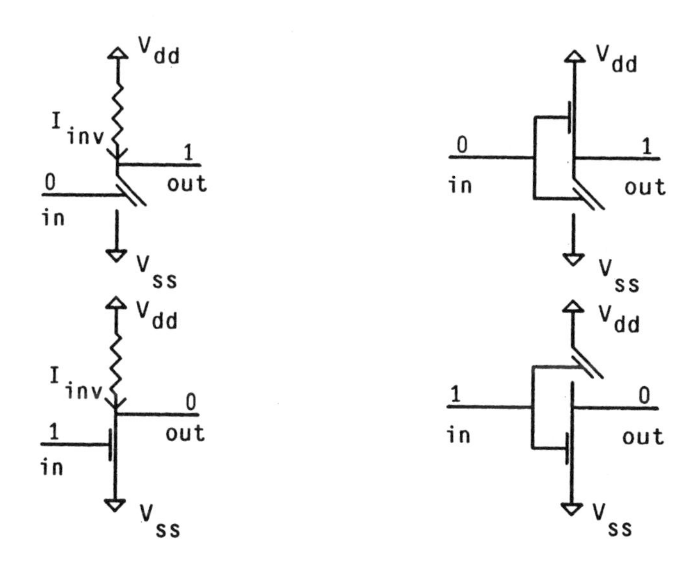

Figure 1-5: nMOS inverter (left) and CMOS inverter: when the output is low, the nMOS inverter consumes power while the CMOS inverter does not.

As far as the future of these technologies goes, CMOS will benefit from scaling in feature size, while nMOS, basically for power dissipation reasons, is much more difficult to scale down. It has been computed that a million gate nMOS chip would require tens of watts [1]. CMOS is also the only available technology suitable to wafer-scale integration.

In addition to its very low power dissipation, excellent noise margin, and scalability, CMOS has already proved to be an excellent subnanosecond technology. In fact, 300ps gate delays have already been achieved. Manufacturers of supercomputers are seriously considering CMOS as an alternative to other promising technologies, like Gallium Arsenide (GaAs).

1.2. CMOS Basic Gates

CMOS provides us with the entire family of logic gates: NAND, NOR, XOR, and so on. Two-input NAND and NOR gates are shown in Fig. 1-6 and Fig. 1-7, respectively. Both figures are accompanied by a table showing the behavior of the four transistors for any combination of inputs.

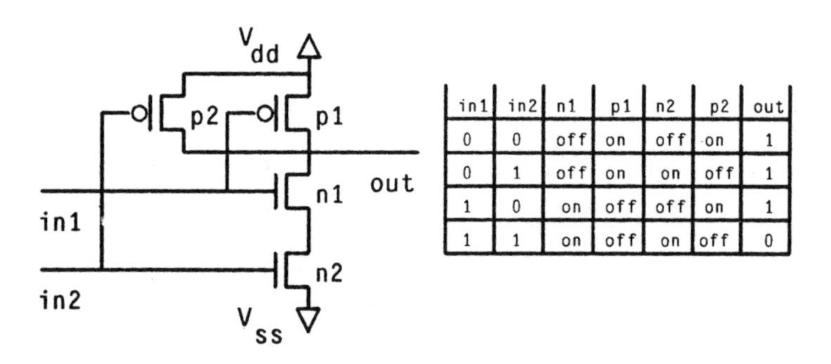

in1	in2	n1	p1	n2	p2	out
0	0	off	on	off	on	1
0	1	off	on	on	off	1
1	0	on	off	off	on	1
1	1	on	off	on	off	0

Figure 1-6: Two-input CMOS NAND gate with logic behavior of the four transistors.

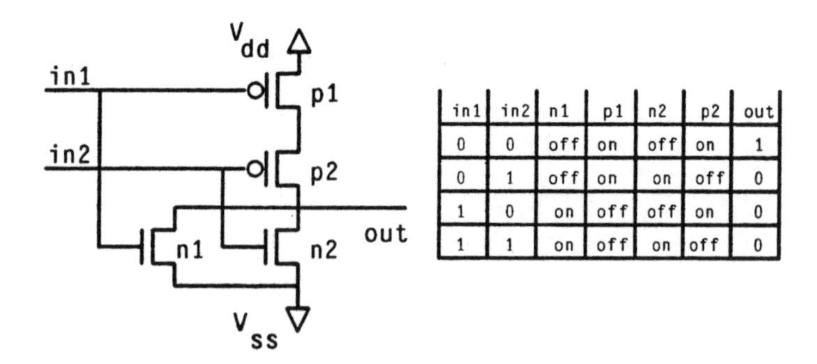

in1	in2	n1	p1	n2	p2	out
0	0	off	on	off	on	1
0	1	off	on	on	off	0
1	0	on	off	off	on	0
1	1	on	off	on	off	0

Figure 1-7: Two-input CMOS NOR gate with logic behavior of the four transistors.

A look at the two gates leads to the following considerations:

- There is a full complementarity between the pull-up and the pull-down stage and, therefore, a "redundancy" in the structures. The pull-up (or the pull-down) stage itself "describes" the boolean equation completely. This will allow implementation of different logics featuring fewer devices (see Chapter 4).

- Each pull-down device has its pull-up counterpart; a five-input NOR gate will have five pull-up transistors and five pull-down transistors. A five-input nMOS NOR would only have six devices.

The last observation leads to the conclusion that nMOS occupies less area than CMOS. Although this kind of judgement is always very risky, given its generality, we can say that *static* nMOS logic is less area-consuming than *static* CMOS logic for processes with the same feature size and number of interconnection layers. The difference in area between the two technologies, when static logic is used, ranges from 30% to even more than 50%.

A CMOS XOR gate is shown in Fig. 1-8 (left). This gate has been designed by simply "translating" into CMOS the conventional XOR TTL implementation shown in the same figure (right). This approach is usually a poor choice in MOS (both nMOS and CMOS), since it does not exploit the intrinsic features of the technology.

The last building-block which is presented is called "transmission gate," and it is shown in Fig. 1-9 with its logic symbol. The transmission gate is the CMOS version of the nMOS "pass transistor." Unlike the latter, it conducts high and low logic levels equally well. Its difference in structure and characteristics from the nMOS "pass transistor" (see Chapter 2) has called for a different name. Note that a transmission gate needs a control signal having *both polarities*, that is, the control signal must also pass through an inverter.

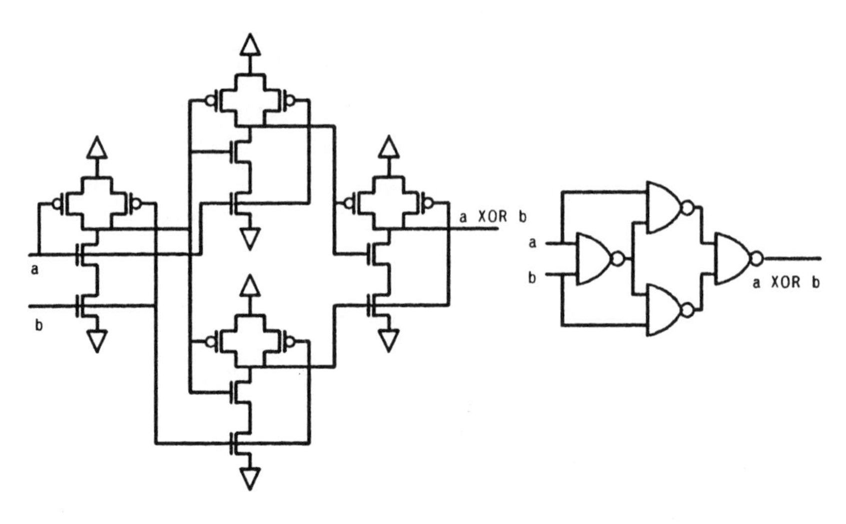

Figure 1-8: A straightforward implementation of a CMOS XOR gate; better approaches are possible, as shown in Fig. 4-6.

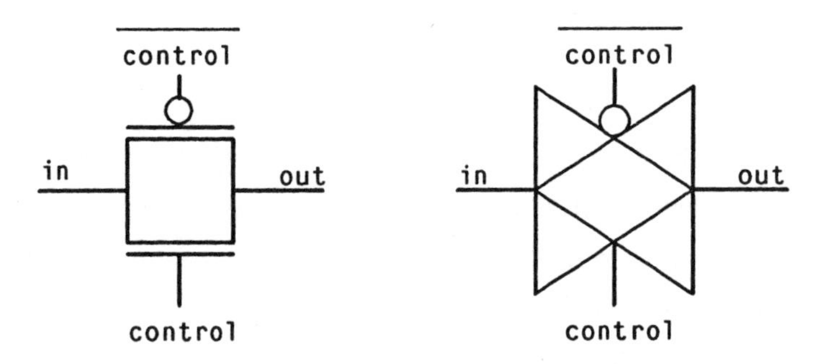

Figure 1-9: Transmission gate: circuit and logic symbol.

Transmission gates, like pass-transistors, are the building-blocks for multiplexer/demultiplexer circuits. Two different structures are presented in Fig. 1-10. Although the circuit on the left seems to be a legitimate implementation, it does not work properly, since both devices are assumed to conduct high and low logic levels equally well. Since this is not the case (see Section 2.4), the structure on the

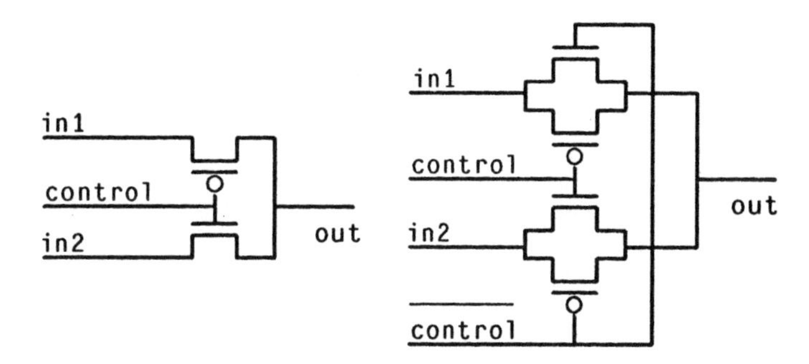

Figure 1-10: Two multiplexer circuits. The circuit on the left does not work properly, and the one on the right must be used.

right of Fig. 1-10 has to be used. This structure consists of two transmission gates that are controlled by two signals having opposite polarity.

Fig. 1-11 shows a CMOS gate that implements a more complex boolean equation. Note again the complementarity between the pull-up and the pull-down stage.

As has already been pointed out, CMOS is a *ratioless* logic, while nMOS is a *ratioed* logic. This holds for static logic: nMOS dynamic and bootstrapping logics are intrinsically ratioless. CMOS logic, both static and dynamic, is always ratioless (with few exceptions, as we shall see in Section 4.1.2 and 4.1.4). Changing the ratio will shift the inverter threshold voltage[1] toward V_{dd} or V_{ss}, *but will not affect the output transient behavior.* Still, careful sizing in CMOS is fundamental not only for performance reasons: unexpected delays due to ratio mismatches might cause race conditions, that, in turn, will not allow the chip to work properly — or not to work at all. Ratio mismatches also affect noise margin.

---

[1]The inverter threshold voltage is defined as the input voltage which makes the inverter produce an identical output voltage, that is $V_{in} = V_{out}$. See Sections 2.5 and 2.6 for details.

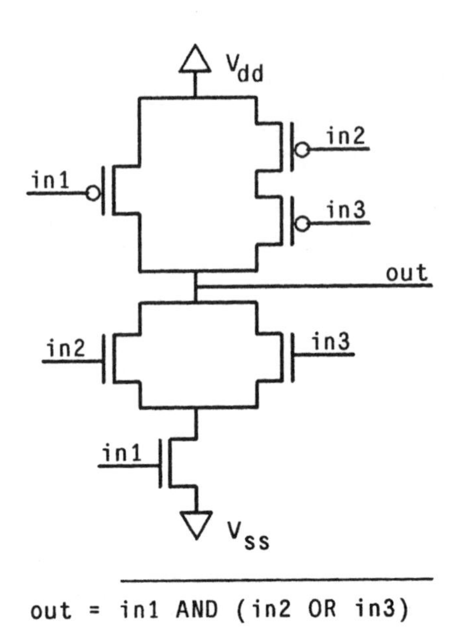

Figure 1-11: A CMOS complex gate.

Another important aspect of CMOS is its operating voltage range: the same CMOS chip can work at lower power supply voltages, with only speed, power dissipation, and noise margin being affected. An increase in the power supply voltage will speed up the chip (at the expense of higher power dissipation) and will improve the noise margin. Typical examples are CMOS operational amplifiers that can work with supply voltages from +5V/-5V up to +15V/-15V and digital CMOS circuits which can operate down to a 2.5V power supply.

CMOS is not free from drawbacks; for instance, it requires more complex fabrication processes than nMOS (see Chapter 3). Its static logic occupies more area than nMOS's, and CMOS is affected by latchup (see Section 3.4). However, the problem of latchup is being solved by the use of more and more sophisticated fabrication processes and layout techniques. With regard to area, the issue is less relevant than might appear, because nMOS cannot attain the same degree of scalability that CMOS has. It is, therefore, somewhat arbitrary to compare these two technologies when one of them cannot be further scaled down, unless extremely sophisticated and expensive cooling procedures are implemented.

References

[1] Iizuka, T.
A Review on CMOS Circuits and Technologies for Digital VLSI.
In *Proc. of the International Symposium on VLSI Technology, Systems and Applications*, pages 30-34. March, 1983.

[2] Mavor, M., M.A. Jack and P.B. Denyer.
Introduction To MOS LSI Design.
Addison-Wesley Publishing Co., Reading, Mass., 1983.

[3] Mead, C. and L. Conway.
Introduction To VLSI Systems.
Addison-Wesley Publishing Co., Reading, Mass., 1980.

[4] Yu, H.N. *et al.*.
1μm MOSFET VLSI Technology: Part I - An Overview.
IEEE Journal of Solid-State Circuits SC-14(2):240-246, April, 1979.

Chapter 2
MOS Transistor Characteristics

The literature on MOS transistor characteristics is extensive. The purpose of this chapter is to review the fundamentals of MOS technology through the use of simplified models. A more accurate model to compute the voltage transfer function of an inverter will be introduced in Section 2.6. Most of the equations presented in this chapter will not be justified. The reader interested in a more comprehensive treatment of MOS physics should refer to references at the end of the chapter [34, 22, 35, 25, 21].

In this chapter we will also analyze the electrical behavior of the transmission gate and the CMOS inverter. Both analyses are under DC conditions, which means that the rate of variation of the input signal is very slow compared to the turn-off or turn-on time of a MOS transistor. Finally, noise margin and power dissipation of the CMOS inverter will be discussed.

2.1. The MOS Transistor

The MOS transistor's most important characteristics are:

- *It is a unipolar device*: conduction in a MOS transistor takes place through the majority carriers only (n-channel device: electrons; p-channel device:

holes). This differs from the behavior of the bipolar transistor where both carriers significantly contribute to the conduction.

- *It is a symmetrical device*: source and drain can be interchanged.

- *It has a high input impedance*: the gate oxide, which will be discussed in the following pages, isolates the gate from both source, drain and substrate. A capacitor is formed between gate and substrate with the oxide as a dielectric and therefore the input impedance is mainly capacitive.

- *It is a voltage-controlled device*: this characteristic, together with a high input impedance, gives the MOS transistor excellent fan-out capabilities.

As discussed in the previous chapter, there are two transistors in CMOS, the n-channel and the p-channel enhancement[1]. Fig. 2-1 shows the voltages between the four terminals for the n-channel (left) and p-channel transistor. The parameters of the n-channel are assumed to be positive, while the parameters of the p-channel, as indicated in the figure, are negative.

A cross-section of the n-channel transistor is shown in Fig. 2-2. The gate is polycrystalline silicon (polysilicon) and a thin oxide (SiO_2) isolates it from the rest of the device and from the substrate, which is p-doped. Source and drain are n-doped regions (doping is much higher than that of an n-type substrate, which is why the symbol n+ is used). These two regions, together with the substrate, make up the n-p-n configuration of the n-channel device. The conduction between the two n-doped regions is controlled by the gate voltage which, when higher than a specified

---

[1] *Four* transistors would actually be possible: *two enhancement transistors* and *two depletion transistors*. Throughout the entire book the two depletion transistors have not been considered, because standard CMOS fabrication processes use only the two enhancement transistors.

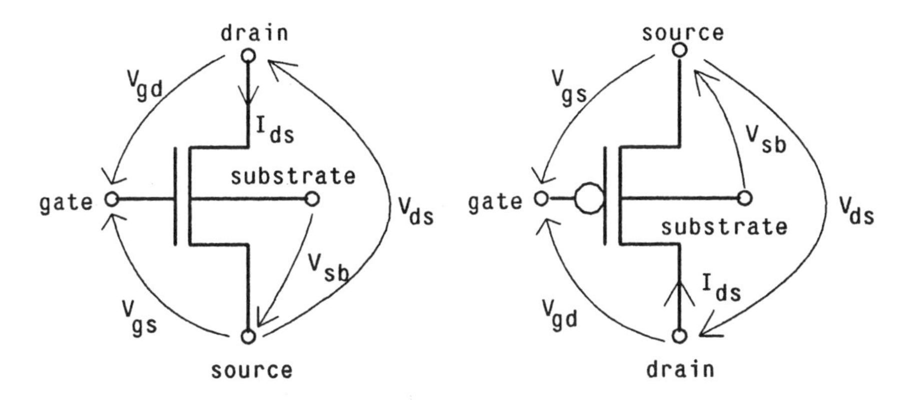

Figure 2-1: Voltage and current for n-channel and p-channel transistor.

voltage V_{Tn} called the *threshold voltage*, creates a *channel* between the two regions allowing conduction to take place. When gate voltage is lower than threshold voltage, no conduction occurs in an ideal device (that is, a device with no *leakage current*). Therefore, the voltage level applied to the gate controls the flow of current between the two n-doped regions (which is the drain-to-source current I_{ds}, shown on the left side of Fig. 2-1).

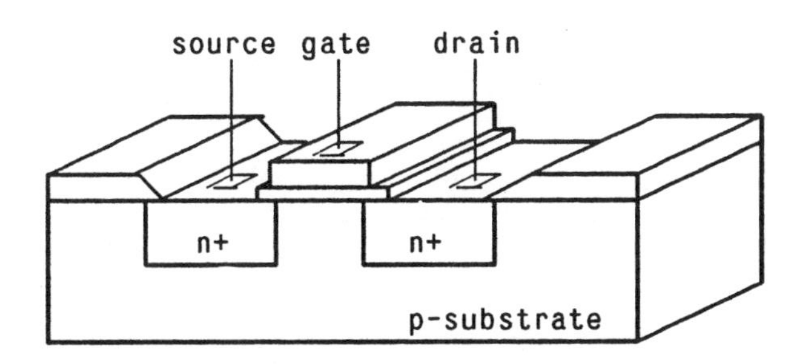

Figure 2-2: n-channel transistor: cross-section.

A cross-section of the p-channel transistor is shown in Fig. 2-3. Again, complementarity is evident: the p-channel transistor is formed over an n-doped

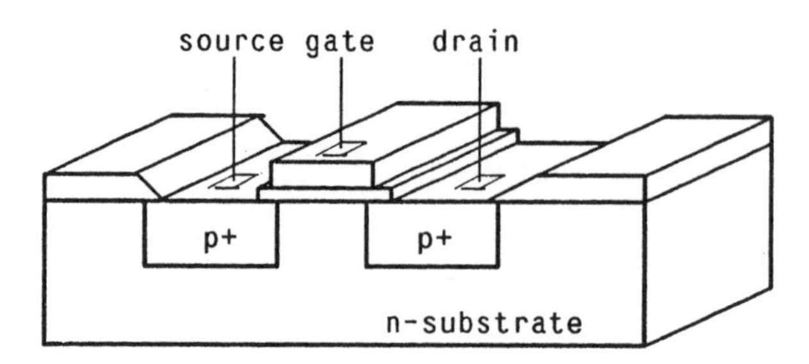

Figure 2-3: p-channel cross-section.

substrate, while source and drain areas are p-doped regions. These two regions, together with the substrate, constitute the p-n-p configuration of the device. Because the doping of the p-type regions is higher than that of a p-type substrate, the symbol p+ is used. The behavior of the p-channel transistor is identical to that of the n-channel transistor, that is, the conduction between the two p-doped regions is controlled by the gate voltage.

In CMOS there are *two threshold voltages*: that of the n-channel transistor, V_{Tn}, and that of the p-channel transistor, V_{Tp}. The threshold voltages cannot be easily controlled by the designer, because they depend mainly on process and physical parameters. An approximate expression for the threshold voltage of an n-channel device is:

$$V_T \approx 2\phi_B + V_F + \phi_{MS} + \sqrt{2\varepsilon_{Si}\varepsilon_o q N_a (2\phi_B + V_{bias})}\ \frac{1}{C_{ox}}, \tag{2-1}$$

where:

- ϕ_B is the *bulk potential*;

$$\phi_B = k\frac{T}{q}\ \ln\frac{N_a}{n_i}, \tag{2-2}$$

where:

- o k: Boltzmann's constant: $k \approx 1.38 \times 10^{-23}$ J/K;
- o q: electronic charge: $q \approx 1.602 \times 10^{-19}$ C;
- o T: temperature (K);
- o n_i: intrinsic carrier concentration in silicon: $n_i \approx 1.45 \times 10^{10}$ cm^{-3};
- o N_a: impurities (acceptor) concentration for p-type substrate (cm^{-3}).

- C_{ox} is the *oxide capacitance* (per unit area):

$$C_{ox} = \varepsilon_o \frac{\varepsilon_{ox}}{t_{ox}} \quad (\text{pF/cm}^2), \qquad (2\text{-}3)$$

where t_{ox} is the *oxide thickness* and is measured in Angstrom (Å). Usually, t_{ox} ranges from 200Å to 500Å.

- ε_{ox} is the *silicon dioxide dielectric constant*: $\varepsilon_{ox} \approx 3.9$;
- ε_o is the *permittivity in vacuum*: $\varepsilon_o \approx 8.86 \times 10^{-14}$F/cm;
- ε_{Si} is the *silicon dielectric constant*: $\varepsilon_{Si} \approx 11.9$.
- V_F takes into account that oxide charges are mainly located in the oxide-substrate interface. They are created during the fabrication process. See [22] for details.
- ϕ_{MS} is the *work function difference* [39]; this parameter will be discussed below.

- V_{bias}: substrate (back-gate) bias is present when the substrate and the source do not share the same potential. Usually, source and substrate are tied together and connected to V_{ss} (n-channel) or V_{dd} (p-channel). However, it is possible to *bias* the substrate: the substrate of an n-channel device could be connected to -2V instead of V_{ss}, for instance. In this case the source would still be at V_{ss} (say, 0V), while the substrate would be at -2V. *Back-gate bias always increases the threshold voltage*; however, sometimes it is useful to apply back-gate bias to eliminate marginal behavior of a chip due to unexpected variations in the process parameters. In fact, back-gate bias is the only way for the designer to modify the threshold voltage, which strictly depends on fabrication process parameters, as we have just seen. This dependency of the threshold voltage on substrate bias is called *body effect.*

The work function difference takes into account the effect of the metal-semiconductor contact (in metal gate technology) or the effect of the contact between differently doped silicon materials (in polysilicon gate technology). In the first case, a cross-section of a gate would show metal (aluminum) on top, oxide in between, and semiconductor (substrate) on the bottom, that is, the "MOS capacitor." This contact between three different materials generates a voltage difference that is called the "work function difference."

When polysilicon gate technology is considered, metal is replaced by "degenerate" — that is, heavily doped — polysilicon. Again, a voltage difference occurs. What we have just described has a significant impact on CMOS, which is largely based on polysilicon gate technology. Indeed, the n-channel transistor and the p-channel transistor have notably different work function differences. To understand why this happens, let us suppose we have degenerate n-type polysilicon (e.g., polysilicon heavily doped with boron). The n-channel transistor's ϕ_{MS} will account for the contact between n-type polysilicon and oxide and between oxide and p-type

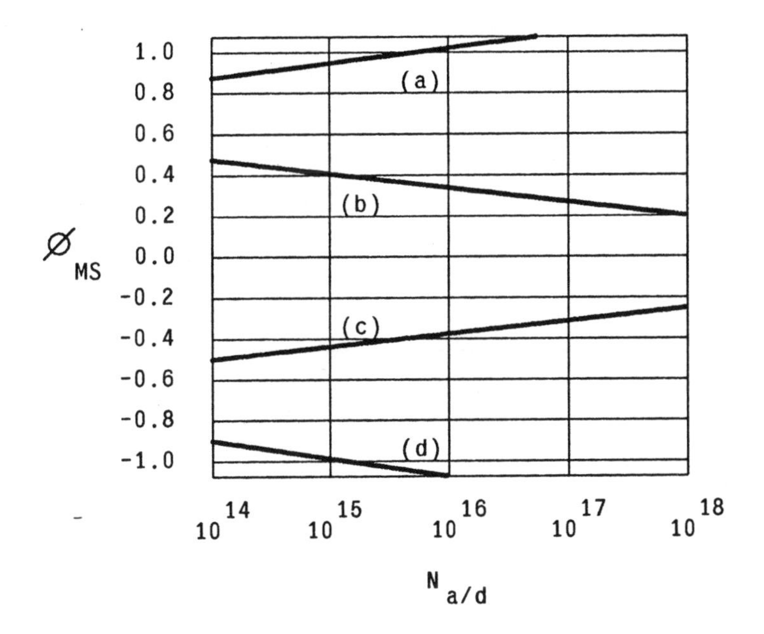

Figure 2-4: Work function difference for different polysilicon gates and substrate types vs. substrate doping concentration (see text). *After Ref. [15].*

substrate. The p-channel transistor's ϕ_{MS} will account for the contact between n-type polysilicon and oxide and between oxide and n-type substrate. Moreover, the doping concentration of the two substrates is different, and the work function difference depends on it. Therefore, the two ϕ_{MS}'s will be different — perhaps by 1 Volt or more. The work function difference is the first parameter we encounter which violates the two transistors' perfect complementarity. Fig. 2-4 shows the work function difference in four cases:

a. degenerate p-type polysilicon over n-type substrate;

b. degenerate p-type polysilicon over p-type substrate;

c. degenerate n-type polysilicon over n-type substrate, and

d. degenerate n-type polysilicon over p-type substrate.

To summarize, the threshold voltage depends mainly on the following parameters:

- Oxide thickness ($t_{ox} \rightarrow C_{ox}$).
- Substrate doping concentration (e.g., $N_a \rightarrow \phi_B$).
- Oxide-semiconductor interface characteristics (e.g., oxide charges $\rightarrow V_F$).
- Gate material (e.g., ϕ_{MS}).
- Temperature.
- Substrate (back-gate) bias.

Second-order effects, such as threshold voltage drift caused by changes in the impurity concentration [15], have not been considered.

The reader will have noted that the two types of transistors require *different substrates.* This is one of the major problems with CMOS, and undoubtedly one of the causes for its slow growth in the 1970's, especially because the presence of two substrates on the same wafer creates several side-effects, the most notable one being latchup (see Section 3.4).

As far as the designer is concerned, two very important parameters are transistor *channel length* (L) and transistor *channel width* (W). They are shown in Fig. 2-5. Channel length is the dimension of the gate which is parallel to the flow of carriers, while its width is perpendicular.

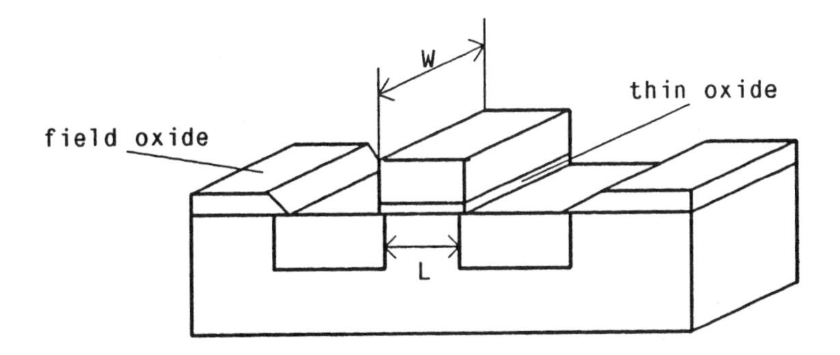

Figure 2-5: Channel length (L) and width (W). Thin oxide isolates the gate from the substrate, while field oxide covers most of the chip.

The same figure also shows the areas where thin oxide (also called "gate oxide") and field oxide are located. Thin oxide (e.g., 200-300Å thick) isolates the gate from the substrate in the bulk process. The field oxide, as its name suggests, covers the entire chip except for the active areas — where gate, source and drain will be formed. Making the channel wider allows more current to flow between source and drain, while making the channel longer increases the source to drain resistance. It should also be noted that the *effective* length and width of the channel are somewhat smaller than the *drawn* channel length and width. Effective length and effective width are process-dependent and should be considered when doing circuit simulation. A typical value for a 3μm drawn channel length is a 2.5μm effective channel length; the shrinkage in this case is about 18%. The electrical characteristics of the device are actually determined by the effective, rather than drawn, channel length and width.

The fundamental characteristics of the n-channel enhancement MOS transistor will now be summarized. The transistor is examined in the configuration shown in Fig. 2-6, with its source and substrate terminals tied to V_{ss}. The same figure shows the I_{ds}-V_{ds} curves and the different regions of operation for increasing values of the gate voltage V_{gs} ($V_{gs1} < V_{gs2} < ... < V_{gsn}$).

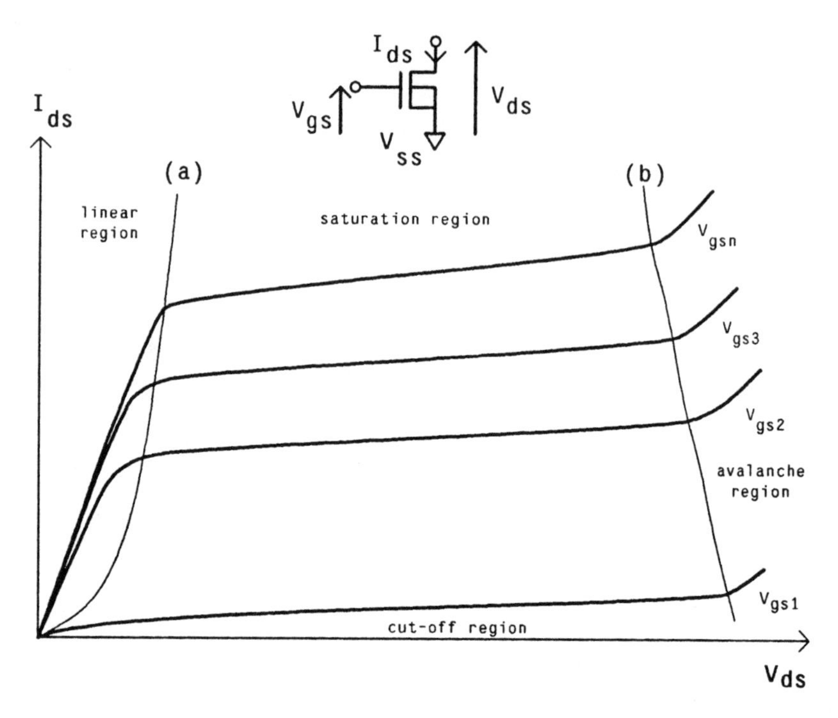

Figure 2-6: Regions of operation for an n-channel enhancement transistor.

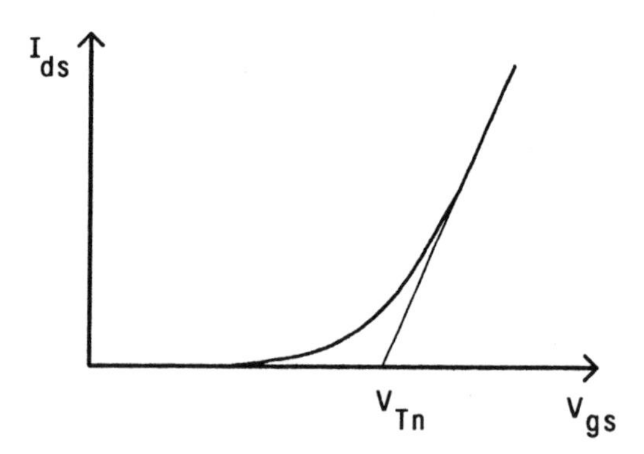

Figure 2-7: For $V_{gs} < V_{Tn}$, the drain-to-source current I_{ds} does not drop drastically to zero; rather, it follows an exponential decay.

Four different regions can be identified in Fig. 2-6:

1. *Linear region*: when V_{ds} increases, I_{ds} increases. The curve has a steep slope; curve (a) is where $V_{ds} = V_{gs} - V_{Tn}$.

2. *Saturation region*: in this region any increase in V_{ds} should *not* increase I_{ds}. However, because of decreasing in channel length — the pinch-off point in Fig. 2-8(c) — I_{ds} slightly increases.

3. *Avalanche region*: when V_{ds} is further increased, the transistor enters a region in which I_{ds} suddenly increases. This increase is almost perpendicular to the V_{ds} axis in metal-gate technology and less drastic in silicon-gate technology — which is the one depicted in Fig. 2-6. Curve (b) is the boundary between the saturation region and the avalanche region.

4. *Cut-off region*: when gate voltage drops below the threshold voltage V_{Tn}, I_{ds} drops to zero. Although we can assume that, for $V_{gs} < V_{Tn}$, $I_{ds} = 0$, the actual behavior is different, as Fig. 2-7 shows: I_{ds} decreases and for $V_{ds} = V_{Tn}$, drain-to-source current still flows. This current is due to leakage effects and it is very important to minimize it, especially in application where very low power dissipation is a critical requirement. The sub-threshold drain-to-source current has been shown to depend on length and width of the transistor, carrier mobility, and other parameters [19]. The doping of the substrate (especially the well for n-channel devices) influences the leakage current [12] significantly.

As has already been pointed out, the MOS transistor is a voltage-controlled device. The voltage applied to the gate controls the conduction. When gate voltage goes above threshold voltage a *channel* is formed between drain and source. Fig. 2-8 shows the various steps of conduction. The channel allows carriers to flow from drain to source; since the type of carriers are determined by the doping of the drain and

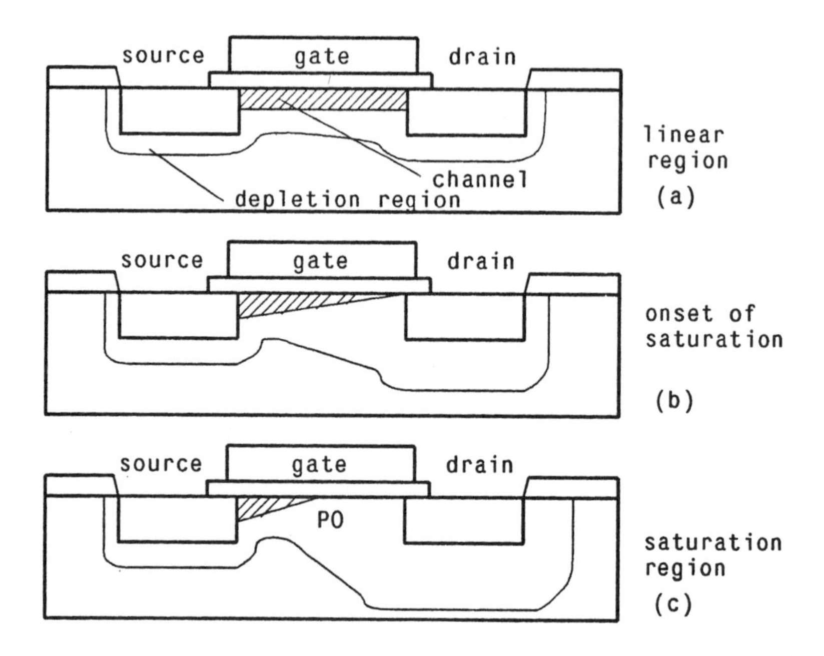

Figure 2-8: The various steps of conduction.

source regions, the carriers in the n-channel transistor are n-type, that is, electrons. The current would leak into the substrate if there were not, in addition to the channel formation, the formation of a *depletion region* below the channel that isolates the carriers from the substrate. This is the behavior of the n-channel enhancement transistor in the linear region which is shown in Fig. 2-8(a). To enter the saturation region, it is necessary to increase the drain voltage. When the drain voltage increases, the drain-to-source current should increase at the same rate. However, this does not happen, and the current stabilizes — although it is still increasing slowly. The physical behavior of the transistor in this region is depicted in Fig. 2-8(b). While the depletion region widens under the drain, the channel starts to shrink. This progressive shrinking of the channel is known as *channel modulation.* If drain-to-source voltage further increases, the channel continues to shrink. Fig. 2-8(c) shows the *pinch-off* point PO. Although it would seem that no current can flow from drain to source, other mechanisms allow conduction between drain and source.

The behavior of the transistor in the three regions is described by the following set of simplified equations:

$$\textit{linear region}: I_{ds} = \beta\left[(V_{gs} - V_{Tn})V_{ds} - \frac{V_{ds}^{2}}{2}\right]; \tag{2-4}$$

$$\textit{saturation region}: I_{ds} = \frac{\beta}{2}(V_{gs} - V_{Tn})^2; \tag{2-5}$$

$$\textit{cut-off region}: I_{ds} \approx 0. \tag{2-6}$$

Eq. (2-6) is actually more complex: I_{ds} depends on β, V_{ds}, V_{gs} and V_{Tn} with overall exponential behavior; see [33] for details. The term β is called the *gain* of the transistor and

$$\beta_n = \mu_n \varepsilon_o \frac{\varepsilon_{ox}}{t_{ox}} \frac{W_n}{L_n}, \tag{2-7}$$

where:

μ_n = electron mobility; measured in cm^2/Vs;
W_n = transistor channel width;
L_n = transistor channel length.

What has been shown up to now can be applied to a p-channel transistor by simply inverting the sign of the parameters, or inverting the direction of the arrows for voltages and currents. In other words, the same equations hold for a p-channel transistor when we assume voltage and current directions as shown in Fig. 2-1.

As has already been pointed out, conduction takes place via one type of carrier, electrons in the n-channel transistor, and holes in the p-channel transistor. Because the mobility μ_n of electrons is higher than the mobility μ_p of holes — by a factor of two or more — the n-channel device is expected to be "faster" than the p-channel transistor. By "faster" we mean that the transition between the cut-off region and the saturation region (and back) will be faster in the n-channel transistor than the same transition in the p-channel transistor. This will have a significant impact on design methodologies and delay minimization procedures, as we shall see in Chapter 4 and 5, respectively.

2.2. Parasitic Parameters

It has already been pointed out that the input impedance of the MOS transistor is mainly capacitive. Determining this capacitance is a topic of fundamental importance, because the logic designer will base the sizing of gates on the output load that each gate has to drive. This output load also consists of the input capacitance of the devices that the gate drives, as we shall see in Chapter 5.

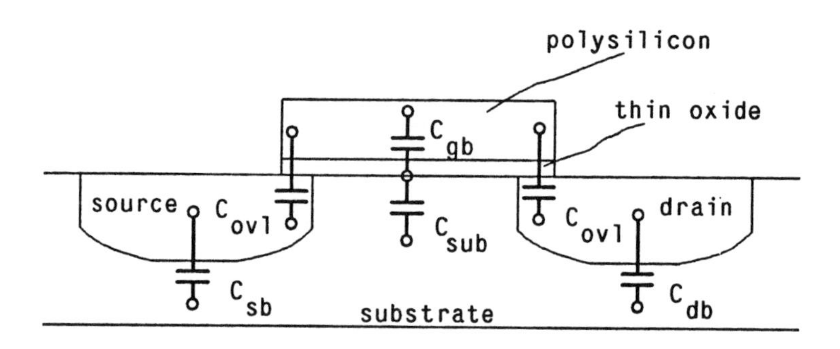

Figure 2-9: MOS transistor parasitic capacitances.

Fig. 2-9 shows a cross-section of a transistor with its associated parasitic capacitances. C_{ovl} is the *overlap capacitance* between gate and either source or drain. C_{sub} accounts for the charge (or lack thereof) under the gate, while C_{db} and C_{sb} are the drain and source *junction capacitances*. The capacitances between the four terminals of the MOS transistor are shown in Fig. 2-10, where C_{sub} is represented by two capacitances C_{gs} and C_{gd} associated with source and drain, respectively. The two overlap capacitances are not shown. In fact, their influence on the overall capacitance is almost negligible, when the fabrication process employs a self-aligned gate technique, that is, when the diffused regions are formed *after* the polysilicon gate has been deposited. It should be noted that Fig. 2-9 presents a *model* for the parasitic capacitances of the MOS transistor. Modeling the behavior of the MOS transistor is an extremely complex task: not only are the capacitances geometry-dependent, but they also depend on process parameters, physical parameters, and transistor voltages. An accurate model is of crucial importance for computer simulation purposes. Here we want to present a simple model to help the designer to determine input and output capacitances. The reader interested in more accurate models should refer to [20, 31, 38, 4, 8].

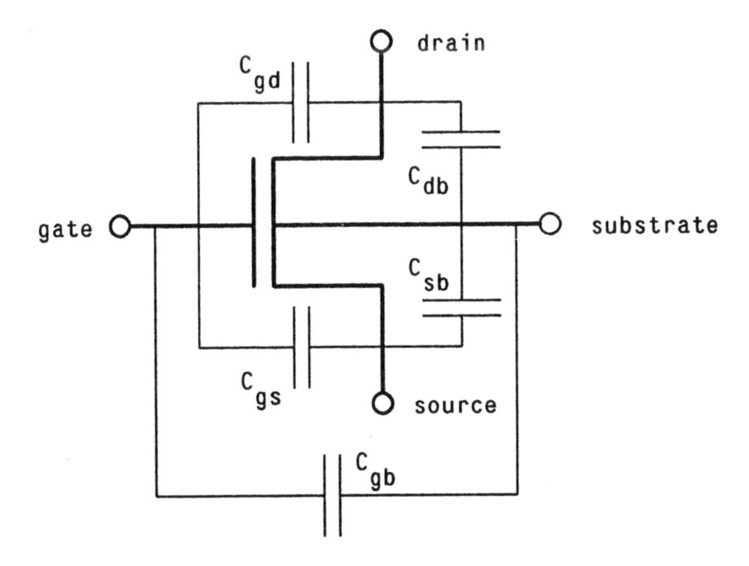

Figure 2-10: Parasitic capacitances and their relationship with the four terminals of a MOS transistor.

The capacitances shown in Fig. 2-10 can be logically divided into three categories: *input capacitance*, *output capacitance* and *feedback capacitance* [11]. Some of these capacitances can be either input, output, or feedback capacitances depending on the configuration in which the transistor is used. For example, if the n-channel transistor has its source and substrate tied to V_{ss} (as in Fig. 2-6), we have:

- *Capacitances contributing to the gate input capacitance*: C_{gb} and C_{gs}.

- *Capacitance contributing to the output capacitance*: C_{db}.

- *Feedback capacitance*: C_{gd}.

C_{sb} is shunted. Both C_{db} and C_{gd} should be kept small. The larger they are, the larger the turn-on time of the transistor will be — although a faster turn-on is also associated with a higher spike at the output [41]. The designer does not have control over C_{gd}, C_{gs} and the overlap capacitances, since they are process dependent. On the other hand, the designer does have control over C_{db} and C_{sb}, which depend on the geometry of the junction (see p. 32).

If we connect the drain to the source, the feedback capacitance C_{gd} effectively contributes to the input capacitance. We can now introduce the *short-circuit input capacitance*, C_G. We have:

$$C_G = C_{gb} + C_{gs} + C_{gd}. \tag{2-8}$$

Let us consider the three capacitances as a function of the gate voltage [30, 38]. When gate voltage is below threshold voltage, there is almost no charge associated with the channel, that is, C_{gs} and C_{gd} are approximately zero (there is only a small amount due to the overlap capacitance). C_{gb} is:

$$C_G \approx C_{gb} = C_o = C_{ox}A = \varepsilon_o \frac{\varepsilon_{ox}}{t_{ox}} A, \tag{2-9}$$

where A is the gate area (LxW). The actual gate capacitance will be smaller than the one shown in Eq. (2-9) shows because of the presence of some charges — and therefore a depletion layer. C_{gb} abruptly drops and then remains constant in the linear and saturation region.

From this point on, C_{gs} and C_{gd} play a dominant role and behave according to channel modulation. When the device is in the linear region we can assume that the charge will be evenly distributed between the source and drain regions (see Fig. 2-8(a)). During saturation the depletion region under the drain is much larger than the one under the source, and C_{gs} will be larger than C_{gd} (see Fig. 2-8(b)), to the extent that C_{gd} can be considered zero for all practical purposes at pinch-off (see Fig. 2-8(c)). Table 2-1 shows the approximate values of the three capacitances during the three operating conditions of cut-off (sub-threshold), saturation, and linear.

Table 2-1: Contributions of the three input capacitances.

Region	C_{gb}	C_{gs}	C_{gd}
cut-off	C_o	0	0
saturation	$0.2C_o$	$0.5\ C_o$	0
linear	$0.2C_o$	$0.4\ C_o$	$0.4\ C_o$

A reasonable assumption for rough sizing of gates is that the gate capacitance is constant and always equal to:

$$C_G = C_o . \tag{2-10}$$

If we assume that the oxide thickness is 400Å, we can compute the actual gate capacitance per square micron:

$$\varepsilon_o \frac{\varepsilon_{ox}}{t_{ox}} \approx 3.9 \frac{8.86x10^{-18}}{400x10^{-4}} \approx 1x10^{-3}\,\mathrm{pF}/\mu\mathrm{m}^2 .$$

Therefore, typical gate capacitances are in the 10^{-3} pF/μm^2 range. As far as C_{db} and C_{sb} are concerned, they do not contribute to the input capacitance but can contribute to the *output load.* These capacitances come from the reverse-biased diode formed by the diffusion layer and the substrate. Fig. 2-11 shows one of these capacitances from a three-dimensional point of view. The capacitance consists of three terms: a "planar" term (region A: C_{jA}), four "cylindrical" terms (regions B, C, D and E: C_{jP}), and four "spherical" terms (regions F, G, H and J: C_{jC}). While A depends on the *area of the junction,* the contribution of the cylindrical regions is proportional to the *perimeter of the junction,* and the spherical contribution is proportional to the *number of corners.* All these capacitances *depend on the voltage applied to the junction.*

For the *planar capacitance per unit area* C_{jA}, we have [34]:

$$C_{jA} = \sqrt{\frac{q\varepsilon_o\varepsilon_{Si}N_{sub}}{2(\phi_b + V)}}\,, \qquad (2\text{-}11)$$

where:

N_{sub} = doping concentration of the substrate;
ϕ_b = built-in junction potential;
V = voltage between the substrate and the diffused region;

and

$$\phi_b = k\frac{T}{q}\ln\left(N_{sub}\frac{N_d}{n_i^2}\right), \qquad (2\text{-}12)$$

where N_{sub} and N_d are the doping concentrations of the substrate and the diffusion region. The *perimetrical capacitance per unit length* C_{jP} (pF/cm) is represented by a transcendental equation:

$$V \approx qx_j^2\frac{qN_{sub}}{4\varepsilon_o\varepsilon_{Si}}\left[1 + \left(\pi\frac{\varepsilon_o\varepsilon_{Si}}{C_{jP}} - 1\right)e^{\varepsilon_o\pi(\varepsilon_{Si}/C_{jP})}\right], \qquad (2\text{-}13)$$

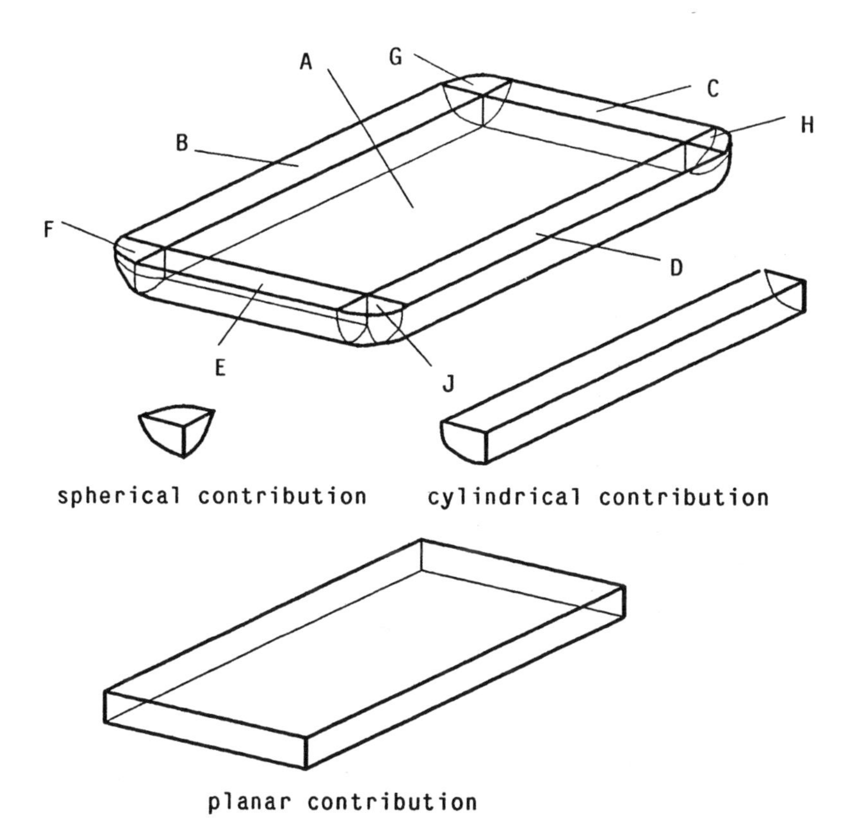

Figure 2-11: Diffusion capacitance.

where x_j is the metallurgical junction depth.

Let $V_{xP} = qN_{sub}\frac{x_j^2}{4\varepsilon_o\varepsilon_{Si}}$ and $x = \pi\frac{\varepsilon_o\varepsilon_{Si}}{C_{jP}}$. We have:

$$f(x) = V - V_{xP}[1 + (x - 1)e^x] \approx 0 ;$$

$$x_{i+1} = x_i + \{V - V_{xP}[1 + (x - 1)e^x]\}\frac{1}{V_{xP}xe^x} . \qquad (2\text{-}14)$$

where Eq. (2-14) is the usual recursive formula to determine x_{i+1} given x_i (by the

Newton-Raphson method) such that $f(x) \to 0$. Similarly, the *spherical capacitance per corner* C_{jC} (pF) is:

$$V \approx V_{xC} \frac{3y - 1}{(y - 1)^3} , \tag{2-15}$$

where:

$$y = \frac{C_{jC}}{32\pi \varepsilon_o \varepsilon_{Si} x_j} , (C_{jC} > 32\pi \varepsilon_o \varepsilon_{Si} x_j) ;$$

$$V_{xC} = qN_{sub} \frac{x_j^2}{6\varepsilon_o \varepsilon_{Si}} .$$

Therefore, as we did above for C_{jP}:

$$f(y) = V - V_{xC} \frac{3y - 1}{(y - 1)^3} \approx 0;$$

$$y_{i+1} = y_k - \left[\left(V(y-1)^3 - V_{xC}(3y - 1) \right) \frac{y-1}{6V_{xC}y} \right] .$$

Let us now consider an example. Suppose we have a 5μm by 4μm junction with a substrate concentration of 1.0×10^{16} cm^{-3} and a metallurgical junction depth of 400Å. The *total capacitance* C_j of the junction can be written as:

$$C_j = AC_{jA} + PC_{jP} + KC_{jC} , \tag{2-16}$$

where A, P, and K are the area of the junction, its perimeter, and the number of corners, respectively. In our example we have:

$$A = 20\mu m^2 ;$$
$$P = 18\mu m ;$$
$$K = 4 .$$

We obtain the following results:

$$C_{jA} \approx 18.3\ \text{nF/cm}^2 = 0.183\text{x}10^{-3}\ \text{pF}/\mu\text{m}^2\ ;$$

$$C_{jP} \approx 720\text{x}10^{-5}\ \text{pF/cm} = 0.72\text{x}10^{-3}\ \text{pF}/\mu\text{m}\ ;$$

$$C_{jC} \approx 4.89\text{x}10^{-4}\ \text{pF/corner}\ .$$

Therefore:

$$C_j = 3.66\text{x}10^{-3} + 12.96\text{x}10^{-3} + 1.95\text{x}10^{-3} = 18.57\text{x}10^{-3}\text{pF}\ (@V = 2V)\ .$$

All the above formulae are easily computed. Finally, Fig. 2-12 shows normalized perimeter capacitance and spherical (corner) capacitance vs. normalized voltage.

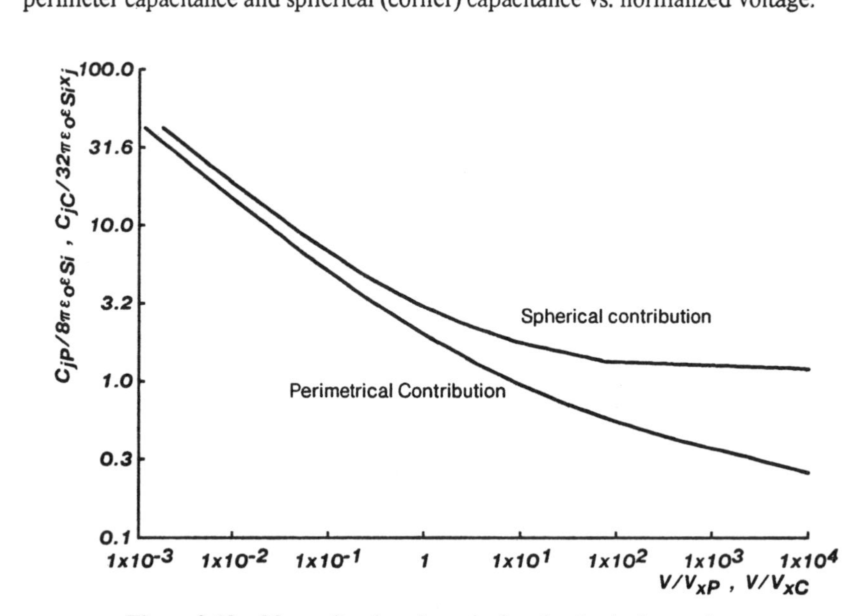

Figure 2-12: Normalized perimetrical and spherical capacitance vs. normalized voltage.

So far, we have analyzed the behavior of the transistor formed on a silicon substrate; if the transistor has been formed on an insulator ("Silicon On Insulator" technology, SOI), the parasitic parameters are significantly different [18]. Moreover, we have considered a transistor which is not affected by small-geometry effects. Small-geometry effects become significant for feature sizes of 3μm and smaller and become predominant at micron and sub-micron feature size.

2.3. Small Geometry MOS Transistor

We now present a brief overview of small geometry effects and their influence on the parameters we have discussed in this chapter. A MOS transistor with a feature size below about 3μm suffers from various effects which influence many of its characteristics, most notably threshold voltage and sub-threshold current. These effects include:

- *Short-channel effects.* They occur in short-channel devices, that is, devices in which the channel length is comparable to the junction depth.

- *Narrow-channel effects.* They occur in narrow-channel devices, that is, devices in which the channel width is comparable to the junction depth.

- *Small-geometry effects.* A small-geometry device is a transistor which has both short-channel and narrow-channel characteristics.

Small-geometry devices are characterized by charge sharing, a phenomenon we shall discuss in Chapter 4, although in a different context. In short-channel devices, source and drain are so close that the charge under the gate redistributes itself among the channel, the source and the drain, and a correct modeling of the parasitic parameters of the transistor becomes more complex, because the transistor is now to be considered as a device with distributed parameters. Although this holds true for long-channel devices as well, the lumped parameter model introduced in the previous sections is sufficient for all practical purposes. However, this model fails to provide an accurate description of small-geometry transistors.

Small-geometry effects affect long-term circuit reliability, tolerance in the fabrication process parameters, long-term circuit performance, etc. We now focus on three major problems: threshold drift, hot-electron emission, and sub-threshold current. Two other important aspects, device isolation and diffusion sheet resistance, will be dealt with later.

Although many of these problems can be solved, if considered separately, global optimization would require satisfying conflicting requirements, and more research is necessary to achieve a solution which satisfies them all. This is especially true for sub-micron devices (0.5μm or shorter effective channel length).

Threshold Voltage Lowering

Charge sharing and drain-induced barrier lowering (DIBL) [37] are mainly responsible for the reduction of the threshold voltage in short-channel devices. The threshold voltage can drop by up to 0.3V for sub-micron channel lengths [5]. Very low threshold voltages are not desirable because they blur the separation between "off" state and conduction state, and make logic circuits more sensitive to noise problems. Several analytical formulations of the threshold voltage for both short- and narrow-channel devices have been presented in the literature, and many are listed in the references at the end of the chapter [40, 3, 16], together with a review of the topic [2].

DIBL characterizes short-channel devices and consists of source injection in the surface of the channel, in the bulk, or in both. This results in lowering the threshold voltage of the transistor and increasing the current in the sub-threshold region. Long channel devices — for instance, 5μm long — feature a constant potential peak under the channel from drain to source; moreover, this potential barrier is largely independent of the drain-to-source voltage. When short-channel devices are considered, the distance between the two diffused regions cannot accommodate the two depletion regions: this proximity effect lowers the potential barrier. Moreover, for drain voltages greater than zero, the peak potential is lowered further. This second effect is caused by unwanted current paths which are formed by the interaction of the electric fields of the two diffused regions, which are now closely coupled. Finally, the peak potential is no longer constant along the channel.

Besides lowering threshold voltage and making it strongly dependent on the drain voltage, DIBL increases sub-threshold current, a critical parameter in some circuits, such as dynamic RAM's and dynamic logic operating at low speed (see Chapter 4). Moreover, higher sub-threshold current increases stand-by power dissipation. A solution to this problem consists of applying a negative voltage to the gate, which significantly reduces the current in the sub-threshold region. Still, the effect of threshold lowering remains.

Increasing the isolation between the two diffused regions by higher substrate concentration reduces the probability of DIBL, because the likelihood of electric fields coupling from source to drain decreases. However, higher doping increases the field at the drain end of the channel, and this results in hot-electron injection, as we shall see below. Decreasing the junction depth can help to reduce DIBL, but favors hot-electron injection occurrence. The same consideration applies to the oxide thickness: decreasing it limits the insurgence of DIBL, but makes hot-electron injection more likely. Therefore, to avoid both DIBL and hot-electron injection, conflicting requirements must be met. Note that the p-channel transistor is more affected by DIBL than the n-channel device is: boron channel implantation used to control the threshold voltage widens the depletion region of p-channel devices with n-type polysilicon gate and this increases short-channel effects [27].

Another important consideration in small-geometry CMOS concerns the work function difference, which becomes a critical issue. Because all tolerances become stricter in small-geometry devices, the fact that polysilicon-gate n-channel and p-channel devices feature different ϕ_{MS} makes it necessary to find a different gate material with reduced difference in ϕ_{MS} between the two devices. Both tungsten [36] and molybdenum [5] are currently being investigated. Lower threshold voltages also require careful analysis of their dependence on temperature; shifts of 100mV or more can be expected over the commercial temperature range (0-70 C).

Narrow-channel effects on the threshold voltage of micrometer and sub-micron devices have been reported in the literature. Expressions for the threshold voltage of micrometer narrow-channel devices [1, 13] and of submicron devices [31, 14] have been presented. The most important narrow-channel effect consists of an *increase in the threshold voltage.* This can be explained by the fact that, while the charge under the gate decreases with the decrease in width, the charge associated with the thick oxide remains approximately constant, and, therefore, its relative contribution increases with narrower widths.

Hot-electron Emission

Hot-electron emission occurs when high fields are present in the surface depletion layer, a typical situation with small-geometry VLSI devices in which the power supply is kept at 5V. This phenomenon is caused by carriers injected from the substrate or from the channel into the gate oxide, where they get trapped. The injection takes place because the electrons gain energy from the high field and, if they are close to the Si-SiO_2 interface, they may have enough energy to surmount the Si-SiO_2 potential barrier. Injection from the substrate is more likely in heavily doped substrates (because the probability of hot-electron emission increases), when charge-pumping from MOS capacitors takes place (e.g., in bootstrap circuits, see Section 4.4), by forward-biased junctions, and by an increase in temperature [6]. This means that a worst-case analysis of hot-electron emission from the substrate should be carried out at the maximum required operating temperature.

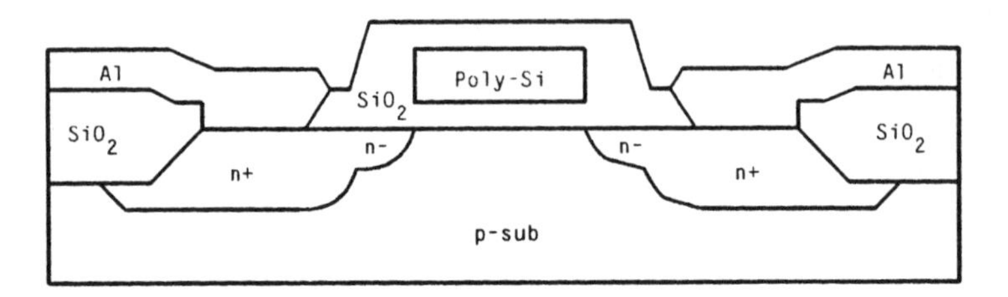

Figure 2-13: Cross-section of n-channel transistor with LDD structure.

Hot-electron emission from the channel is induced by large drain-to-source voltages, which generate high fields close to the drain junction. Hot-electron emission from the channel imposes limits on the maximum V_{ds}, which for small-geometry devices — for instance, $0.5\mu m$ line width — should be kept well below 5V, or, alternatively, lightly doped drain-source structures (LDD) [24] could be used. The cross-section of an n-channel transistor with LDD is shown in Fig. 2-13. The structure consists of narrow, self-aligned, lightly doped n-type regions (n-) which are introduced between the source (drain) and the channel. Electric fields which cause hot-electron injection are spread into the lightly doped regions, and their effect is thus reduced. LDD structures allow higher power supply voltages or shorter channel lengths. Note that the two lightly doped regions increase the drain-source resistance, and some performance degradation is to be expected. The reader is referred to Chapter 11 of [22] and to [6, 10, 9] for a detailed analysis of hot-electron carrier injection.

Sub-threshold Voltage Conduction

The ability to drastically reduce the I_{ds} current below the threshold voltage by a small decrease in gate voltage is essential to obtain a device which very closely resembles an ideal switch. The sub-threshold current depends on the channel doping concentration and other parameters, some of which have already been mentioned. To achieve a drastic reduction in current for small variations in the gate voltage, low doping concentration is necessary. However, this contradicts the requirement for high channel doping concentrations to minimize DIBL.

2.4. CMOS Transmission Gate

As was shown in Chapter 1, the CMOS transmission gate requires two devices: a p-channel and an n-channel device. On the other hand, the CMOS transmission gate does not have the drawbacks that the nMOS pass-transistor has; that is, it does not require any level restoring logic at its output. In this section we will analyze in detail the reason why the n-channel transistor is not able to pass a "1" logic level (V_{dd}) without degrading the signal, and, conversely, why the p-channel transistor does not pass well "0" (V_{ss}) logic levels.

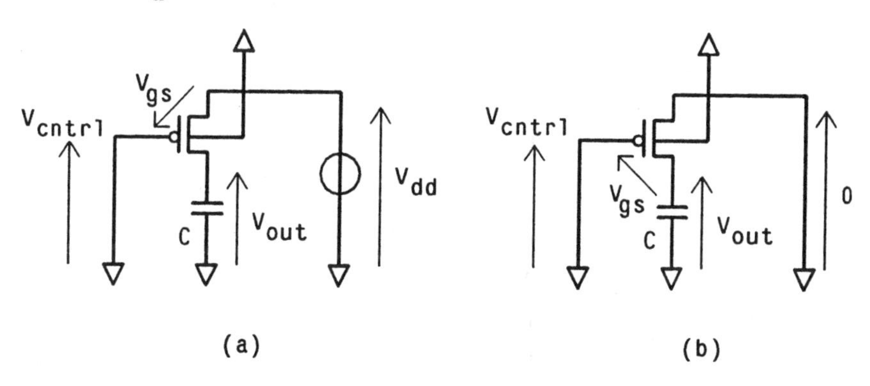

Figure 2-14: p-channel pass-transistor.

Fig. 2-14 shows a p-channel transistor acting as a pass-transistor. The gate of the transistor is driven by the V_{cntrl} signal. When V_{cntrl} is "1," the p-channel is "off," and no conduction takes place. Let V_{cntrl} go to "0," and suppose (see Fig. 2-14(a)) that the load capacitor C is initially uncharged, that is, $V_{out} = 0$. We have $V_{gs} = V_{dd} \gg |V_{Tp}|$. Since the input of the pass-transistor is V_{dd}, C will start to charge, and, when $V_{out} = V_{dd}$, no current will flow, i.e., we have reached the equilibrium. As V_{out} approaches V_{dd}, V_{ds} decreases, and less current flows between drain and source.

Fig. 2-14(b) presents the opposite case: the capacitor C is charged and will discharge through the p-channel into V_{ss}, which is to say that the p-channel is passing "0" (V_{ss}) to the output (V_{out}). To start ($V_{cntrl} = 0$), we have $V_{gs} = V_{out} \gg |V_{Tp}|$. However, in this case, V_{gs} does not remain constant, but follows V_{out}. The output voltage V_{out} across the capacitor decreases until it reaches the threshold voltage of the p-channel transistor ($V_{out} = V_{gs} = |V_{Tp}|$). From then on the p-channel device is off and the capacitor cannot completely discharge, thus making it impossible to deliver a V_{ss} voltage into V_{out}. V_{out} can at most reach $|V_{Tp}|$. Therefore, the p-channel transistor can transfer high logic levels with no degradation, but it cannot transfer low logic levels without degrading the output logic level[1].

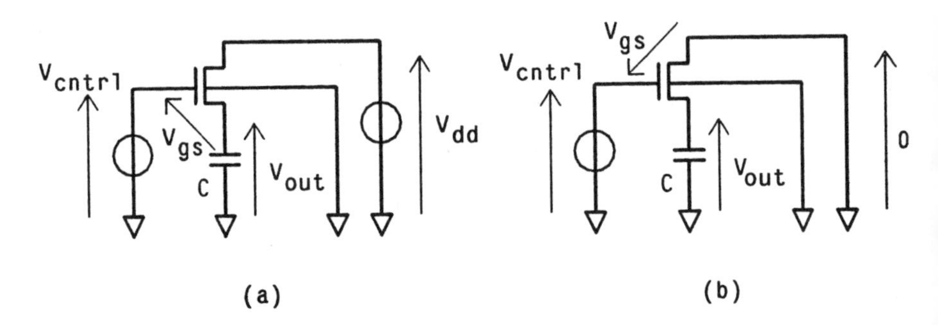

Figure 2-15: n-channel pass-transistor.

Fig. 2-15 shows the other case, with the n-channel transistor used as a pass-

[1]Note the inversion of source and drain between Fig. 2-14(a) and (b). As we said in Chapter 1, source and drain are interchangeable terminals. Usually, for a p-channel device, the drain's potential is lower than that of the source. The same consideration applies to the n-channel transistor; in this case the source is at lower potential.

transistor. The device is controlled by the voltage V_{cntrl} applied at its gate. When V_{cntrl} is low, no conduction takes place between drain and source; the transistor is in the cut-off region. Let us now consider Fig. 2-15(a) and assume that the load capacitor C is initially uncharged. We have $V_{gs} = V_{cntrl}$. When V_{cntrl} is high ($V_{gs} \gg V_{Tn}$), conduction takes place, the capacitor starts to charge, and V_{gs} decreases. However, the voltage across the capacitor, V_{out}, never reaches V_{dd} because when V_{gs} reaches V_{Tn} the device will turn off. Therefore, V_{out} can only reach (V_{dd} - V_{Tn}). The n-channel transistor does not conduct high logic levels well. In the case depicted in Fig. 2-15(b), the capacitor is charged, and, when V_{cntrl} is high, it will discharge through the device up to when the voltage V_{out} will be zero. No degradation of the signal through the transistor occurs. Therefore, the n-channel device passes a low voltage level well.

The transmission gate features the two devices in parallel which are controlled by two signals having opposite polarities. When the transmission gate is "off," both the n-channel and the p-channel devices are "off." When it is "on," both of them are "on." In this case, good conduction for high logic levels — through the p-channel device — and good conduction for low logic levels — through the n-channel device — coexist. The output signal will always be an exact replica of the input signal. Note that this is also true in the n-channel pass-transistor if we make sure that only low logic levels will be passed through the device. If this is the case, no level restoring logic following the pass-transistor is necessary.

Although the transmission gate is extensively used in CMOS design, it also has some drawbacks that preclude its use in some situations. The transmission gate does not require V_{dd} and V_{ss} signals, and this leads to smaller area when it is used to implement logic functions. However, the transmission gate lacks the capability to provide the current necessary to drive an output load. The transmission gate is a "passive" device from the viewpoint of current capability. Therefore, when the output load is significant — like in tri-state pads and tri-state drivers for buses — the use of a transmission gate is not recommended. The "on" resistance of a transmission gate, although usually small, can cause significant delays when it is coupled with a large capacitive output load because of the increased RC constant.

When we need to drive large output loads, or we need to tri-state an output, we have to use *active transmission gates*, such as the one shown in Fig. 2-16.

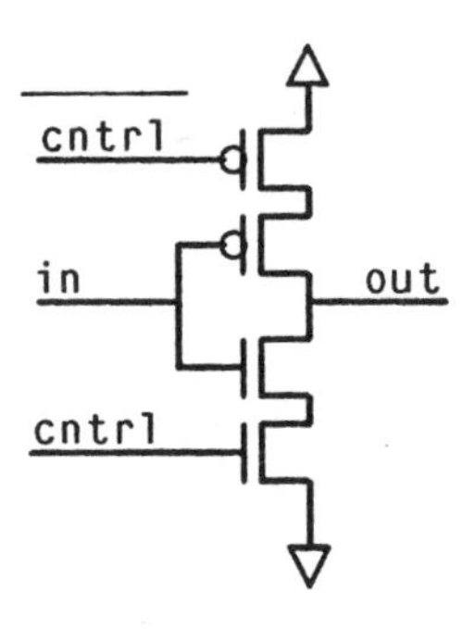

Figure 2-16: Active transmission gate.

The driving capability of this gate can be increased or decreased by changing the widths of the four transistors. Because the use of this gate is not completely straightforward, the gate will be discussed in detail in Chapter 5.

2.5. CMOS Inverter

We now analyze the behavior of the CMOS inverter in order to determine its DC voltage transfer characteristic. This section presents the analysis of the inverter according to Eqs. (2-4), (2-5), and (2-6). These equations are the simplest analytical formulation of the behavior of the MOS transistor. As we said, I_{ds} depends on V_{ds} in saturation, but this dependence is not shown in Eq. (2-5). Therefore, more complex equations are necessary to accurately model the MOS transistor and obtain a precise DC voltage transfer characteristic of the inverter . Section 2.6 will present one of these models. In this section we present the results derived from the simplest theory of the MOS transistor behavior as anticipated in Section 2.1.

When we study the behavior of a CMOS gate we have to introduce a new parameter, that is, the *inverter threshold voltage*, V_{Ti}. There are two different definitions for V_{Ti}:

1. V_{Ti} is the input voltage at which the output voltage *starts* to change;

2. V_{Ti} is the input voltage such that:

$$V_{Ti} = V_{out}. \tag{2-17}$$

We will adopt the second definition. When Eq. (2-17) holds and, moreover, $V_{Ti} = V_{dd}/2$, we say that the gate is *balanced*. We shall see in Section 2.6 that a balanced inverter optimizes the noise margin.

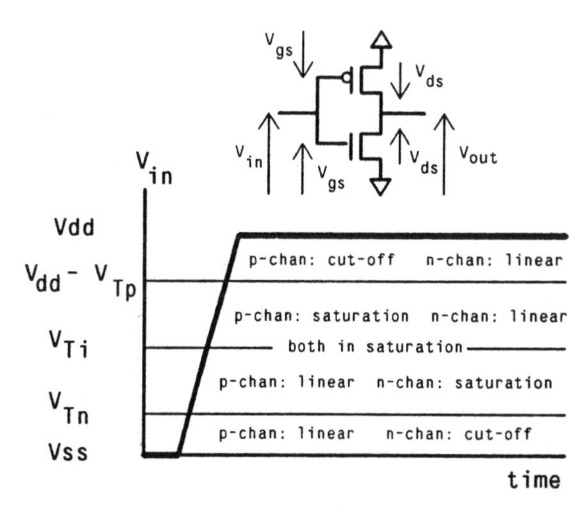

Figure 2-17: Input signal and different operation regions for the two transistors.

Let us suppose that the input voltage starts at V_{ss} and then increases up to V_{dd}. Fig. 2-17 depicts the five different phases the CMOS inverter will pass through, distinguished by the different regions of operation for the two transistors.

1. When the input voltage is equal to V_{ss}, the p-channel transistor is strongly conducting. The n-channel transistor is in its cut-off region, and its drain-to-source current is zero. The output voltage is equal to V_{dd}. We saw in Section 2.4 that high logic levels are passed through the p-channel with no signal degradation. Although the drain-to-source current of a MOS device in the cut-off region is not exactly zero, it is so small when compared with the current of the same device in conduction — a factor of about 10^{-7} — that we can assume it is zero. We can conclude that there is no power dissipation in this situation.

2. The input voltage is now above the threshold voltage of the n-channel transistor; the p-channel device is in the linear region while the n-channel transistor enters the saturation region. Current is flowing between the V_{dd} and the V_{ss} terminals. The currents generated by the p-channel and n-channel devices are (see Eq. (2-4) and (2-5)):

$$I_{dsp} = \beta_p \left[(V_{gsp} - V_{Tp}) V_{dsp} - \frac{1}{2} V_{dsp}^2 \right];$$

$$I_{dsn} = \frac{\beta_n}{2} (V_{gsn} - V_{Tn})^2.$$

Assuming that $V_{gsp} = (V_{in} - V_{dd})$, $V_{dsp} = (V_{out} - V_{dd})$, $V_{gsn} = V_{in}$ and $V_{dsn} = V_{out}$, the two above equations can be rewritten as:

$$I_{dsp} = \beta_p \left[(V_{in} - V_{dd} - V_{Tp})(V_{out} - V_{dd}) + \right.$$

$$\left. - \frac{1}{2} (V_{out} - V_{dd})^2 \right];$$

$$I_{dsn} = \frac{\beta_n}{2} (V_{in} - V_{Tn})^2.$$

We can now equate the two currents, that is, $I_{dsn} = I_{dsp}$, to compute the output voltage V_{out}. We have:

$$V_{out} = \left[(V_{in} - V_{Tp})^2 - 2(V_{in} - \frac{1}{2}V_{dd} - V_{Tp}) + \right.$$

$$\left. - (V_{in} - V_{Tn})^2 \frac{\beta_n}{\beta_p} \right]^{1/2} + (V_{in} - V_{Tp}).$$

3. The input voltage increases and reaches the inverter threshold voltage. At this point, both devices are in saturation. The value of V_{Ti} depends on the relative β of the two transistors and their threshold voltages.

   By using the Eq. (2-5) for both devices we obtain (after substitution):

$$I_{dsn} = \frac{\beta_n}{2}(V_{in} - V_{Tn})^2;$$

$$I_{dsp} = \frac{\beta_p}{2}(V_{in} - V_{dd} - V_{Tp})^2.$$

   By equating I_{dsp} with I_{dsn} we find that V_{in} is equal to $V_{dd}/2$, provided that $\beta_n = \beta_p$ and the two threshold voltages have the same absolute value.

4. With the input voltage still increasing, we reach a point which is complementary to point 2: the n-channel device is in the linear region and the p-channel device is in saturation. We have:

$$V_{out} = (V_{in} - V_{Tn}) - \left[(V_{in} - V_{Tn})^2 + \right.$$

$$\left. - (V_{in} - V_{dd} - V_{Tp})^2 \frac{\beta_p}{\beta_n} \right]^{1/2}.$$

5. When the input voltage is between ($V_{dd} - V_{Tp}$) and V_{dd}, the n-channel transistor is conducting while the p-channel is off.

The result obtained at point 3. is very important. The fact that V_{Ti} is equal to $V_{dd}/2$ means that the noise margin is largely independent of the supply voltage and is optimized, so long as the two betas and threshold voltages are the same. We shall see later that the noise margin for actual inverters is somewhat lower than $V_{dd}/2$.

The inverter voltage transfer function is plotted in Fig. 2-18 for different values of $\beta_r = \beta_n/\beta_p$. We notice that V_{Ti} can be controlled by the beta ratio β_r. When $\beta_r = 0.1$, for instance, the inverter will switch at about $V_{in} = 3.5V$. This ability to easily control the inverter threshold voltage can be of some interest, as we shall see in Section 6.4.

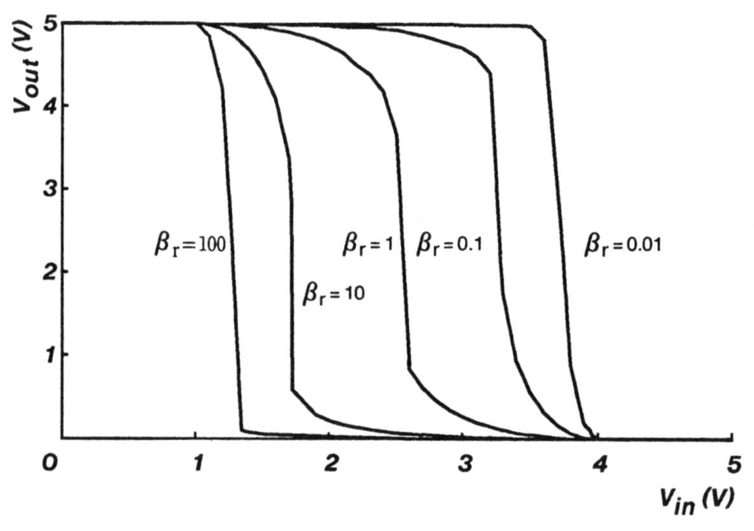

Figure 2-18: Inverter voltage transfer function for different values of β_r.

2.6. A More Accurate Model for the CMOS Inverter

Eqs.(2-4), (2-5), and (2-6) do not take into account several parameters that are important for accurately modeling the behavior of the MOS transistor. Moreover, they do not distinguish between the n-channel and the p-channel transistor. In addition to different mobilities, we know that the influence of the substrate cannot be neglected.

In this section we present a very accurate model of the MOS transistor that has been developed by Brews [4]. The Brews model has an accuracy which is comparable to the accuracy of the Pao-Sah model [26], which is the most rigorous model known for long channel MOS transistors. Moreover, while the Pao-Sah model consists of a two variable integral, the Brews charge-sheet model is a simple analytical formula. The basic assumption of the Brews's model considers the channel as a very thin layer of charge. A comparison between the two models is presented in [23]; this paper also explains why the Brews model so accurately matches the Pao-Sah model. Not only is Brews's charge-sheet model more economical than the Pao-Sah model from a computational point of view, but it is, like Pao-Sah's, valid throughout the entire range of operation, that is, saturation, sub-threshold, and linear regions.

Fig. 2-19 shows the MOS transistor. The drain current I_{ds} in the Brews model is:

$$I_{ds} = \frac{1}{\eta}\mu\frac{W}{L}\Big[C_o(1+\eta V_{gs})(\phi_s-\phi_{so}) - \frac{\eta}{2}C_o(\phi_s^2-\phi_{so}^2) + \\ -\frac{2}{3}\sqrt{2}\,qN_aL_b[(\eta\phi_s-1)^{3/2}-(\eta\phi_{so}-1)^{3/2}] + \\ +\sqrt{2}\,qN_aL_b[\sqrt{\eta\phi_s-1}-\sqrt{\eta\phi_{so}-1}]\Big], \tag{2-18}$$

where η is equal to q/kT and L_b is the extrinsic Debye length:

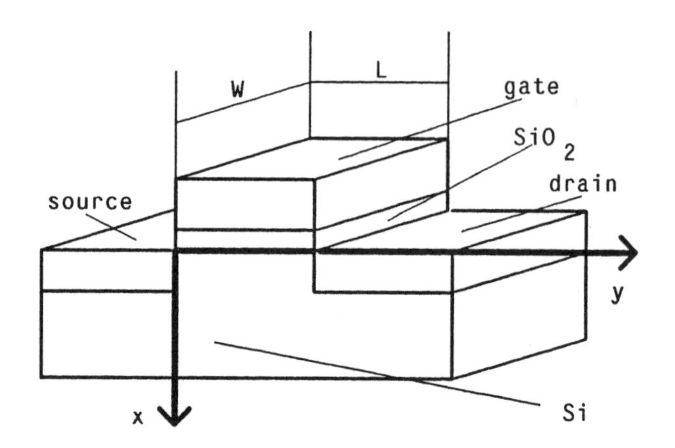

Figure 2-19: MOS transistor.

$$L_b = \frac{1}{q}\sqrt{\varepsilon_{Si}\frac{kT}{N_a}} . \tag{2-19}$$

Eqs. (2-18) and (2-19) are for an n-channel device. Similar equations can be written for a p-channel transistor. The formula ignores fixed-charge and work function difference effects. However, both of them are simple additive terms for V_{gs} in Eq. (2-18). ϕ_s is the potential along the oxide-silicon surface, and it should be more rigorously written as $\phi_s(y)$. Similarly, ϕ_{so} should be written as $\phi_s(0)$. Because we are interested in the current at the drain end of the channel, we have to substitute $\phi_s(L) = \phi_{sL}$ for ϕ_s in Eq. (2-18). In fact, ϕ_{so} and ϕ_{sL} are the potential at the source end of the channel and the potential at the drain end, respectively. We have:

$$I_{ds} = \frac{1}{\eta}\mu_n\frac{W}{L}\Big[C_o(1+\eta V_{gs})(\phi_{sL}-\phi_{so}) - \frac{\eta}{2}C_o(\phi_{sL}{}^2-\phi_{so}{}^2) +$$

$$-\frac{2}{3}\sqrt{2}\,qN_aL_b[(\eta\phi_{sL}-1)^{3/2} - (\eta\phi_{so}-1)^{3/2}] +$$

$$+\sqrt{2}\,qN_aL_b[\sqrt{\eta\phi_{sL}-1} - \sqrt{\eta\phi_{so}-1}\,]\Big] . \tag{2-20}$$

Eq. (2-20) can be solved together with the following set of equations:

$$C_o(V_{gs} - \phi_{so}) = \sqrt{2}\, qN_a L_b \sqrt{\eta\phi_{so} - 1 + (\frac{n_i}{N_a})^2 e^{\eta\phi_{so}}} ,$$

and

$$\eta\phi_{sL} = \eta\phi_{so} + \eta V_{ds} + \ln\left[\frac{N(L)}{N(0)}\right] ;$$

$$qN(0) = C_o(V_{gs} - \phi_{so}) - \sqrt{2}\, qN_a L_b \sqrt{\eta\phi_{so} - 1} ;$$

$$qN(L) = C_o(V_{gs} - \phi_{sL}) - \sqrt{2}\, qN_a L_b \sqrt{\eta\phi_{sL} - 1} .$$

Given V_{gs} and V_{ds}, it is possible to determine the two potentials N(0) and N(L) which are the carrier density per unit area at the source and at the drain end, respectively. Eq. (2-18) can be implemented on a pocket calculator, and, given all the physical parameters, it can be used to compute the I-V characteristics of a MOS transistor. Fig. 2-20 shows the voltage transfer function of a CMOS inverter obtained from Eq. (2-18) for five different pull-up|pull-down ratios. The same figure also shows the parameters of the fabrication process.

Note that the curve that comes closest to a balanced inverter is shown with a dotted pattern and corresponds to $W_n = 5\mu m$ and $W_p = 10\mu m$. An interesting characteristic of these curves is their sharp transition: this is a peculiar feature of CMOS technology. The sharpness of the transition is due to the fact that having both devices in saturation is a very unstable configuration, because the two transistors, acting as current generators, are in series. A very small increase — or decrease — in input voltage causes a drastic change in output voltage. Moreover, *the sharpness of the transition does not depend on the pull-up/pull-down ratio*, as is the case with an nMOS inverter with depletion load. Finally, the reader should be aware of the fact that *no*

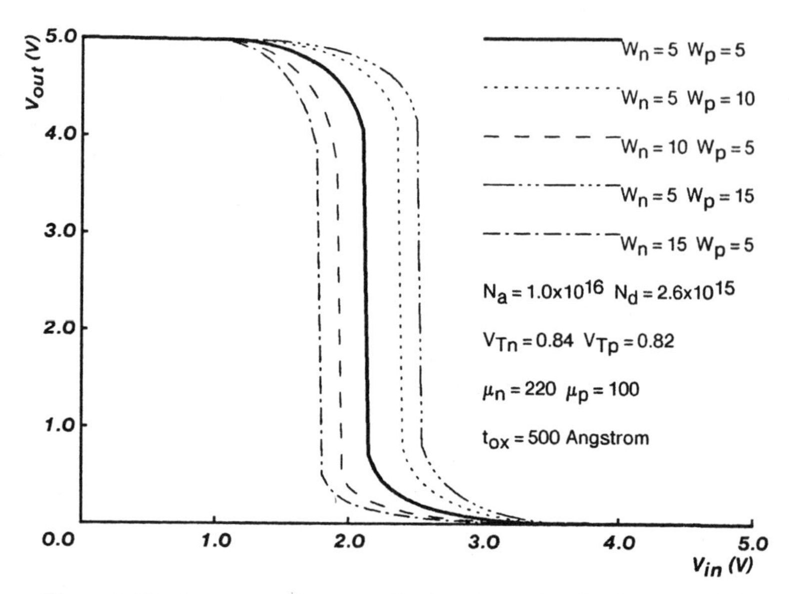

Figure 2-20: Inverter voltage transfer function using the Brews model.

temporal information is contained in Fig. 2-20. Again, what has been shown in Section 2.5 and in this section is only the DC analysis of the CMOS inverter. Delay has not been discussed yet and will be dealt with in Chapter 5.

Eq. (2-18) has also been used to determine the drain current of the CMOS inverter for the same pull-up|pull-down ratios of Fig. 2-20; the results are shown in Fig. 2-21. From a qualitative point of view, the shape of the curve was expected. When V_{in} is close to either V_{dd} or V_{ss}, one of the two transistors is in the cut-off region, and the current has to be very small. The maximum of the curve corresponds to the situation when both transistors are in saturation and also corresponds to the sharp transition in Fig. 2-20. Also note the difference between the curve for $W_n = 5\mu m$, $W_p = 10\mu m$ and the curve for $W_n = 10\mu m$, $W_p = 5\mu m$. The two peaks have different values, showing once again that the two transistors are not perfectly "complementary."

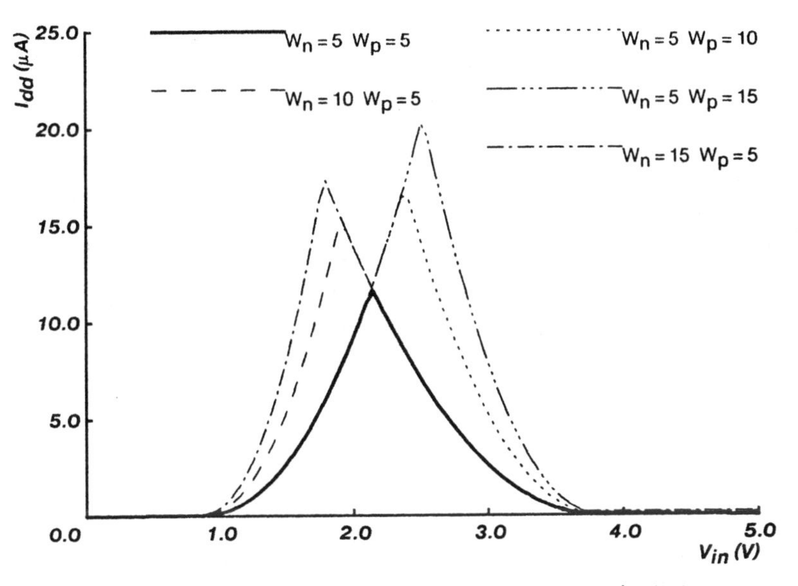

Figure 2-21: Inverter drain current for different pull-up|pull-down ratios using the Brews model.

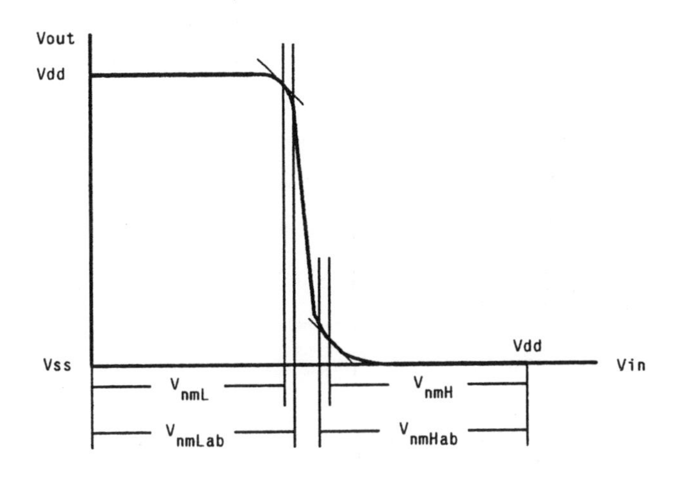

Figure 2-22: Noise margin definition (see text).

The noise margin characterizes the circuit's "noise insensitivity". In the case of the CMOS inverter, an input of either V_{ss} or 1.0V generates a V_{dd} output value. Therefore, the "low" noise margin should be at least 1V. The same considerations apply to the "high" noise margin. An input of either V_{dd} or 4V generates the same output — that is, V_{ss}. Therefore, the "high" noise margin should be at least 1V. (V_{dd} = 5V and V_{ss} = 0V have been assumed.)

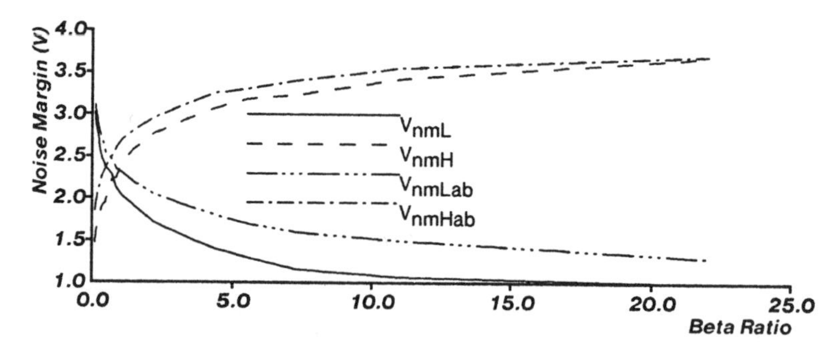

Figure 2-23: "Low" and "high" noise margins for a CMOS inverter with the process parameters shown in Fig. 2-20.

The "low" and "high" noise margins can be computed in several different ways, and we present two of them. The CMOS inverter voltage transfer function is shown in Fig. 2-22. V_{nmL} and V_{nmH} are the "low" and "high" noise margins, respectively. They are computed from the voltage transfer function when the tangent to it has a 45 degree angle. Therefore, any input voltage between V_{ss} and V_{nmL} generates a V_{dd} output, while any input voltage between V_{dd} and V_{nmH} generates a V_{ss} output. A different way of computing the "low" and "high" noise margins is shown in Fig. 2-22. V_{nmLab} and V_{nmHab} are the input voltages that correspond to the abrupt change in the output voltage — that is, when the gate switches. This way of computing the noise margin is more optimistic than the previous one. The two contributions to the noise margin as a function of the beta ratio are reported in Fig. 2-23. The *noise margin* is the average of the "low" and "high" contributions. We have:

$$V_{nm} = (V_{nmL} + V_{nmH})/2 ;$$

$$V_{nmab} = (V_{nmLab} + V_{nmHab})/2 .$$

$V_{nm} \approx 2.25V$ and $V_{nmab} \approx 2.45V$ for any beta ratio. As anticipated in Section 2.5, the actual noise margin is somewhat lower than $V_{dd}/2$.

2.7. CMOS Power Dissipation

If we neglect leakage effects, CMOS does not have static power dissipation in the sense in which it was used for nMOS technology. Basically, CMOS power dissipation is the sum of two terms, both depending on the utilization rate of the gate, that is, its switching frequency:

1. *Short-circuit power dissipation*: as we just saw in the previous section, there is a dynamic short-circuit between V_{dd} and V_{ss}, when both transistors are in saturation.

2. *Charge and discharge of parasitic capacitors*: all the parasitic capacitors, such as junction capacitances, interconnect capacitances, etc., cause power dissipation by releasing/storing energy.

During a single cycle, the short-circuit power dissipation is proportional to the integral of one of the curves shown in Fig. 2-21 or, with a simpler expression,

$$P_{s\text{-}c} = I_{mean} \times (V_{dd} - V_{ss}) . \quad (2\text{-}21)$$

If the gate has to charge an output load C_L, the total *energy* dissipated can be expressed as:

$$E_L = C_L \times V_L^2, \quad (2\text{-}22)$$

where V_L is the voltage across the load. The right-hand side of Eq. (2-22) should be multiplied by a frequency factor F to compute the power dissipation due to charge/discharge of parasitic capacitances, i.e.:

$$P_{dyn} = E_L F \,. \tag{2-23}$$

The switching frequency F depends on the gate input, and cannot be estimated "a priori" without knowing the exact function driving the logic. If we assume that the state of the circuit changes every 60ns, for instance, the worst-case would be a frequency of about 16MHz. However, the gate might not switch every cycle, and the actual power dissipation may be much less than the worst-case. Power dissipation considerations are very important when gates capable of delivering a large amount of current must be designed. Typical examples are pad drivers or bus drivers. Note that, because there is no static power dissipation, the process can be shrunk without requiring sophisticated cooling systems. On the other hand, because smaller feature size means higher switching frequency, power dissipation does increase due to the linear dependency on the switching frequency. However, the actual increase is not linear because the values of the parasitic capacitors decrease with the decrease in feature size. All this assumes that the power supply voltage is not scaled down with the feature size.

References

[1] Akers, L.A., M.M.E. Beguwala and F.Z. Custode.
A Model of a Narrow-Width MOSFET Including Tapered Oxide and Doping Enchroachment.
IEEE Trans. on Electron Devices ED-28:1490-1495, December, 1981.

[2] Akers, L.A. and J.J. Sanchez.
Threshold Voltage Models of Short, Narrow and Small Geometry MOSFET's: A Review.
Solid-State Electronics 25(7):621-641, July, 1982.

[3] Bandy, W.R. and D.P. Kokalis.
A Simple Approach for Accurately Modeling the Threshold Voltage of Short-Channel MOSTs.
Solid-State Electronics 20(8):675-680, August, 1977.

[4] Brews, J.R.
A Charge-Sheet Model of the MOSFET.
Solid-State Electronics 21(2):345-355, February, 1978.

[5] Chatterjee, P.
Device Design Issues for Deep Submicron VLSI.
In *Proc. of the 1985 International Symposium on VLSI Technology, Systems and Applications - Taiwan*, pages 221-226. May, 1985.

[6] Cottrell, P.E., R.R. Troutman and T.H. Ning.
Hot-Electron Emission in N-Channel IGFET's.
IEEE Trans. on Electron Devices ED-26(4):520-532, April, 1979.

[7] El-Mansy, Y.
MOS Device and Technology Constraints in VLSI.
IEEE Journal of Solid-State Circuits SC-17(2):197-203, April, 1982.

[8] Guebels, P.P. and F. van de Wiele.
A Small Geometry MOSFET Model for CAD Applications.
Solid-State Electronics 26(4):267-273, April, 1983.

[9] Hess, K.
Review of Experimental Aspects of Hot Electron Transport in MOS Structures.
Solid-State Electronics 21(1):123-132, January, 1978.

[10] Hoefflinger, B., H. Sibbert and G. Zimmer.
Model and Performance of Hot-Electron MOS Transistors for VLSI.
IEEE Trans. on Electron Devices ED-26(4):513-520, April, 1979.

[11] Ihantola, H.K.J. and J.L. Moll.
Design Theory of a Surface Field-Effect Transistor.
Solid-State Electronics 7:423-430, 1964.

[12] Jeppson, K.O. and J.L. Gates.
The Effects of Impurity Redistribution on the Subthreshold Leakage Current in CMOS n-Channel Transistors.
Solid-State Electronics 19:83-85, January, 1976.

[13] Ji, C.R. and C.T. Sah.
Two-Dimensional Numerical Analysis of the Narrow Gate Effect in MOSFET.
IEEE Trans. on Electron Devices ED-30:635-647, June, 1983.

[14] Ji, C.R. and C.T. Sah.
Analysis of the Narrow Gate Effect in Submicrometer MOSFET's.
IEEE Trans on Electron Devices ED-30(12):1672-1677, December, 1983.

[15] Kim, O.H. and C.K. Kim.
Threshold Voltage Shift due to Change of Impurity Type of Polysilicon in Heavily Doped Polysilicon Gate MOSFET.
In *Proc. of the International Symposium on VLSI Technology, Systems and Applications*, pages 170-173. March, 1983.

[16] Lai, P.T. and Y.C. Cheng.
An Analytical Model for the Narrow-width Effect in Ion-Implanted MOSFETs.
Solid-State Electronics 27(7):639-649, July, 1984.

[17] Lewis, E.T.
Design and Performance of 1.25μm CMOS for Digital Applications.
Proc. IEEE 73(3):419-432, March, 1985.

[18] Lim, H.K. and J.G. Fossum.
A Charge-Based Large-Signal Model for Thin-Film SOI MOSFET's.
IEEE Journal of Solid State Circuits SC-20(1):366-377, February, 1985.

[19] Masuhara, T. and J. Etoh.
Low-Level Currents in Ion-Implanted MOSFET.
IEEE Trans. on Electron Devices ED-21(12):799-807, December, 1974.

[20] Meyer, J.E.
MOS Models and Circuit Simulation.
RCA Review 32:42-63, March, 1971.

[21] Milne, A.D.
MOS Devices.
John Wiley & Sons, 1983.

[22] Nicollian, E.H. and J.R. Brews.
MOS (Metal Oxide Semiconductor) Physics and Technology.
John Wiley & Sons, New York, 1982.

[23] Nussbaum, A. *et al.*.
The Theory of the Long-Channel MOSFET.
Solid-State Electronics 27(1):97-106, 97-106, 1984.

[24] Ogura, S. *et al.*.
Design and Characteristics of the Lightly Doped Drain-Source (LDD) Insulated Gate Field-Effect Transistor.
IEEE Trans. on Electron Devices ED-27(8):1359-1367, August, 1980.

[25] Ong, DeW.G.
Modern MOS Technology.
McGraw-Hill Book Co., 1984.

[26] Pao, H.C. and C.T. Sah.
Effects of Diffusion Current on Characteristics of Metal-Oxide (Insulator) - Semiconductor Transistors.
Solid-State Electronics 9:927-937, 1966.

[27] Pfiester, J.R., J.D. Shott and J.D. Meindl.
Performance Limits of CMOS ULSI.
IEEE Journal of Solid-State Circuits SC-20(1):253-263, February, 1985.

[28] Pierret, R.F. and J.A. Shields.
Simplified Long-Channel MOSFET Theory.
Solid-State Electronics 26(2):143-147, February, 1983.

[29] Poorter, T. and J.H. Satter.
A D.C. Model for an MOS-Transistor in the Saturation Region.
Solid-State Electronics 23(7):765-772, July, 1980.

[30] Sah, C.T. and H.C. Pao.
The Effects of Fixed Bulk Charge on the Characteristics of Metal-Oxide-Semiconductor Transistors.
IEEE Trans. on Electron Devices ED-13(4):393-409, April, 1966.

[31] Sah, C.T.
The Narrow Gate Effect in Silicon VLSI MOSFET.
In *Proc. of the International Symposium on VLSI Technology, Systems and Applications*, pages 165-169. March, 1983.

[32] Sodini, C.G., T.W. Ekstedt and J.L. Moll.
Charge Accumulation and Mobility in Thin Dielectric MOS Transistors.
Solid-State Electronics 25(9):833-841, September, 1982.

[33] Swanson, R.M. and J.D. Meindl.
Ion-implanted Complementary MOS Transistors in Low Voltage Circuits.
IEEE Journal of Solid-state Circuits SC-7(2):146-153, April, 1972.

[34] Sze, S.M.
Physics of Semiconductor Devices.
John Wiley & Sons, New York, 1969.

[35] Sze, S.M. (ed.).
VLSI Technology.
McGraw-Hill Publishing Co., 1983.

[36] Takeda, E., G.A.C. Jones and H. Ahmed.
Constraints on the Application of 0.5-μm MOSFET's to ULSI Systems.
IEEE Journal of Solid-State Circuits SC-20(1):242-247, February, 1985.

[37] Troutman, R.R.
VLSI Limitations from Drain-Induced Barrier Lowering.
IEEE Trans. on Electron Devices ED-26(4):461-469, April, 1979.

[38] Ward, D.E. and R.W. Dutton.
A Charge-Oriented Model for MOS Transistor Capacitances.
IEEE Journal of Solid State Circuits SC-13(5):703-707, October, 1978.

[39] Werner, W.M.
The Work Function Difference of the MOS-System with Aluminium Field Plates and Polycrystalline Silicon Field Plates.
Solid-State Electronics 17:769-775, 1974.

[40] Wu, C.Y. *et al.*.
An Analytic and Accurate Model for the Threshold Voltage of Short Channel MOSFETs in VLSI.
Solid-State Electronics 27(7):651-658, July, 1984.

[41] Zahn, M.E.
Calculation of the Turn-on Behavior of MOST.
Solid-State Electronics 17(8):843-854, August, 1974.

Chapter 3

Fabrication Processes

CMOS processes are generally more complex and expensive than nMOS processes because extra steps and extra masks are required in the fabrication process. Although many different processes are available, all the CMOS processes fall into either one of the two following classes:

- *Bulk processes*: the substrate is doped silicon. Examples of bulk processes are p-well, n-well, and twin-tub. Newer bulk processes [18, 32] can also have bipolar transistors on the same wafer, which has positive effects on driving capability — which is not as high in MOS as in bipolar technology — and when digital/analog applications — such as sense amplifiers in memory design [17] — are considered.

- *Silicon-on-insulator (SOI) processes*: the substrate is an insulator, such as sapphire ("silicon-on-sapphire," SOS) or silicon dioxide (SiO_2).

All bulk processes have to solve the problem of having both p- and n-channel transistors on the same substrate. This is accomplished by forming a "secondary" substrate into a "primary" substrate. This procedure characterizes both p-well and n-well processes. The SOI processes do not have this problem, because the two transistor types are isolated by the substrate.

Plates referred to in this chapter are found in the PLATES section located between pages 172 and 173.

The "double-substrate" structure of bulk processes has serious side-effects. As we shall see in Section 3.4, this structure is the principal cause of latchup, an often destructive phenomenon. Moreover, the double substrate degrades the performance of the transistors on the secondary substrate.

This chapter will present the most common bulk processes and the SOS process. Although more expensive and delicate, SOS CMOS outperforms the bulk process both in terms of speed and circuit density. Finally, the chapter will introduce the "design rules" of a hypothetical CMOS process. These design rules will be used in the layout examples shown in the plates.

3.1. The p-well Fabrication Process

The earliest CMOS fabrication process was a p-well bulk process, because it is compatible with the pMOS fabrication process — the first MOS process used commercially. The two different transistor types are isolated by a "secondary substrate," the *well or tub*, which is formed into a "primary substrate," as shown in Fig. 3-1. In the p-well process the primary substrate is n-doped, and, therefore, capable of accepting p-channel devices, while the well is p-doped and accepts n-channel transistors. The p-channel transistors are only formed on the n-substrate, while the n-channel transistors must be in the p-doped well.

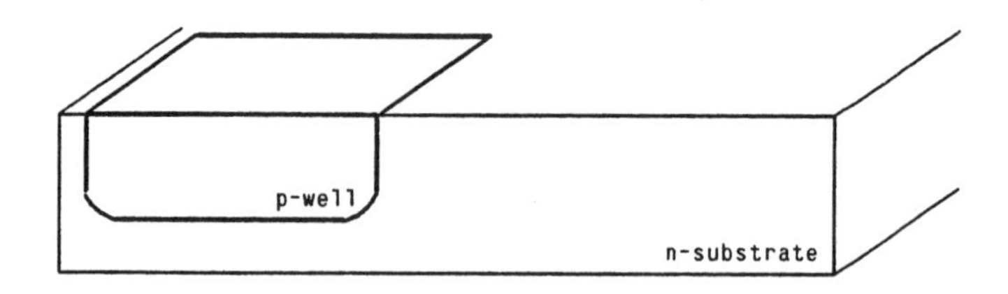

Figure 3-1: p-doped well (p-well) is formed into an n-doped substrate.

The well has a much higher doping concentration than that of the n-substrate to achieve better control on the threshold voltage. This produces negative side-effects — such as increased parasitic capacitances, particularly junction capacitance — in the n-channel transistors. Moreover, this difference in substrate doping makes the threshold voltage and the work function difference (i.e., the threshold voltage, again) of the two devices different. Special techniques — such as boron implantation — are used to make V_{Tn} and V_{Tp} as equal as possible in absolute value.

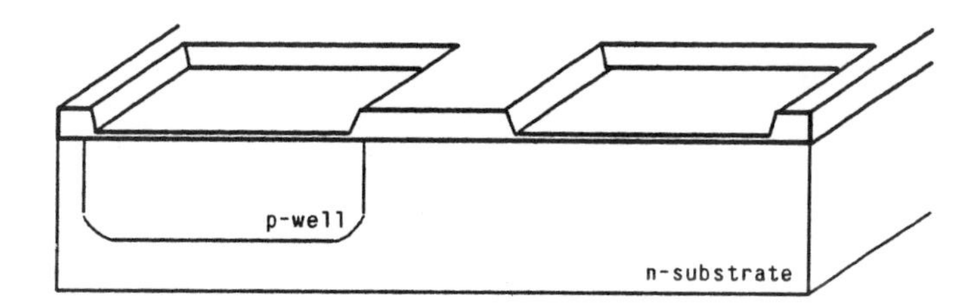

Figure 3-2: Thick and thin oxide are grown: thin oxide covers the active areas, while thick oxide covers the rest of the substrate.

A p-well process is now presented. The steps that will be shown are only the most representative. Some intermediate steps are missing. The process shows the fabrication of an inverter. After creating the well, oxide is grown on the substrate, as shown in Fig. 3-2. Field (thick) oxide will cover the entire substrate except for the *active areas,* that is, the areas where the n- and p-channel transistors will be formed. In the active areas only a thin oxide is present, so that the entire wafer will be covered by oxide, either thick or thin. This would not be true if the process featured *buried contacts,* which allow a contact between polysilicon and active area. If the process does feature buried contacts, a window into the thin oxide would be patterned where the polysilicon makes contact with the active area, as shown in Fig. 3-3.

The next step is the deposition and doping of polysilicon, shown with bold lines in Fig. 3-4. Polysilicon is deposited undoped and then doped either by diffusion or ion

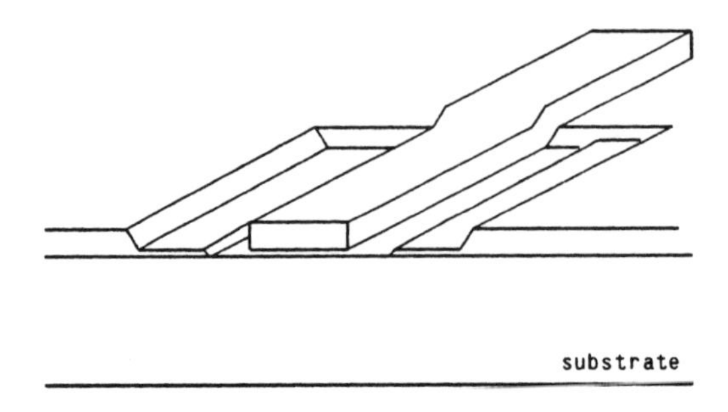

Figure 3-3: Buried contact: a window is patterned into the thin oxide to allow the polysilicon to make a contact with the substrate.

implantation[1]. An alternative approach is doping during deposition by using dopant gases. Diffusion results in polysilicon with the lowest resistivity [28].

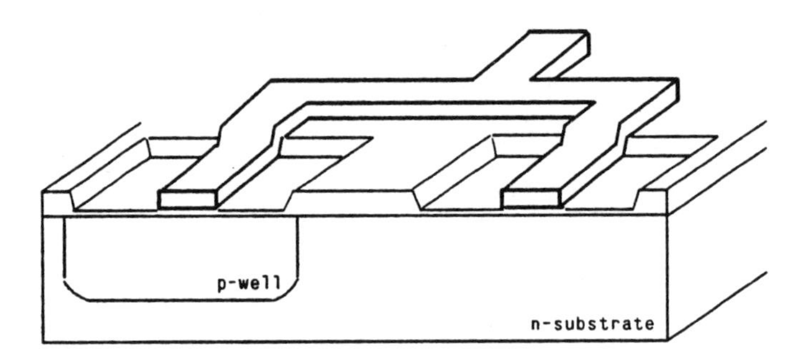

Figure 3-4: Polysilicon is deposited.

The source and drain areas are then implanted (vertical arrows in Fig. 3-5). The thickness of the polysilicon and field oxide is sufficient to protect the area of the substrate below them, while the thin oxide allows the implantation to penetrate the

---

[1] Both diffusion and ion implantation are used to introduce impurity atoms in silicon. While diffusion is a thermal process, ion implantation is a kinetic process. High temperatures are necessary to diffuse impurities in silicon, while implantation achieves doping by providing the ionized atoms with enough energy to penetrate the surface of the material. Ion implantation allows us to obtain more accurate doping concentration and distribution profiles. See [28] for details.

substrate. This procedure allows self-alignment between a gate and its drain and source. This in turn reduces the overlap between gate and source/drain thereby minimizing one of the parasitic capacitances discussed in the previous chapter (i.e., C_{ovl}).

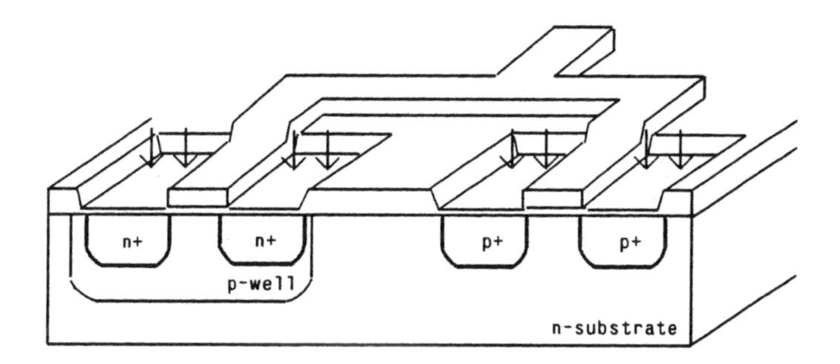

Figure 3-5: The source and drain areas are implanted.

The next step involves the deposition of a dielectric on the entire chip to isolate the polysilicon from the next layer — usually, the metal layer.

Before depositing metal, windows ("contacts") are etched into the dielectric to allow contact between metal and polysilicon or active area. Then, metal — usually aluminum — is deposited. Fig. 3-6 shows the aluminum in bold lines. The dielectric between poly and metal is not shown. Finally, the circuit is covered with a protective material, and windows are etched for bonding pads. This overcoat material is known as "overglass" (SiN).

To make the inverter, the source of the n-channel transistor is connected to V_{ss}, and the source of the p-channel transistor is connected to V_{dd}. Normally, the p-well is connected to V_{ss}, while the n-substrate is connected to V_{dd}. It is important to have the two substrates connected to their respective potentials to avoid performance degradation or, worse, latchup. The threshold voltage of an n-channel or p-channel transistor depends on the voltage between substrate and source terminal (V_{bias} in

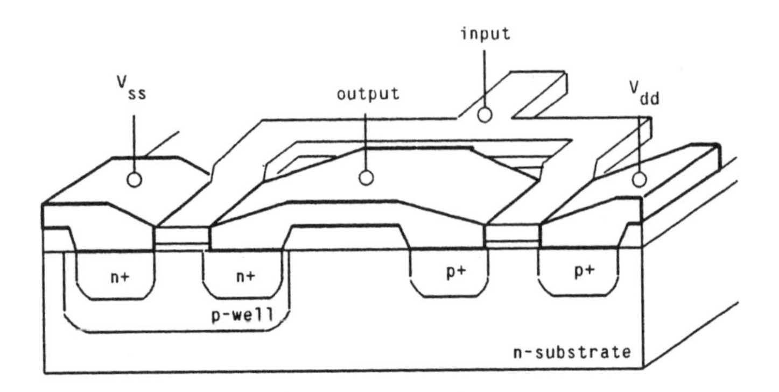

Figure 3-6: After windows are etched (contacts), metal is deposited (dielectric between metal and poly is not shown).

Eq. (2-1)), and a non-zero V_{bias} increases the threshold voltage and the turn-on time of the transistor.

Note that, unlike in nMOS technology, the output of the gate is only available in metal. If poly is required, a poly-metal contact has to be laid out.

3.2. The n-well Fabrication Process

The n-well process is complementary to the p-well process, that is, an n-type well is created into a p-doped substrate. All p-channel devices will be inside the well, and all n-channel transistors will be on the p-doped substrate. Fig. 3-7 shows a cross-section of the inverter in the n-well fabrication process. The well is normally tied to V_{dd} while the p-doped substrate is tied to V_{ss}. The fabrication steps are similar to the ones presented in the last section for a p-well process.

Is the n-well process significantly different from the p-well in terms of performance? The well, a heavily doped substrate, increases the parasitic capacitances. Therefore, for either a p- or n-type well, some loss in performance is to be expected for n-channel transistors or p-channel transistors, respectively. In terms

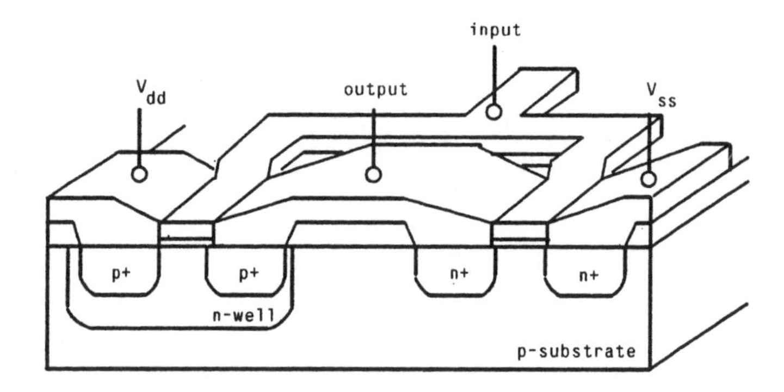

Figure 3-7: n-well process: CMOS inverter.

of overall speed, however, there seems to be no appreciable difference between the two processes.

Although a clear superiority of one process over the other cannot be shown, the n-well process does have an important feature: because its substrate is p-doped, the bulk n-well CMOS process is compatible with the bulk nMOS process. This means that on the p-doped substrate we can form n-channel enhancement and depletion transistors. This allows the designer to include nMOS circuitry, which is usually more compact than a CMOS implementation with the same functionality. Moreover, CMOS circuits — such as memories — feature a much larger number of n-channel transistors than p-channel transistors, and the n-well process affects the performance of p-channel devices. Note that the p-well process would be compatible with a bulk pMOS process, but this compatibility is far less attractive than the compatibility between n-well and nMOS because of the poor performance of the pMOS transistor (both enhancement and depletion).

Both p-well and n-well processes may feature two — or more — metal layers and/or polysilicon layers. This requires multiple dielectric depositions to isolate the different layers in multi-metal and/or multi-poly fabrication processes. Multi-metal layer fabrication processes are used extensively in gate-array technology, while memories need processes with multi-poly layers — usually polycide. Polycide, which is polysilicon with metal silicide on top, lowers the sheet resistance of the conductor by up to an order of magnitude.

3.3. LOCMOS Technology

"Local Oxidation of Silicon" (LOCOS) [3] is a technique which allows tighter spacing between regions of different doping than is the case for the conventional processes presented in the last sections. In this technique, which can be applied to both p-well and n-well fabrication processes, the silicon substrate is coated with a layer of silicon nitride (Si_3Ni_4), which is then used as a mask during the oxidation of silicon to form SiO_2.

The application of LOCOS technique to CMOS is called LOCMOS [4, 24, 35] and will be presented in detail now. Fig. 3-8 shows the various steps for a p-well process:

a. The silicon wafer is coated with a thin layer of silicon nitride, which is then removed where the thick oxide has to be formed (silicon nitride is shown with a bold line). The oxide sinks into the silicon, thus providing good isolation between the different regions.

b. The silicon nitride is etched from the areas where the well has to be formed.

c. Silicon nitride is etched from the other areas, and thin oxide is thermally grown. Polysilicon is then deposited and doped.

d. p-type regions are formed in the substrate.

e. n-type regions are formed in the well.

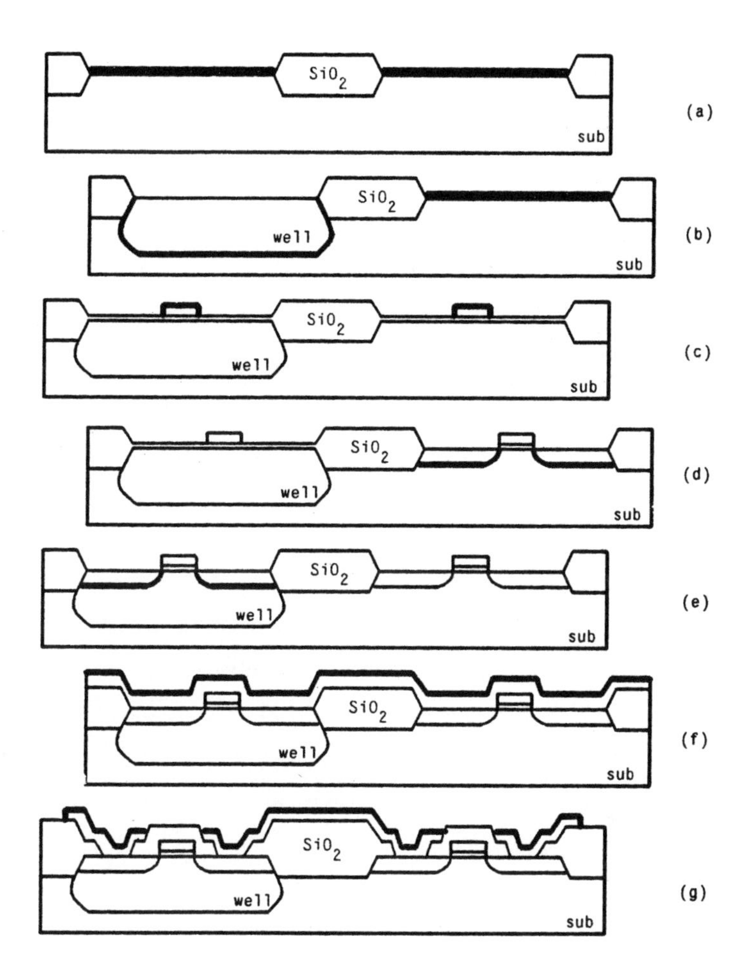

Figure 3-8: LOCMOS fabrication process. *After [4].*

f. Insulator (silicon dioxide) is deposited.

g. Openings for metal contacts are etched, and metal is deposited — through vaporization — and patterned.

Fig. 3-9 shows a comparison of a conventional bulk CMOS process and a

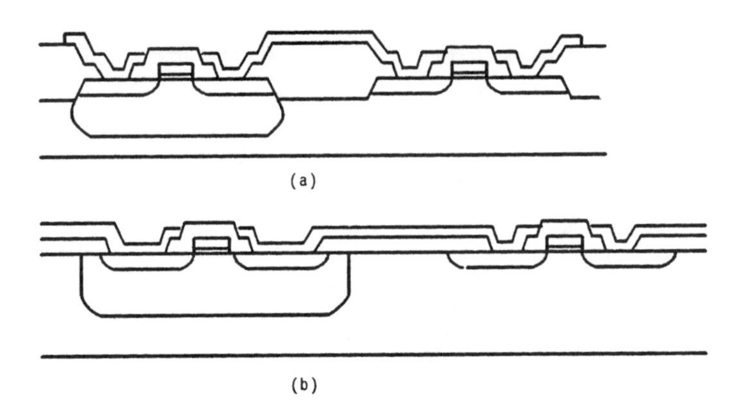

Figure 3-9: LOCMOS process (a) vs. conventional process (b): inverter cross-section.

LOCMOS process. The figure shows a cross-section of an inverter. The LOCMOS technique can halve the area of a circuit, as compared to conventional techniques [4]. Extra process steps — such as field implantation in the active areas of the substrate, which increases the doping concentration — help reduce punch-through[1] effects between the well and outside regions of similar type (e.g., between p-well and p+ regions). More than one step of a fabrication process can be of LOCOS type. An example of this is found in [35], where a fabrication process for dynamic memories is presented.

---

[1]Punch-through in a MOS transistor takes place when the depletion regions of the drain and source merge together and injection of majority carriers from the source to the drain takes place. Punch-through is more likely to happen in short-channel devices and/or when high resistivity substrates are used. As a general term, punch-through is a mechanism where two geometrically separated regions become electrically connected.

3.4. Latchup

The formation of *lateral and vertical* bipolar, parasitic transistors is intrinsic to the bulk processes. These transistors contribute to the effect known as "latchup," which is undoubtedly the most dangerous event that can take place in a CMOS circuit: latchup can easily destroy a chip. Protection against latchup is carried out in various ways, including steps during the fabrication process and by using appropriate layout techniques. In either method it is important to understand the phenomenon and identify the precautions that have to be taken by the designer to avoid its occurrence. This last topic will be dealt with in Section A.2, where layout techniques for latchup avoidance are presented.

Fig. 3-10 shows a cross-section of an inverter. p-well technology is shown, but the same considerations hold for n-well technology. The same figure also shows the four bipolar transistors which are intrinsic to the fabrication process. The four transistors are two vertical n-p-n devices (Q1 and Q4) and two lateral p-n-p devices (Q2 and Q3); each n-p-n/p-n-p pair (e.g., Q1 and Q2) is configured as a Semiconductor-Controlled-Rectifier (SCR or thyristor). Under certain conditions, the SCR exhibits a positive feedback loop inside its structure. The net effect of latchup is a static short-circuit between V_{dd} and V_{ss} that cannot be broken unless either V_{dd} or V_{ss} terminals are disconnected *very quickly*. If this does not happen, the current between V_{dd} and V_{ss} will increase to the point where metal lines vaporize or junction punch-through takes place, thus overheating the devices and destroying the gate. Eventually, the whole chip stops working.

For the sake of simplicity in the discussion, the simplified model shown in Fig. 3-10 is used. However, important parasitic parameters not shown in the figure contribute to latchup. Some of these are listed below:

- Substrate-well capacitance [30].

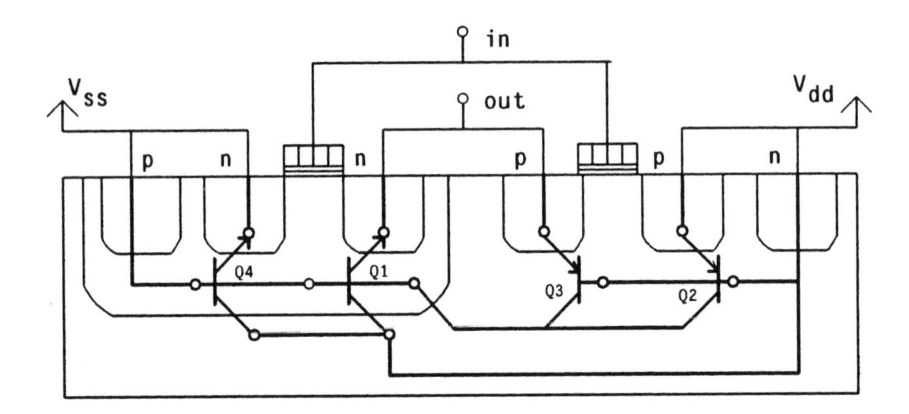

Figure 3-10: Inverter cross-section with the four bipolar parasitic transistors. Q1 and Q4 are vertical; Q2 and Q3 are lateral.

- Contact resistances between the metal line and the various junctions.
- Emitter (junction) resistances.
- Base resistances which depend on the characteristics (doping concentration, depth, etc.) of the well (Q1, Q4) and of the n-substrate (Q2, Q3).
- Resistances between the collectors of the transistor pairs (Q1 and Q4, Q2 and Q3).

Most of the resistances listed above depend on fabrication process characteristics — such as substrate and diffused region doping concentration — and also depend on the *tri-dimensional geometry* of the various regions. This makes latchup modeling extremely difficult. Finally, both resistances and transistors are characterized by non-linear behavior, and, therefore, a model such as the one shown in Fig. 3-10, although useful to understand latchup, does not accurately model the actual physical situation. This occurs because all parasitic elements are not lumped together but, instead, are distributed throughout the 3-D space. The literature shows that a 2-D analysis suffices to model latchup accurately [9]. This analysis is usually performed using finite element techniques.

Let us suppose that the inverter output voltage goes below V_{ss} - 0.7V. As shown in Fig. 3-11(a), the emitter-base junction of Q1 becomes forward-biased, and majority carriers are injected into the base. The electrons reach the collector (n-substrate) and migrate toward the V_{dd} terminal. If the current is high enough and is combined with a significant resistance between the V_{dd} contact and the p+ source (which corresponds to the emitter of Q2), a voltage drop occurs that lowers the substrate potential below that of the p+ source. If this voltage drop is 0.7V or more, the emitter-base junction of Q2 becomes forward-biased and holes are injected through the collector into the well, as shown in Fig. 3-11(b). If there is sufficient resistance between the V_{ss} contact and the n+ source (which corresponds to the emitter of Q4), yet another voltage drop takes place. The emitter-base junction of Q4 becomes forward-biased and injects majority carriers into the substrate. The carriers add to those injected by the emitter-base junction of Q1, more current is injected into the n-substrate, the voltage drop at the p+ source becomes larger, and this process continues indefinitely . A closed-loop is established with the current reinforcing itself every cycle. Disconnection of the input to the gate cannot stop the build-up of current, because the current delivered to the input of the inverter plays no role in the closed-loop, at this point. Only disconnecting V_{dd} or V_{ss} can stop the process.

We saw that transistor Q1 starts the latchup process when the output goes below V_{ss} by an amount which causes the emitter-base junction of the n-p-n transistor to become forward biased. The same thing can happen when the output goes above V_{dd} by the same amount; in this case transistor Q3 will fire latchup.

By examining how latchup occurs, we can identify four factors that cause this phenomenon [26]:

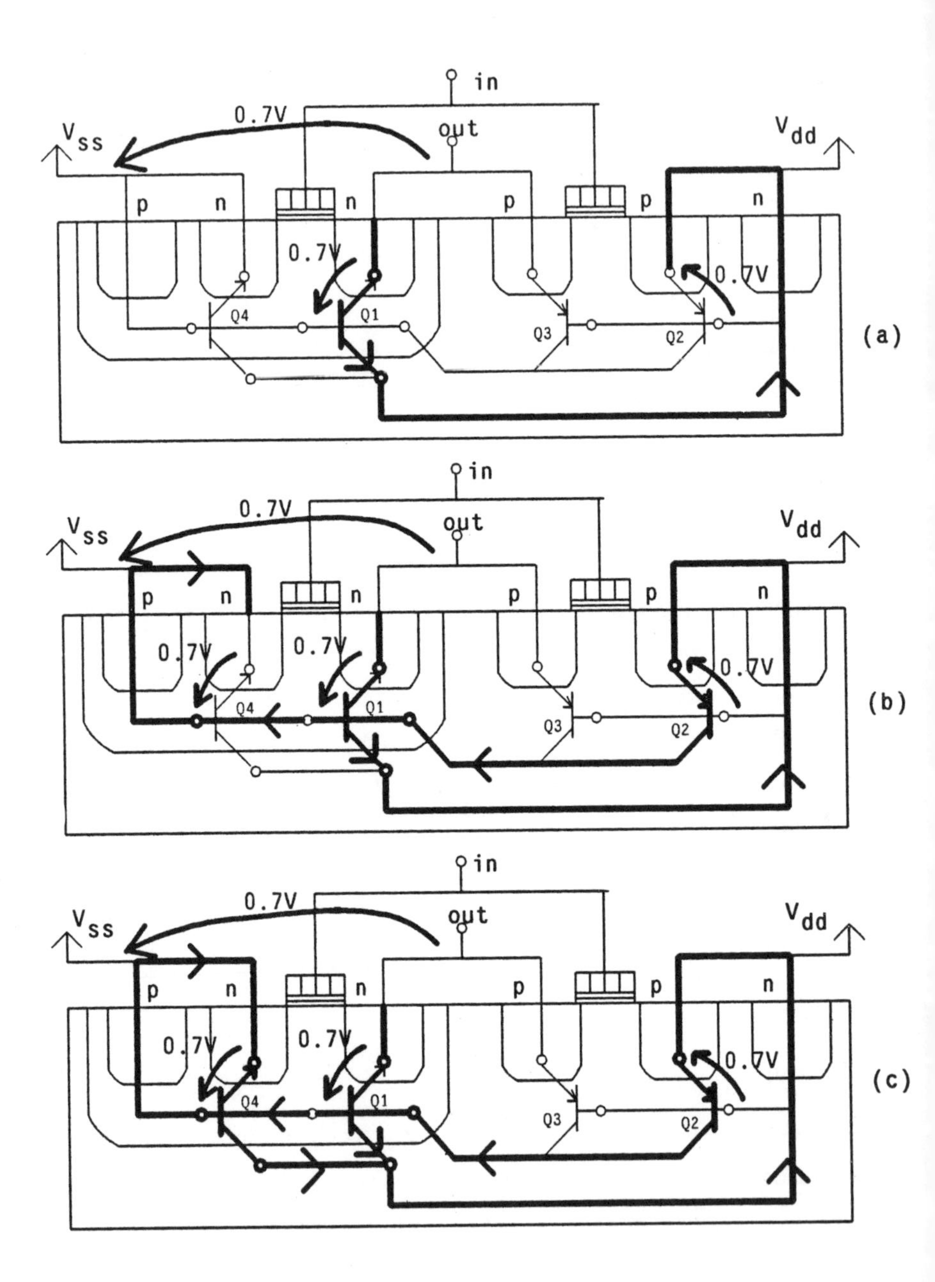

Figure 3-11: Majority carrier flow and junction voltages during latchup.

- The sum of the common base current gains (α) of the vertical n-p-n and lateral p-n-p transistor pair is such that:

$$\alpha_{Q1} + \alpha_{Q2} \geq 1,$$

or, equivalently, the product of the common emitter current gains (h_{FE}) of the same transistors is:

$$h_{FE\text{-}Q1} h_{FE\text{-}Q2} \geq 1.$$ [1]

- The emitter-base junctions are forward-biased, and the transistors are working in their active region. This happens when the output goes below V_{ss} by more than 0.7V or when the output goes above V_{dd} by more than 0.7V. Note that under normal conditions these junctions are always reverse-biased.

- The voltage is greater than the holding voltage.

- The current supplied to the terminals is greater than the holding current. Holding current and voltage are characteristic parameters of an SCR. When the current and the voltage exceed the holding current and voltage, a very small increase in the voltage applied to the SCR will cause a massive increase in the current flowing through [27].

The output of a CMOS gate can go below V_{ss} or above V_{dd} for various reasons:

---

[1] β is used widely to identify the common emitter current gain. h_{FE} has been preferred here to avoid confusion with the *transistor gain factor* β:

$$\beta = \mu\varepsilon_0 \frac{\varepsilon_{ox}}{t_{ox}} \frac{W}{L},$$

which is used in this book.

overshoot (particularly in bootstrapped gates, see Section 4.4), electrostatic discharges, and noise spikes. Also, latchup can occur during power up, when a signal is applied to a gate without power supply connected. A typical example is in battery-powered circuits (e.g., back-up memories), when the main voltage supply is turned off, and the batteries take over with some delay. Alternatively, a voltage drop in a power line inside the circuit can fire latchup. Bus reflections can cause signal overshoot, and fire latchup as well. Noise problems especially affect the I/O pad section, and special care must be taken in the layout and design of I/O pads (see Chapter 7). The reader is referred to references [9, 21, 30, 34, 23] at the end of the chapter for further information on physical aspects of latchup.

Several solutions have been proposed to reduce or eliminate latchup. They can be divided into three different approaches:

1. Reduction of the gain of the parasitic transistors to ensure that the h_{FE} product (α sum) is always less than one [10, 2]. This can be achieved by increasing the depth of the well. Unfortunately, this also increases the well area and, therefore, decreases circuit density. An alternative solution is to introduce doping that decreases the life-time of the carriers. This approach negatively influences the performance of the MOS transistor. An alternative approach to reduce the minority carrier life-time is through neutron irradiation of bulk substrate of the wafer [1, 25]. Decreasing the resistance of the well by using a buried epitaxial layer [14, 8] has proved to be an effective solution.

2. Elimination of the SCR structure either through modifications of the bulk process [26], or through the use of SOI.

3. Layout methodologies [22]. They do not eliminate latchup but are effective in most circuits.

Among layout methodologies, one solution consists of increasing the base resistance of the bipolar transistors by separating the p-channel transistors from the well — in the p-well process — as much as possible. This fsdecrease the closed-loop gain of the SCR. To be really effective, this distance should be greater than 75μm [19], which is unacceptable for dense circuitry.

A much more effective approach uses *guard rings* surrounding the n-channel and the p-channel devices [22]. Fig. 3-12 shows the cross-section of an inverter with guard rings (bold) surrounding the p-channel and the n-channel transistors.

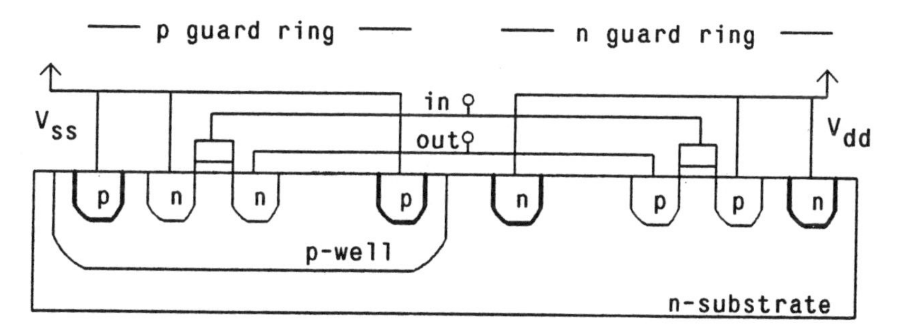

Figure 3-12: Guard rings (bold) are an effective approach to reduce the possibility of latchup.

Guard rings have been used extensively in bipolar technology [28, 7] together with Schottky clamp diodes to keep the transistors far from their saturation region. The basic effect of guard rings applied to bulk CMOS technology is to collect the carriers injected during the latchup process. During latchup onset the n-p-n transistor will inject majority carriers through its collector into the n-substrate. The guard ring surrounding the p-channel transistor will provide a "discharge path" for these electrons. Therefore, the voltage drop caused by the carriers, in conjunction with the resistance between the V_{dd} contact and the p-n-p emitter, will decrease. The p-n-p emitter-base junction will be less likely to become forward-biased. Even if the junction does become forward-biased, the minority carriers injected into the well through the p-n-p collector will be "discharged" through the well guard ring. Conversely, the guard ring does not allow a build-up of the voltage drop that might make the emitter-base junction of Q4 forward-biased, which closes the positive feedback loop. Note that, to be effective, guard rings should be placed as close as possible to the edge of the well — or just outside it. We will return to this topic in Section A.2.

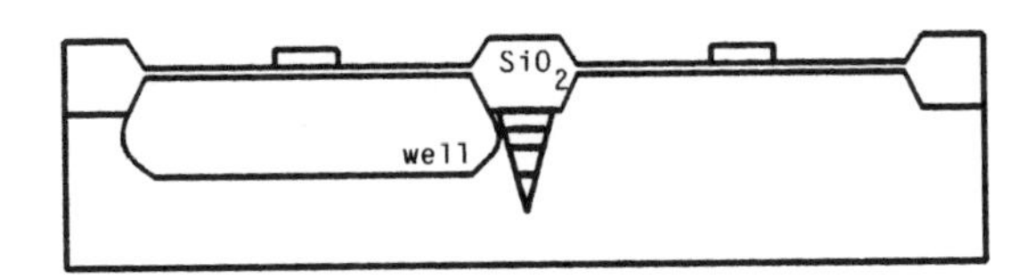

Figure 3-13: Cross-section of CMOS inverter with trench isolation technique.

Micron and sub-micron fabrication processes make the problem of isolating the two complementary devices more critical. Usually, the separation between the n+ and p+ regions has been about 2.5 times the feature size or line width [5]. A 2μm processes will have a 5μm separation between a p+ region and an n+ region — for instance, the drains of the p-channel and n-channel devices in the inverter. Scaling this dimension linearly with the scaling of the line width makes latchup more likely. Because the current which triggers latchup is directly proportional to the depth of the well, and because the well depth is also scaled down, the likelihood of latchup increases. However, by using proper fabrication techniques — such as epitaxial-layers [36] — latchup does not represent a major obstacle, at least down to 1μm feature size. Sub-micron processes will require different isolation techniques, and many of them are currently being investigated. It is clear that a deeper well would be desirable, because latchup triggering current would be increased. However, a deeper well occupies more area, thus increasing the separation between n+ and p+ regions and decreasing circuit density. The problem can be successfully tackled by using "trench" isolation [36] between the n-channel and p-channel devices. Fig. 3-13 shows the inverter cross-section with trench isolation. The process consists of etching the silicon substrate and then refilling the trench with SiO_2 and polysilicon. Excellent isolation between the complementary devices can be achieved, still maintaining a deep well.

An interesting use of the lateral and vertical transistors created by the bulk CMOS process is presented in [31]; their use allows us to have both CMOS and bipolar transistors on the same chip without modifying the fabrication process in any way. Although these transistors have many limitations, most notably poor frequency response, they can be successfully used in low-noise, low-frequency applications — such as DC operational amplifiers. Finally, new dynamic RAM cells have been proposed which combine a MOS device together with a "parasitic" bipolar transistor, the base-collector of which is used as the storage junction [35].

3.5. The Twin-tub Fabrication Process

The twin-tub process [20] is an effective solution to the latchup problem. Moreover, it does not decrease circuit density. In the twin-tub process the two transistor types are formed over two different wells. The wells are created in a lightly doped epitaxy over a heavily doped n-substrate. Fig. 3-14 shows the cross-section of a CMOS inverter.

The major advantage of the twin-tub process is that the concentrations of the two wells can be tuned separately: overcompensation — the high doping necessary to characterize the well — is no longer necessary, and more precise tuning of the process parameters is possible. Twin-tub processes do not have heavily doped wells below the two transistors, and parasitic capacitances are smaller. Moreover, twin-tub processes are far less prone to latchup, although they are not completely immune as are the SOI processes [23].

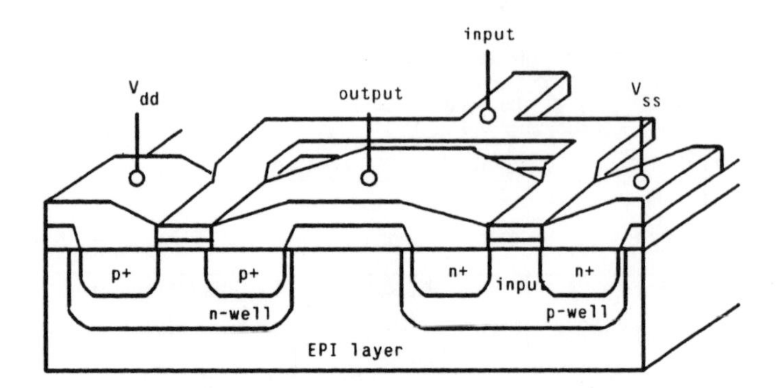

Figure 3-14: Twin-tub process: CMOS inverter.

3.6. The SOS Fabrication Process

The cross-section of an SOS inverter is shown in Fig. 3-15. Silicon is deposited over the sapphire substrate and then etched to form the transistors. The p- and n-type regions are generated by diffusion. Because of the insulating substrate, the process is intrinsically latchup-free. Since no wells are necessary and no guard rings are required, the circuit density of SOS is much higher than that of a bulk process. Moreover, since there is no well, there are none of the parasitic capacitances associated with a well. SOS parasitic capacitances are small compared to those in bulk processes. Wire capacitances (see Chapter 5) are also significantly smaller, and drain and source capacitances are close to zero.

Small parasitic capacitances and high circuit density (shorter wires) lead to higher speed. In 1981, for instance, it was possible to fabricate 4K-bit static RAM's in SOS technology with access time of 18ns. [11] — an outstanding performance for that time, especially when the speed/power product is considered. Three years later, subnanosecond gate delays could be achieved in gate-array technology using SOS [29], thus making SOS a direct competitor of Emitter-Coupled Logic (ECL), with the advantages of low power dissipation and scalability. Gate delays as low as a

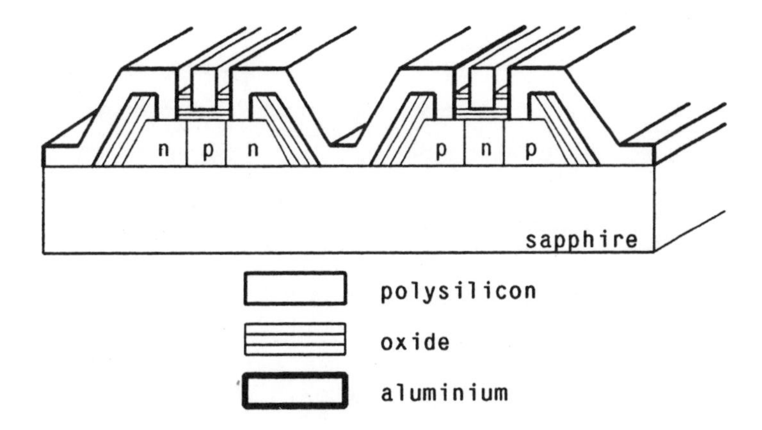

Figure 3-15: Inverter cross-section in an SOS process.

few tens of picoseconds have been achieved recently [15, 6].

SOS is not the only SOI process currently available, although it is the oldest and the only one for which (few) commercial chips have been mass-produced. Extensive research is under development in this field. One promising alternative to SOS is Silicon-on-SiO_2 [12, 13], where the transistors are formed over silicon dioxide — or amorphous polysilicon — that acts as insulator. However, it is too early to say whether this technology will be successful from a marketing point of view. The major drawbacks of SOS are its fabrication process, which is more complex than bulk's, and fabrication costs. Early on, the process was also more prone to cosmetic defects, but lately the situation has improved in this respect. Sapphire wafers are more expensive than silicon wafers by about a factor of four.

3.7. Bulk vs. SOI

Bulk CMOS processes have dominated the market in the past years, while SOS has not been as successful. There are many reasons, not only technical and/or commercial, behind the difficulties that SOS has encountered in capturing a significant part of the CMOS market. Although SOS has been shown to be superior in many aspects, some problems still remain. This section summarizes the major advantages and drawbacks of bulk vs. SOS. For a more complete analysis of this topic the reader is referred to [33].

Gate delay is undoubtedly smaller in SOS than in bulk. Moreover, SOS's higher circuit density provides even better performance at the chip level. However, for sub-micron technologies this advantage seems less notable. In fact, bulk and SOS feature about the same speed, because limiting factors other than the transistor parasitic capacitances start to play a dominant role, most notably interelectrode capacitances — such as the coupling capacitance between parallel conductors, which is not significantly influenced by the fabrication technology of the substrate.

Another drawback that SOS has always had, when compared to bulk CMOS, is leakage currents. The sapphire-silicon interface is far from ideal, and source-to-drain leakage currents are larger than in a bulk process with the same minimum line width. This makes the use of dynamic logic in SOS extremely difficult (see Chapter 4), since nodes which store charge are more likely to discharge through leakage paths than in bulk CMOS. This is not strictly a design consideration, but does have a significant influence on marketing issues. In fact, it means that dynamic memories cannot be produced in SOS, and it is well-known how important the DRAM market is. Finally, the SOS feature of being radiation hardened is of interest only in special fields — such as in military and space applications, and nuclear control systems.

3.8. Design Rules

Table 3-1: Design rules for a hypothetical p-well, CMOS process (1).

First-Metal Rules	
metal width	2λ
metal spacing	2λ
Second-Metal Rules	
metal width	4λ
metal spacing	4λ
Well Rules	
well width	4λ
well spacing	5λ
well overlap of active area	2λ
Active Area	
active area width	2λ
active area to active area spacing	3λ
active area in n-substrate to well edge	4λ

We now present the design rules of the "generic" process that will be used in the layout examples. They are shown in Tables 3-1 and 3-2. This process is not necessarily a real one; in fact, actual processes usually require a much more detailed set of design rules. We show a p-well process with one polysilicon and two metal layers. All the dimensions are in λ units [16].

Plate I shows the patterns that are used for each layer. Only the P-type mask is

Table 3-2: Design rules for a hypothetical p-well, CMOS process (2).

Rule	Value
Poly Rules	
poly width	2λ
poly spacing	2λ
poly gate extension over active area	2λ
poly to active area spacing	2λ
poly gate to active area spacing	2λ
N+/P+ mask Rules	
mask overlap of active area	1λ
mask overlap of poly gate	1λ
First Contact Rules: poly - first-metal or active area - first-metal	
minimum contact size	2λx2λ
maximum contact size	2λx4λ
poly overlap of contact	1λ
first-metal overlap of contact	1λ
contact to active area spacing	1λ
first contact to first contact spacing	2λ
Second Contact Rules: first-metal - second-metal	
minimum contact size	2λx2λ
maximum contact size	2λx2λ
first-metal overlap of contact	2λ
second-metal overlap of contact	2λ
second contact to second contact spacing	2λ
first contact to second contact spacing	2λ

explicitly shown. Where there is no P-type mask, an N-type mask has to be assumed. Devices "covered" by the P-type mask are p-channel transistors. The contact, labeled "Cut" in Plate I, is a metal to polysilicon contact. The second contact is a second-metal to first-metal contact. There is no second-metal to polysilicon contact.

Each new signal layer, for example a third-metal layer, would require at least five new design rules: minimum width, minimum spacing, minimum size of third-metal to second-metal contact (third contact), third-metal overlap of third contact and second-metal overlap of third contact. More sophisticated processes may also feature a second-metal to polysilicon contact.

Plates II through V show the layout of basic gates, including an inverter, a two-input NAND, a two-input NOR, and a transmission gate. Note that the p-channel device in the inverter shown in Plate II is wider than the n-channel device. How much wider the p-channel transistor has to be involves an important delay minimization problem. The same argument applies to the widths of the devices shown in Plate III, IV and V. The topic will be dealt with in Chapters 5 and 7.

Finally, note that circuit protections, such as guard rings or V_{dd}/V_{ss} substrate contacts, are not shown. In reality, the well should be connected to V_{ss}, while the n-type substrate should be connected to V_{dd}. In these layouts the n-type substrate is defined as the area which is not p-well. These layouts should be considered as a simple reference, because many variations are possible depending on the aspect ratio of the cell and the location of the input/output terminals and power/ground lines. An example of this is presented in Plate VI, where the layout of an inverter with aspect ratio and terminal locations different from those of Plate II is shown.

References

[1] Adams, J.R. and R.J. Sokel.
Neutron Irradiation for Prevention of Latch-up in MOS Integrated Circuits.
IEEE Trans. on Nuclear Science NS-26(6):5069-5073, 1979.

[2] Anagnostopoulos, C.N. *et al.*.
Latch-Up and Image Crosstalk Suppression by Internal Gettering.
IEEE Journal of Solid State Circuits SC-19(1):91-97, February, 1984.

[3] Appels, J.A. and M.M. Paffen.
Local Oxidation of Silicon.
Philips Res. Reports 26:157-165, 1971.

[4] Brandt, B.B.M., W. Steinmaier and A.J. Strachan.
LOCMOS, a New Technology for Complementary MOS Circuits.
Philips Technical Review 34(1):19-23, 1974.

[5] Chatterjee, P.
Device Design Issues for Deep Submicron VLSI.
In *Proc. of the 1985 International Symposium on VLSI Technology, Systems and Applications - Taiwan*, pages 221-226. May, 1985.

[6] Chen, M.-L., B.C. Leung and B. Lalevic.
A High-performance CMOS/SOS Device with a Gradually Doped Source-Drain Extension Structure.
IEEE Electrons Device Letters EDL-4(10):372-374, October, 1983.

[7] Elmasry, M.I.
Digital Bipolar Integrated Circuits.
John Wiley & Sons, 1983.

[8] Estreich, D.B., A. Ochoa and R.W. Dutton.
An Analysis of Latch-up Prevention in CMOS IC's Using an Epitaxial Buried Layer Process.
In *IEDM Tech. Dig.*. December, 1978.

[9] Genda, J.H.
An Improved Model for Latch-up in CMOS Structures.
In *Proc. of the 1983 International Symposium on VLSI, Technology, Systems and Applications - Taiwan*, pages 99-103. April, 1983.

[10] Hashimoto, K. *et al.*.
Counter-doped Well Structure for Scaled CMOS.
In *IEDM Tech. Dig.*. December, 1982.

[11] Isobe, M. *et al.*.
An 18 ns CMOS/SOS 4K Static RAM.
IEEE Journal of Solid State Circuits SC-16(5):460-465, October, 1981.

[12] Lam, H.W.
Silicon on Insulating Substrates - Recent Advances.
In *IEDM Tech. Dig.*, pages 348-351. 1983.

[13] Lim, H.K. and J.G. Fossum.
A Charge-Based Large-Signal Model for Thin-Film SOI MOSFET's.
IEEE Journal of Solid State Circuits SC-20(1):366-377, February, 1985.

[14] Manoliu *et al.*.
High-Density and Reduced Latchup Susceptibility CMOS Technology for VLSI.
IEEE Electron Device Letters EDL-4(7):240-245, July, 1983.

[15] Mayer, D.C. *et al.*.
A Short-Channel CMOS/SOS Technology in Recrystallized 0.3-μm-Thick Silicon-on-Sapphire Films.
IEEE Electron Device Letters EDL-5(5):156-158, May, 1984.

[16] Mead, C. and L. Conway.
Introduction To VLSI Systems.
Addison-Wesley Publishing Co., Reading, Mass., 1980.

[17] Miyamoto, J.I. *et al.*.
A High-Speed 64K CMOS RAM with Bipolar Sense Amplifiers.
IEEE Journal of Solid-State Circuits SC-19(5):557-563, October, 1984.

[18] Momose, H. *et al.*.
1.0μm n-Well CMOS/Bipolar Technology.
IEEE Journal of Solid State Circuits SC-20(1):137-143, February, 1985.

[19] Ong, DeW.G.
Modern MOS Technology.
McGraw-Hill Book Co., 1984.

[20] Parrillo, L.C. *et al.*.
Twin-Tub CMOS - A Technology for VLSI Circuits.
In *IEEE Int. Electron Device Meeting*. 1980.

[21] Pattanayak, D.N. *et al.*.
Switching Conditions for CMOS Latch-Up Path with Shunt Resistances.
IEEE Electron Device Letters EDL-4(4):116-119, April, 1983.

[22] Payne, R.S., W.N. Grant and W.J. Bertram.
Elimination of Latch Up in Bulk CMOS.
In *Proc. of the International Electron Device Meeting*, pages 248-251. IEEE, 1980.

[23] Pinto, M.R. and R.W. Dutton.
Accurate Trigger Condition Analysis for CMOS Latchup.
IEEE Electron Device Letters EDL-6(2):100-102, February, 1985.

[24] Preckshot, N.E. *et al.*.
Design Methodology of a 1.2-μm Double-Level-Metal CMOS Technology.
IEEE Journal of Solid State Circuits SC-19(1):81-90, February, 1984.

[25] Soden, J.M., H.D. Stewart and R.A. Pastorek.
ESD Evaluation of Radiation-hardened, High Reliability CMOS and MNOS ICs.
In *Proc. EOS/ESD Symposium*, pages 134-146. Reliability Analysis Center, September, 1983.

[26] Sugino, M., L.A. Akers and M.E. Rebeschini.
Latchup-Free Schottky-Barrier CMOS.
IEEE Trans. on Electron Devices :110-118, 1983.

[27] Sze, S.M.
Physics of Semiconductor Devices.
John Wiley & Sons, New York, 1969.

[28] Sze, S.M. (ed.).
VLSI Technology.
McGraw-Hill Publishing Co., 1983.

[29] Tanaka, S. *et al.*.
A Subnanosecond 8K-Gate CMOS/SOS Gate Array.
IEEE Journal of Solid State Circuits SC-19(5):657-662, October, 1984.

[30] Troutman, R.R. and H.P. Zappe.
A Transient Analysis of Latchup in Bulk CMOS.
IEEE Trans. on Electron Devices ED-30(2):170-179, February, 1983.

[31] Vittoz, E.A.
MOS Transistors Operated in the Lateral Bipolar Mode and Their Application in CMOS Technology.
IEEE Journal of Solid State Circuits SC-18(3):273-279, June, 1983.

[32] Walezyk, I. and J. Rubinstein.
A Merged CMOS/Bipolar VLSI Process.
In *IEDM Technical Digest*. December, 1983.

[33] White, M.H.
Characterization of CMOS Devices for VLSI.
IEEE Journal of Solid-State Circuits SC-17(2):208-214, April, 1982.

[34] Wieder, A.W., C. Werner and J. Harter.
Design Model for Bulk CMOS Scaling Enabling Accurate Latchup Prediction.
IEEE Trans. on Electron Devices ED-30(3):240-245, March, 1983.

[35] Wu, C.Y. and M.Z. Lin.
The Modified Bimos Dynamic RAM Cell Using LOCOS and N-Well Technologies.
In *Proc. of the International Symposium on VLSI Technology, Systems and Applications*, pages 267-269. March, 1983.

[36] Yamaguchi, T. *et al.*.
Process and Device Performance of 1 μm-Channel n-Well CMOS Technology.
IEEE Journal of Solid State Circuits SC-19(1):71-80, February, 1984.

Chapter 4

Logic Design

Several logic design techniques can be used in CMOS; however, all of them belong to one of the following logic disciplines:

- Static logic.
- Dynamic logic.
- Bootstrap logic.

Static logic is the simplest and most straightforward. It is called "static" because the information is permanently stored — that is, so long as the circuit is powered — and any gate output node is strongly connected to either V_{dd} or V_{ss}. Moreover, if we exclude switching periods, the output of the gate at all times assumes the value of the boolean function implemented by the circuit.

Dynamic logic is based on the concept of "precharging," which consists of pulling a gate output node up (to V_{dd}) or down (to V_{ss}) either to charge or discharge the parasitic capacitance associated with that node. Therefore, it is possible to have either precharge-low or precharge-high logic. Any gate implemented with dynamic logic passes through two different phases: *precharge* and *evaluation*. In the precharge

---

Plates referred to in this chapter are found in the PLATES section located between pages 172 and 173.

phase the output assumes a value which is independent of both the input values and the boolean function which the gate implements, depending only on the characteristics of the precharging technique used — output low for precharge-low, output high for precharge-high. In the evaluation phase, the gate behaves like a static gate, that is, the output depends on the input values and the boolean function. If the inputs and the boolean function generate the output value driven during precharge, no change in the output node occurs. Otherwise, the node is strongly pulled down if precharge was high, or pulled up if precharge was low. During precharge, the information is stored in parasitic capacitors, such as the gate capacitance, and disappears in a few milliseconds if the node is not "refreshed," that is, if the information stored in the node is not updated. This phenomenon is caused by leakage currents, which have been discussed in Chapter 2, or by charge sharing, which will be dealt with in Section 4.3.

Bootstrapping techniques use the charge previously stored in parasitic capacitors to "overdrive" some nodes in a circuit, thereby speeding up the circuit with a small increase in power dissipation. Overdriving consists of driving gates with voltages higher than V_{dd}. Boostrapped gates are different from dynamic logic gates because an explicit precharge signal is not always required. Note, however, that, for reasons such as protection against latchup, the use of bootstrap logic is not recommended in a bulk CMOS process. If such a fabrication process is used, only static or dynamic logic should be used. Therefore, we shall concentrate our attention on these two logic disciplines.

The reasons benind the choice of a logic family are somewhat different when comparing CMOS and nMOS design; in fact, dynamic nMOS logic mainly serves the following purposes:

- It allows us to decrease power dissipation, because, as compared to nMOS static logic, dynamic logic never creates a static path between V_{dd} and V_{ss}.

- It allows us to implement a *ratioless* logic (see below), with the benefit of not requiring particular pull-up|pull-down ratios to achieve correct logic behavior.

- It allows us to decrease the area in some cases. Since dynamic logic is a ratioless logic, it is possible to implement multi-input NAND gates without laying out very long channel depletion devices.

- It allows us to increase the circuit's speed in some cases. Depending on the system clocking strategy, nMOS dynamic logic can be faster than its static counterpart. We shall return to this topic later on, when a comparison between static and dynamic logic is carried out. Although the discussion will refer specifically to CMOS design, much of the discussion can be applied to nMOS design as well.

We have already mentioned that static CMOS is a ratioless logic and has no static power dissipation. For this reason, the rationale for using dynamic logic cannot be to reduce power dissipation. CMOS dynamic logic features the following advantages over its static counterpart:

- The area of dynamic logic gates is usually smaller than the area of static logic gates implementing the same function.

- Dynamic CMOS circuits can be faster than static CMOS circuits.

As we shall see, what we have been calling "static logic" in Chapter 1 is actually a subclass of static logic circuits that from now on will be referred to as "complementary static logic." In fact, although dynamic logic usually occupies less area and can be faster than complementary static logic, this is not *always* true. Other static logic subclasses allow us to decrease area at the expense of power dissipation, to decrease area at the expense of speed, or to increase speed at the expense of power dissipation.

Dynamic logic suffers from a phenomenon called "charge sharing." This phenomenon is associated with the storage of information into a parasitic capacitor. Therefore, no dynamic logic is completely immune from charge sharing, and the designer always has to be very careful when dynamic logic is being used. Back in the 1970's, the rationale for using nMOS dynamic logic was to speed up the circuit and decrease the power dissipation. At that time, parasitic capacitances were significant, and the process of storing some charge — that is, information — into these capacitors was robust and reliable. More recently, the shrinkage of feature sizes has decreased the size of these parasitic capacitors, greatly increasing circuit speed. This increased speed is making it more difficult to control the storage of information into these parasitic capacitors, since they can be discharged — or charged — more easily by noise spikes or, to a lesser extent, by radiation effects, such as alpha-particles. Moreover, small feature size processes have higher sub-threshold (leakage) currents, which decrease the lifetime of charge stored in the parasitic capacitors, endangering the correct behavior of dynamic logic. Although dynamic memories have been designed with small feature sizes, the fabrication processes are specially designed for this purpose — for instance, by using "trench capacitors" [1] — and it is not clear whether a "general purpose" fabrication process will allow reliable use of dynamic logic techniques for submicron line widths. Other processes, for example SOS, have such high leakage currents and small parasitic capacitances that dynamic logic is seldom used. Moreover, low-speed functional testing can become extremely difficult when dynamic logic is used with submicron processes, unless proper modifications to the gates are carried out.

Finally, static logic allows the designer to control small-geometry effects — such as hot-electron injection — more easily. As we saw in Chapter 2, injection from the channel is mainly induced by drain-to-source voltage, which is very well controlled in static logic: it cannot be larger than $V_{dd} - V_{ss}$. This is not true in dynamic logic where local, small-scale bootstrap effects can bring the signal to levels higher than V_{dd}. Theoretical analysis of MOS transistor scalability limits has shown that static logic will likely be used down to 0.15μm feature size, while dynamic logic is limited to 0.3μm [10]. This is based on the power supply voltage dropping to 2V. Below these feature sizes, poor noise margin, leakage currents, and very small parasitic capacitances make it impossible to achieve reliable circuit operation.

4.1. Static Logic

Static logic design is the simplest form of logic design and includes four techniques: *complementary, nMOS-like, transmission gate intensive*, and *cascode logic*. It is clear that this subdivision is somewhat arbitrary, and it is often difficult to decide whether a certain logic is complementary or transmission gate intensive (while nMOS-like and cascode logic are much more easily recognizable).

4.1.1. Complementary Logic

Some examples of complementary logic have already been presented in Chapter 1. This logic features the same number of p-channel and n-channel transistors. For an *n* input gate, there will be *2n* devices. The logic is ratioless, and there is no static power dissipation — we can ignore leakage currents. Fig. 4-1 shows an And-Or-Invert (AOI) circuit implemented in complementary logic.

The major disadvantages of this logic are low density and gate delay. There are many "redundant" transistors, and the gate of each n-channel device is connected to the gate of the corresponding p-channel device. Therefore, the gate capacitances are in parallel, increasing the input load. Moreover, area and speed are no longer orthogonal parameters, because the length of the interconnect becomes the limiting factor in achieving high speed; larger areas can lead to longer interconnections and therefore to lower speed.

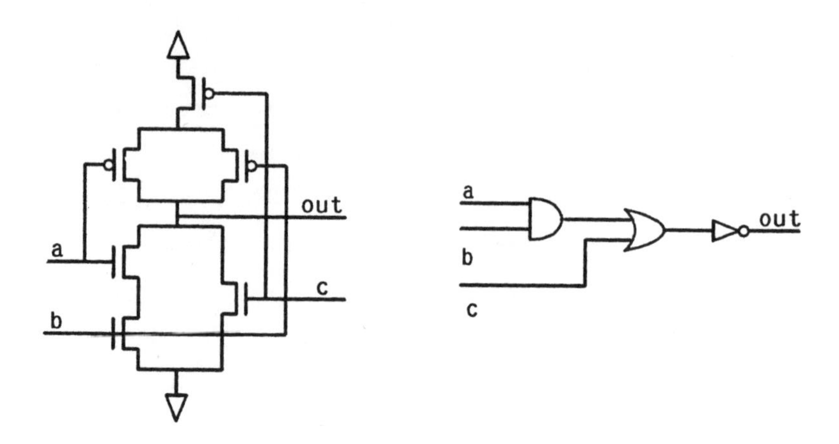

Figure 4-1: AOI implemented in complementary logic (left) and its representation in terms of basic gates.

4.1.2. nMOS-like Logic

The nMOS-like implementation of an AOI gate is shown in Fig. 4-2. Fig. 4-2(a) shows the n-channel based version, while Fig. 4-2(b) shows the p-channel based version. Each input is connected to the gate of only one device — instead of two as in complementary logic. Moreover, if the gate has n inputs, only $(n+1)$ devices are necessary. This logic occupies less area than complementary logic. Concerning circuit speed, nMOS-like logic was faster than complementary logic [12] in the early days of CMOS technology, when the gate capacitance of the MOS transistor was indeed a limiting factor. With fabrication processes of about 3μm and smaller, this is no longer true. Although the input capacitance is smaller than in complementary logic, because the input is connected to only one device, the pull-up section in Fig. 4-2(a) is a p-channel transistor used as a resistor, and the current which is delivered suffers from energy dissipation. We conclude that the two effects — decreased input capacitance but also decreased driving capability — balance each other.

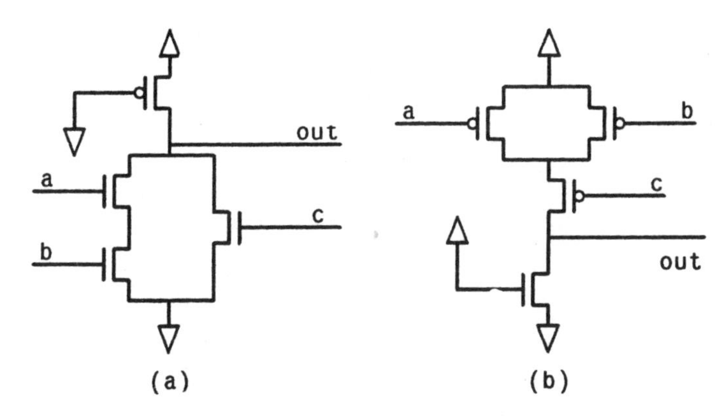

Figure 4-2: CMOS, nMOS-like AOI gates using n-channel devices (a) or p-channel devices (b).

nMOS-like logic is a *ratioed* logic. The pull-up and pull-down sections have to satisfy a certain range of ratios, otherwise the output voltage might not be able to switch the next gate(s) and/or the noise margin might be adversely affected, as shown in Figs. 4-4 and 4-5. Fig. 4-4 shows the input to an nMOS-like inverter (see Fig. 4-3 (left)) and six output curves. These curves refer to the same inverter with different pull-up|pull-down ratios. The two transistors have the same width, and the inverter drives a load of 10^{-3}pF. Curve 1 refers to the inverter with a pull-up|pull-down ratio of 1|1; curve 2 refers to the inverter with a pull-up|pull-down ratio of 2|1 (i.e., the p-channel device is twice as long as the n-channel device and both have the same width), and so on. Ratios of 1|1 and 2|1 are unacceptable. Besides providing poor noise margin, the output low level is close to 0.6V. When this output is connected to similar nMOS-like gates, it keeps the n-channel device(s) weakly off and drastically increases static power dissipation. In order to improve the noise margin and strongly turn off n-channel transistors in other gates, a ratio of either 4|1 or 5|1 is normally used. This is done at the expense of output rise-time, which increases when the ratio increases.

Figure 4-3: Inverter configurations for nMOS-like logic.

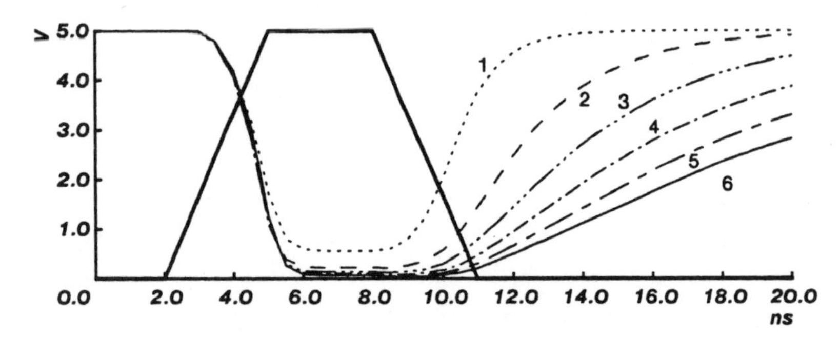

Figure 4-4: nMOS-like inverter of Fig. 4-3 (left): input signal (bold) and six output signals for different pull-up|pull-down ratios.

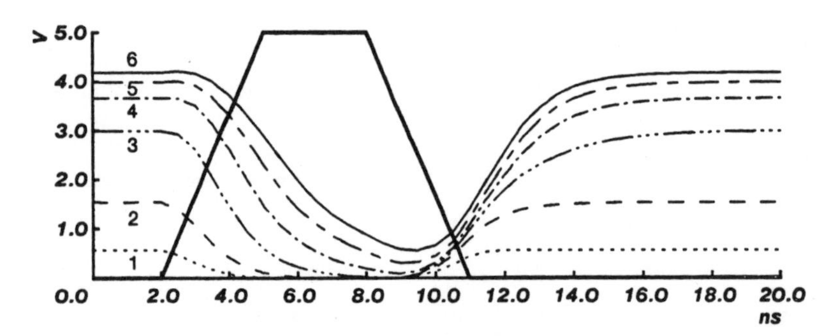

Figure 4-5: nMOS-like inverter of Fig. 4-3 (right): input signal (bold) and six output signals for different pull-up|pull-down ratios.

Fig. 4-5 shows the input signal to an inverter with a "complementary" configuration (see Fig. 4-3 (right)) and six output waveforms. The gate of the n-channel device is connected to V_{dd}, and the input signal is connected to the gate of the p-channel device. Both transistors have the same length. Curve 1 shows the output waveform for a pull-up|pull-down ratio of 1|1; curve 2 refers to a pull-up|pull-down ratio of 1|2 (i.e., the n-channel device is twice as wide as the p-channel transistor), and so on. Note that performance degradation is much more significant, and a ratio of 6 has to be considered the minimum requirement. Therefore, only a logic like the one shown in Fig. 4-2(a) is usually implemented.

Finally, as in nMOS design, NOR-based gates are always preferred to NAND-based gates, because a NAND gate in this logic requires a long p-channel device to satisfy the necessary pull-up|pull-down ratio, making the layout more complex and slowing down the circuit. The most serious drawback of this logic is that a direct path between V_{dd} and V_{ss} exists when the input is high in the circuit in Fig. 4-3 (left). nMOS-like logic suffers from static power dissipation.

There are some applications where this logic can be useful, such as decoders with severe area limitations and no clock availability. As a general guideline, this logic can be implemented when power dissipation is not a major concern, circuit density is of fundamental importance, and no clock is available — either because there is no clock or because it cannot be used. Finally, note that the difference in power dissipation between nMOS-like and complementary logic *decreases* with higher frequency of utilization of the complementary gates, because dynamic power dissipation is proportional to the frequency of operation of the circuit, as shown in Eq. (2-23).

4.1.3. Transmission Gate Intensive Logic

Transmission gate intensive logic is unique to CMOS design. In fact, not only do we mean extensive use of transmission gates but also an interesting utilization of them that Fig. 4-6 helps to explain. Fig. 4-6 shows an exclusive-OR (XOR) logic circuit implemented with a transmission gate intensive technique. The figure shows an inverter at the input and a transmission gate at the output. The two transistors in the center of the circuit act as an inverter *which is conditionally activated.* When **a** is high, the circuit behaves like an inverter; when **a** is low, both transistors are off, and the output is in the high impedance state. This technique can be implemented in CMOS, because the output signal makes a full swing between V_{dd} and V_{ss}, and the logic is ratioless. Fluctuations in the power supply do not affect the correct logic behavior of the circuit, because the logic levels in CMOS are largely independent of the supply voltage — a typical example being the inverter threshold voltage — and only the noise margin is affected. In other words, a circuit which functions with a 5V power supply will function at 4V as well — with a decrease in power dissipation (see Eqs. (2-21) and (2-22)), and also a decrease in speed.

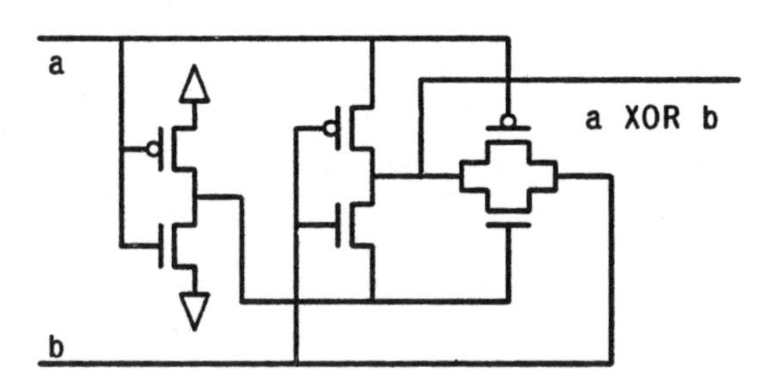

Figure 4-6: Transmission gate intensive XOR gate.

Normally, transmission gate intensive logic is a good choice when inputs are available in both polarities. An example is shown in Fig. 4-7, where a very compact OR-gate is implemented with only three transistors (vs. six in complementary logic and five in nMOS-like logic). The major advantage of this logic is small area [11], which benefits from the fact that few V_{dd} and V_{ss} lines — if any — have to be routed through the circuitry. At the same time, this can be also a drawback: the logic features poor driving capability and should never be used extensively on large areas of the chip. This would introduce a significant delay. The delay grows quadratically with the length of the chain of transmission gates, rather than linearly as in complementary logic, because both capacitance and transistor "on" resistance have to be considered. Finally, gates implemented with this logic should never drive long lines and/or large loads.

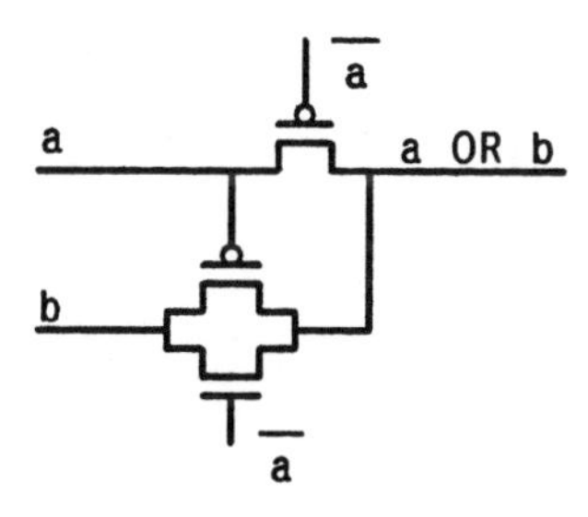

Figure 4-7: Transmission gate intensive OR gate.

Transmission gate intensive logic belongs to the class of switching network logics [16], and switching network theory can be used to help the designer in the synthesis and implementation of logic functions. Significant research has been carried out to formalize a design methodology based on switching logic. Although the MOS switch is far from being ideal (with particular reference to the "on" resistance), switches made of superconductor materials closely resemble ideal switches, and transmission gate intensive logic may become an attractive design methodology in the future.

4.1.4. Cascode Logic

This logic features cross-coupled p-channel device loads and n-channel devices used in a differential configuration. Moreover, a signal from one gate to the next is always transferred together with its complement. This logic includes two different schemes: *cascode voltage switch logic*, CVSL [4] and *differential split-level logic*, DSLL [9].

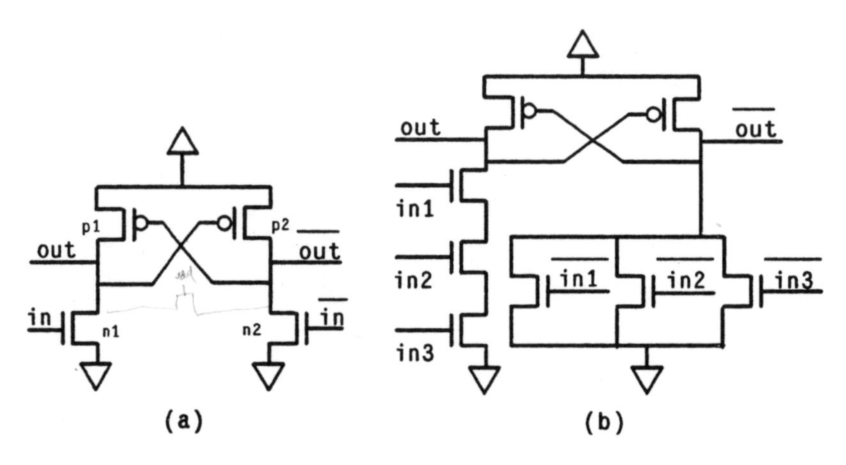

Figure 4-8: Cascode voltage switch inverter (a) and NAND/AND gate (b).

A CVSL inverter gate is shown in Fig. 4-8(a). Let us assume that in is low. Node $\overline{\text{out}}$ is low, and pl is conducting: node out is high. When node in goes high and $\overline{\text{in}}$ goes low, pl is turned off, and p2 is turned on. Note that when in goes high, pl is still on, and, therefore, CVSL is a ratioed logic. Moreover, a direct path is created by nl and pl — or n2 and p2 — between V_{dd} and V_{ss}. During switching a current spike takes place. This current spike is usually larger than that of a complementary gate [9]. A NAND/AND gate is shown in Fig. 4-8(b). Both out and $\overline{\text{out}}$ are sent to other gates. Note that this logic is fully compatible with complementary logic, because the output makes a full swing between V_{dd} and V_{ss}.

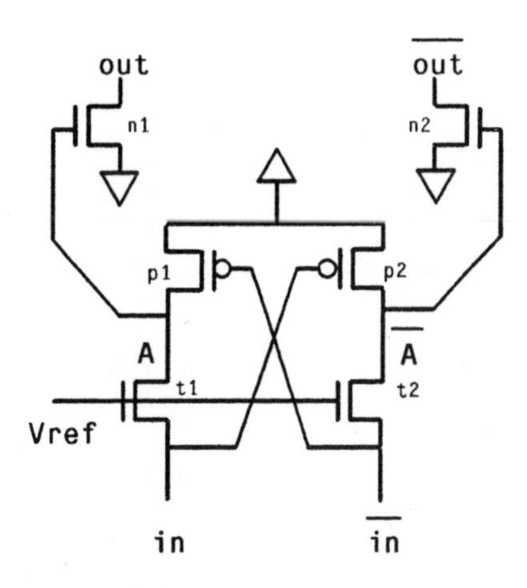

Figure 4-9: Differential split-level inverter.

A DSLL inverter is shown in Fig. 4-9. DSLL is an aggressive logic which provides high speed at the expense of static power dissipation. Two extra n-channel devices are present, and they are driven by a reference voltage. DSLL achieves high speed because it keeps some devices close to their switching threshold and limits the voltage swing at the output. The output does not reach V_{dd}, but settles at about $V_{dd}/2$. The concept of reducing the voltage swing is appealing. Given an inverter threshold voltage of $V_{dd}/2$, it is clear that, at least in theory, we could drive the gate with low voltages of about 2V and high voltages of about 3V (V_{dd} = 5V), still achieving correct logic behavior. In fact, an input voltage level of 2V makes the output of a balanced CMOS inverter switch to V_{dd}, while an input voltage of 3V makes the inverter output switch to V_{ss}. The evident drawback is that both devices conduct all the time, and power dissipation drastically increases. Moreover, the noise margin is significantly affected, because a small variation of the input signal can make the gate switch. However, the example discussed above is an extreme case. In fact, as we shall see below and in Section 7.8, limiting the voltage swing can be actually implemented without introducing unacceptable side-effects.

The DSLL "inverter" shown in Fig. 4-9 closely resembles a CVSL gate, with the only difference being two extra n-channel devices driven by a reference voltage. Its electrical behavior differs significantly, however. The value of the reference voltage should be equal to $V_{dd}/2$ plus the n-channel transistor threshold voltage. Let in be low and $\overline{\text{in}}$ high — approximately, $V_{dd}/2$. Therefore, p2 is conducting, node $\overline{\text{A}}$ is pulled to V_{dd}, and $\overline{\text{out}}$ is low. Node A is *not* at V_{ss}, but rather at a few hundred millivolts. Transistor n1 is off, and, so far as the gate shown in the figure is concerned, its drain voltage has an indeterminate value. Note that the value of $\overline{\text{in}}$ is such that p1 is *not* strongly turned off, and, at the same time, t1 is on. Therefore, there is some static power dissipation in the p1-t1 path. At the same time, t2 is off while p2 is strongly on. Therefore, there is no static power dissipation in the p2-t2 path — except for leakage currents.

We now let in switch to high and $\overline{\text{in}}$ to low. Because p1 is not completely off, it will turn on very fast, pulling node A high faster than in a conventional CMOS inverter. At the same time, t2, which was just inside the sub-threshold region, starts to conduct immediately, while p2 is turned off. As soon as node $\overline{\text{A}}$ drops to the threshold voltage of the n-channel device, n2 is turned off.

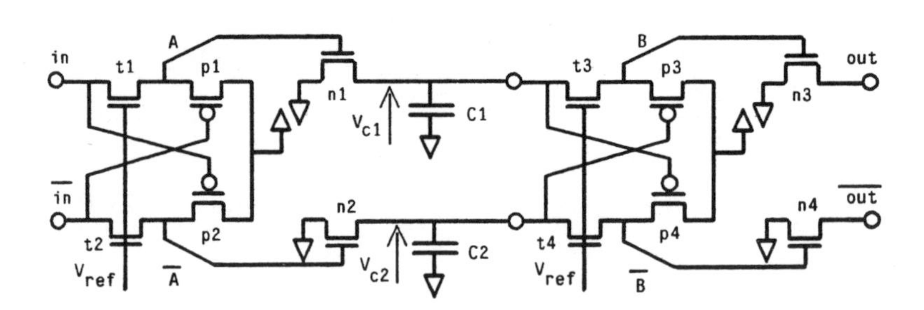

Figure 4-10: Cascade of two DSLL gates.

A complex gate is obtained by substituting n1 and n2 with the logic network that executes the required boolean function. Therefore, the computing section of the gate includes n-channel devices only. When we cascade two gates like the one shown in Fig. 4-9 (see Fig. 4-10), the voltages V_{c1} and V_{c2} across the stray capacitances are limited to values between V_{ss} and $V_{dd}/2$. The open drain n-channel transistor which was off in the previous example has its drain pulled up by the next gate, and, therefore, the drain voltage is no longer indeterminate. An important side-effect of this logic is that the drain-to-source voltages of n1, ..., n4 are always limited to at most $(V_{dd}/2 - V_{ss})$. This decreases the device sensitivity to hot-electron injection, which strongly depends on the drain-to-source voltage, and allows the use of short-channel devices in the n-channel device evaluation network without resorting to complex fabrication processes. This only holds true up to a certain point: DIBL and hot-electron emission affect submicrometer devices to such an extent that new fabrication processes are indeed necessary.

Finally, this logic is compatible with any other logic, static or dynamic. When a DSLL gate has to be connected to a gate implemented in a different logic, it is sufficient to use the A and $\bar{A}$ outputs shown in Fig. 4-9, because they execute a full swing between V_{dd} and V_{ss} (actually, V_{ss} + 100-200mV).

4.2. Dynamic Logic

Dynamic logic design is considerably more difficult than static logic design. Problems such as charge sharing do not exist in static logic design. Dynamic logic overcomes some of the problems of static logic:

- *It occupies less area*: the area of a CMOS dynamic circuit is comparable to the area of an nMOS circuit or to the area of a CMOS circuit using nMOS-like logic.

- *It has higher speed*: not only is the input capacitance of a dynamic gate smaller than that of the corresponding complementary static implementation — no transistors in parallel are needed to implement a logic function — but the switching threshold of the gate depends on the switching threshold of the *device itself*, rather than on $V_{dd}/2$, as is the case with a complementary static gate. We will return to this topic in Section 4.5.

Dynamic logic allows us to design circuits that have low power dissipation, high speed, and occupy little area. The drawbacks of dynamic logic can be summarized as follows:

- Dynamic logic can be affected by charge sharing (also known as "charge redistribution" or "charge splitting").

- Dynamic logic always requires clocks. This is not necessarily a drawback so long as the clocks fit smoothly into the clocking discipline at the system level.

- Dynamic logic makes it difficult — if not impossible — to operate at very low speed (i.e., in the KHz range and below). This can be a serious drawback when performing functional testing, which is commonly carried out at low operating speed. Moreover, some products, such as portable systems for data acquisition or those oriented to the low-end of the consumer market (e.g., toys), require circuits with very low power dissipation, because they are battery-powered. As we saw in Section 2.7, power dissipation depends linearly on the operating frequency of the circuit, hence there is a need for low frequency operation. In this case, either static logic must be used or some modifications to the basic dynamic gate have to be performed.

- Dynamic logic cannot be fully utilized. This is also true for static logic, because of the intrinsic latency that all circuits have. However, all dynamic logics use precharging techniques that lower the *availability* of the circuit. During precharge, the logic cannot be utilized. Again, this is not necessarily a drawback so long as the designer is aware of the problem.

The last point brings us directly to the critical issue of which logic to use, given a certain set of requirements. The first choice is to decide whether static logic or dynamic logic shall be used. Before tackling the problem, however, let us consider the most frequently used dynamic logics and discuss the problem of charge sharing.

4.2.1. Ripple-through Logic

The simplest form of dynamic logic is called ripple-through. Its n-channel based implementation is shown in Fig. 4-11. The gate performs an And-Or-Invert (AOI) function. The clocked device is a p-channel transistor. The clock cycle can be divided into two phases:

1. *Clock low*: the output of the gate is precharged high, and we assume that no path from the output to V_{ss} exists, that is, both a — or b — and c must be low in the gate shown in the figure. When the clock is low, the output is pulled high and the parasitic capacitor associated with the output node is charged. This capacitor consists of the gate capacitance of all inputs connected to the output of this gate and of three drain-substrate junction capacitances: those of the two n-channel transistors whose inputs are a and c, and that of the p-channel device.

2. *Clock high*: the p-channel does not conduct, and the output node is either left floating at a high logic level or, depending on the input pattern, is strongly pulled down. This can happen if c is high or a and b are high. This phase is known as the *evaluation* phase.

The same function can be implemented by a gate based on p-channel devices, as shown in Fig. 4-12. When the clock is high, precharging takes place, and the output

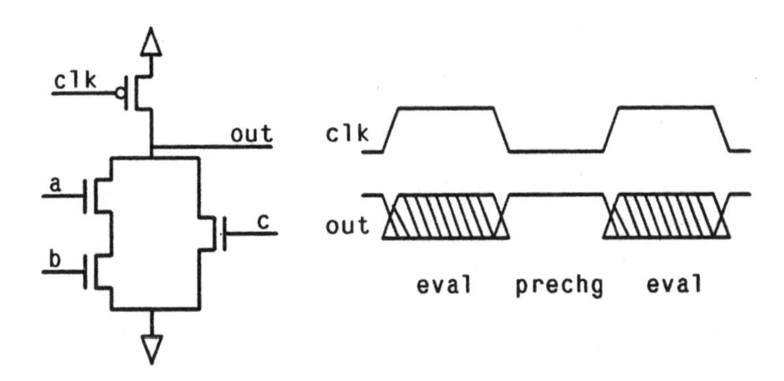

Figure 4-11: n-channel based, ripple-through dynamic logic gate.

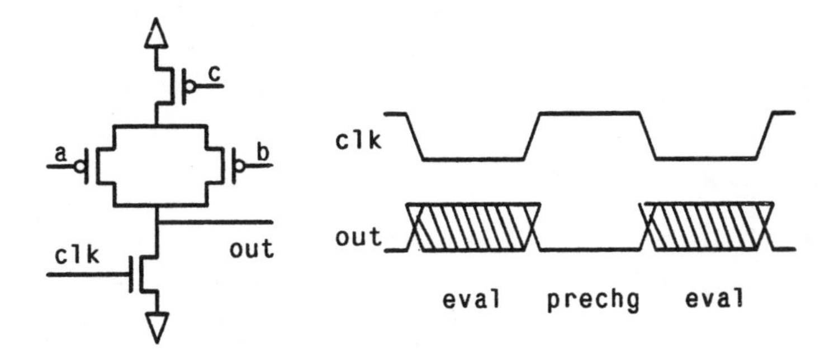

Figure 4-12: p-channel based, ripple-through dynamic gate.

node is strongly pulled down. Again, we assume that the input pattern does not create a direct path to V_{dd} — which would happen if b and c are low, for instance — and the parasitic capacitor associated with the output node is discharged. When the clock goes low, the n-channel transistor turns off, and evaluation takes place. Depending on the input pattern, the output node remains at a low logic level or is pulled up — when a and c are both low, for example.

Other combinations are possible, such as circuits consisting of n-channel only or p-channel only devices. They are shown in Fig. 4-13(a) and (b). These gates are not recommended because they exhibit poor noise margin. A full output swing is

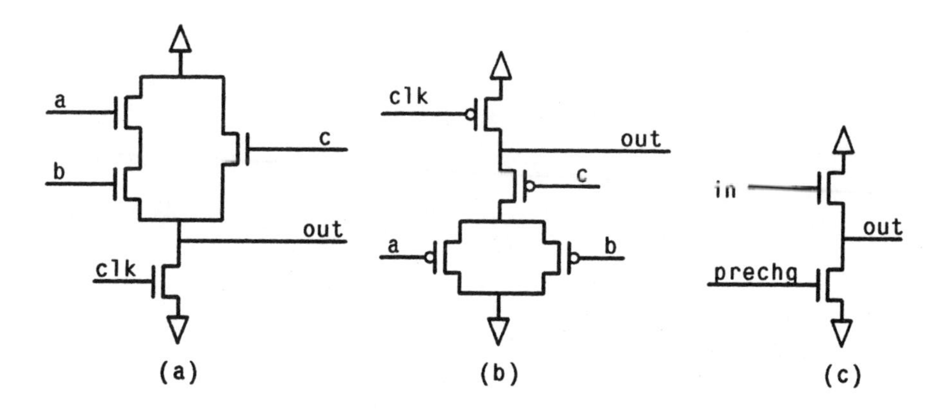

Figure 4-13: Alternative ripple-through gates: their use is not recommended.

impossible, because the n-channel device does not conduct high logic levels well and the p-channel device does not conduct low logic levels well; both degrade the output signal. This is shown in Fig. 4-14, where the performance of the precharged inverter in Fig. 4-13(c) is shown.

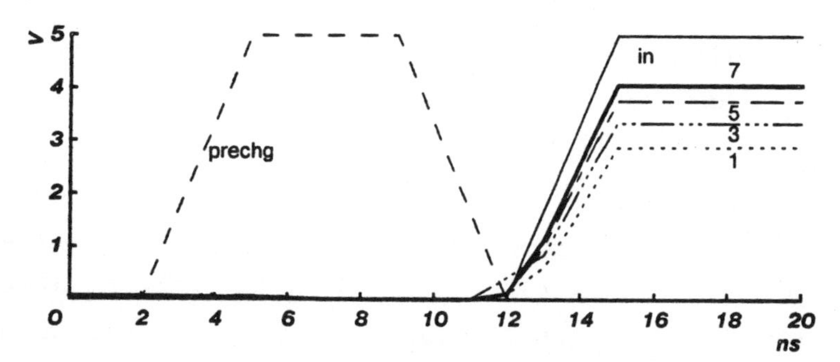

Figure 4-14: Output signals with different transistor ratios for a gate like the one in Fig. 4-13(c).

Different pull-up|pull-down ratios can help improve the noise margin (such as curve 7 in the figure), but they increase the circuit area and the input gate capacitance. Four output signals are shown in Fig. 4-14. The numbers of the curves are the ratios that have been used. These ratios *are different* from the ratios used in Section 4.1.2. Curve 3, for instance, refers to the inverter whose pull-up n-channel device is *three times wider* than the pull-down n-channel device, and both have the same length. We can increase the output voltage by making the pull-up transistor wider; however, the threshold voltage of the n-channel device will always limit the output swing. At best, the output voltage can equal $V_{dd} - V_{Tn}$.

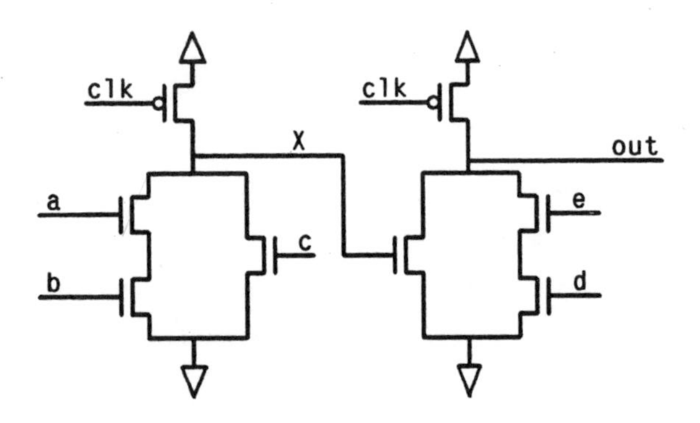

Figure 4-15: It is impossible to cascade ripple-through gates of the same kind.

A serious problem in dynamic logic arises when different gates are combined to implement more complex functions. It is impossible to cascade gates of the same kind, say n-channel based, as Fig. 4-15 shows. When one gate is being precharged (`clk` low), its output **X** is high, and this turns on the transistor on the left side of the second gate. This makes it impossible to correctly precharge the gate, because of the direct path between V_{dd} and V_{ss}. For this reason, `out` has an indeterminate value which depends on the charge distribution among the parasitic capacitors, the diffusion resistance, etc..

A different approach consists of cascading gates in an alternate fashion. As an example, let us consider cascading two gates of the kind shown in Fig. 4-11 and 4-12. This situation is shown in Fig. 4-16. An n-channel based ripple-through gate is followed by a p-channel based ripple-through gate. The first thing to notice is that we

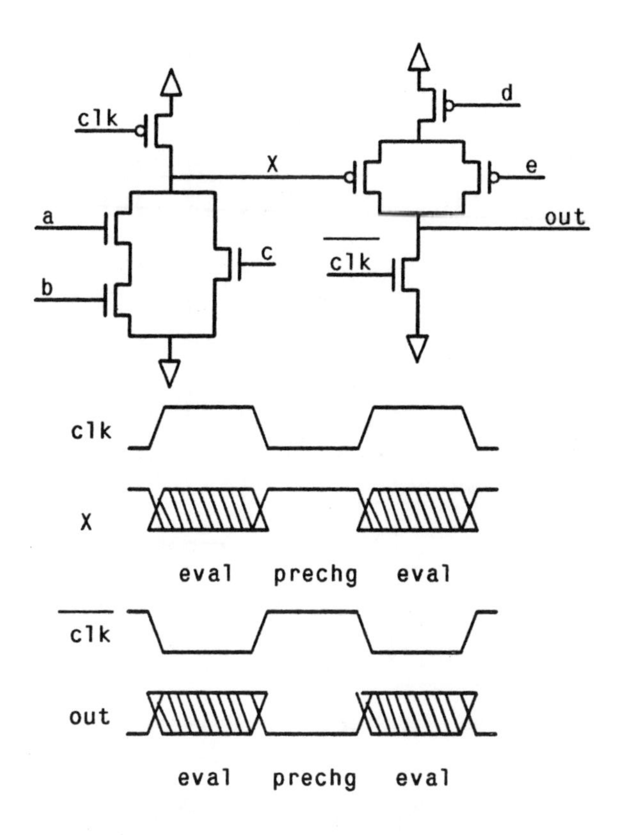

Figure 4-16: Cascading dynamic gates of different kinds.

need two clocks, that is, `clk` and $\overline{\texttt{clk}}$. When the n-channel based gate is being precharged, its output X goes high and turns off the transistor in the p-channel based gate. This gate can be precharged correctly because all its p-channel devices are off. This implies that both d and e inputs are coming from n-channel based gates.

However, so far as clocks are concerned, Fig. 4-16 can be considered an ideal case. In reality, `clk` and $\overline{\texttt{clk}}$ *might* not be so perfectly aligned. What happens if the two clocks are skewed? Fig. 4-17 shows the same clocks, but with $\overline{\texttt{clk}}$ skewed. During the interval t_1-t_2 the first gate is precharging and the node X goes high. However, because of skews, the second gate is still evaluating and the sudden change in one of

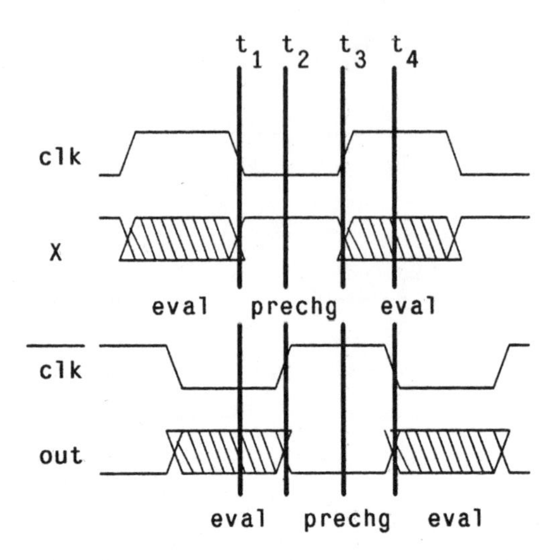

Figure 4-17: When there are clock skews, the circuit shown in Fig. 4-16 does not always work reliably.

its inputs can cause either a glitch in the output or even an incorrect output.

The same situation happens during the interval t_3-t_4. The second gate is precharging while the first one is still evaluating. The output will either be forced low or set to an indeterminate state — depending on the logic equation the gate is performing. In other words, clock skews can create a situation similar to the one shown in Fig. 4-15. Although the skews are usually not so large, even a very small overlap would cause spikes that do not allow correct precharging or that can discharge precharged nodes. Note that a spike which discharges a precharged node may cause irreparable damage to the output of the gate.

Finally, it has been assumed that all the inputs arrive perfectly synchronized with the clocks. This is not necessarily true. The e and d inputs in Fig. 4-16 might be delayed, and, while precharging takes place in the second gate, their value could still be low. The effect of this is a short-circuit between V_{dd} and V_{ss} with static power dissipation. For even larger skews, the result is similar to clock skews, that is, incorrect precharging can occur. Finally, if only d is delayed and the corresponding transistor is conducting when precharging takes place, the transistor whose input is X has V_{ds} equal to *or exceeding* ($V_{dd} - V_{ss}$). This makes hot-electron injection from the channel more likely to happen.

The example we just considered made us aware of the problems with using dynamic logic and pinpointed the major sources of problems: clock skews, race conditions, output spikes, and so on. Several of the dynamic logics that have been presented in the literature try to overcome some of these problems. We now analyze some of them in detail.

4.2.2. P-E Logic

P-E logic stands for "Precharge-Evaluation Logic": the two actions of precharging and evaluation are performed by one clock and *two separate devices.* Fig. 4-18 shows the implementation of P-E gates using n-channel (a) and p-channel devices (b). The behavior of the gate is as follows (the n-channel based gate of Fig. 4-18(a) is used as an example):

1. *Clock low*: the gate is precharged, and the output node is pulled up to a high logic level. Regardless of the input pattern, no direct path between V_{dd} and V_{ss} exists because the n-channel transistor on the bottom ("evaluation transistor") is not conducting.

2. *Clock high*: the evaluation transistor conducts and, if the result of the logic function in the block is to connect the output to the n-channel transistor, the output is also connected to V_{ss}. That is, the output node is discharged and goes to a low logic level. If this is not the case, the output node remains charged.

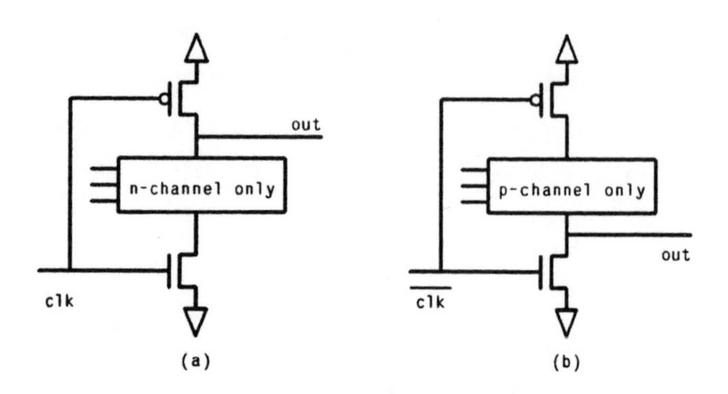

Figure 4-18: P-E logic gates.

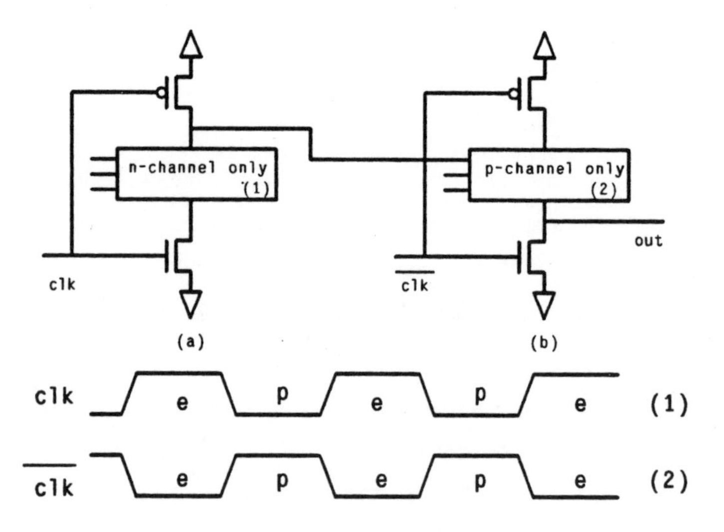

Figure 4-19: Timing for P-E logic gates.

P-E logic allows the use of both gates shown in Fig. 4-18. This is shown in Fig. 4-19. All the n-channel based gate outputs must be connected to p-channel based gate inputs to allow correct precharging. Both ripple-through and P-E logic present the same problems, so far as clock skews are concerned, but P-E logic, with its "precharge" and "evaluation" transistors, offers more reliable behavior. When precharging takes place, the output node is physically disconnected from V_{ss} — in the n-channel based gate — or V_{dd} — in the p-channel based gate — regardless of input pattern, race conditions, and so on.

4.2.3. Clocked CMOS Logic

A clocked CMOS logic (C^2MOS) inverter is shown in Fig. 4-20(a) [14]. When the clock is high, the gate behaves like a normal inverter. When the clock is low, the output value is *stored* in the gate capacitance of the inputs connected to the output node and — depending on the output node's value — either in the p-channel or in the n-channel drain junction capacitance. Therefore, the gate shown in Fig. 4-20(a) can be considered an element of dynamic storage. In fact, this gate can be considered functionally similar to the one shown in Fig. 4-20(b), which is a typical building-block for shift registers. The difference between the two is that C^2MOS requires less area, as Plate VII and Plate VIII show. C^2MOS needs fewer contacts and well-crossings. Both plates show a basic cell which can be cascaded to obtain a shift-register.

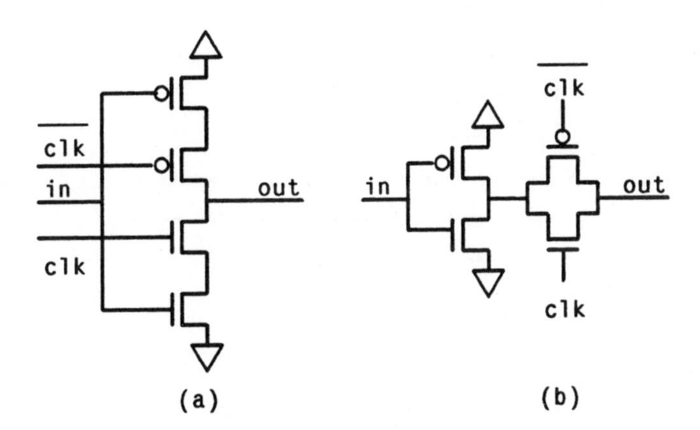

Figure 4-20: Inverter scheme in C^2MOS logic (a) and a functionally equivalent topology (b).

This logic features very compact layout because the area is naturally partitioned between n-channel and p-channel transistors without crossings between the two sections. Different gates can be obtained by substituting the inverter in Fig. 4-20(a) with the appropriate set of transistors. Note that since precharging is not used and the gate maintains complementarity (equal number of n-channel and p-channel devices), the output of a C^2MOS gate can be connected to any other gate — either p-channel or n-channel based.

4.2.4. Domino Logic

A domino logic gate [5] consists of two basic elements: a P-E gate and a static inverter at the output. Two domino logic gates are shown in Fig. 4-21: on the left, the n-channel based version; on the right, the p-channel based gate. The output of an n-channel based domino gate can only go to a similar gate, and the output of a p-channel based gate can only go to a p-channel based domino gate. The output of the gate is taken from the output of the static inverter. The dynamic section of a domino gate works essentially as a P-E gate. The behavior of an n-channel based domino gate is as follows:

- Precharge phase: the output of the gate is precharged high and, therefore, the output of the buffer is low.

- Evaluation phase: the output of the buffer remains low, or, if a path between the output of the dynamic gate and V_{ss} has been created, the buffer output goes high.

Domino gates have significant advantages over conventional P-E or ripple-through gates. First, the availability of a static buffer increases the driving capability of the gate. In fact, both high and low logic levels are provided by connecting the output to either V_{dd} or V_{ss}, while the previously mentioned dynamic logics generate one of the two logic levels by means of the charge stored into a parasitic capacitor. The dynamic section of the gate always has a fan-out of one, because it is only connected to the inverter — this makes the sizing of the transistors easier. Moreover, during

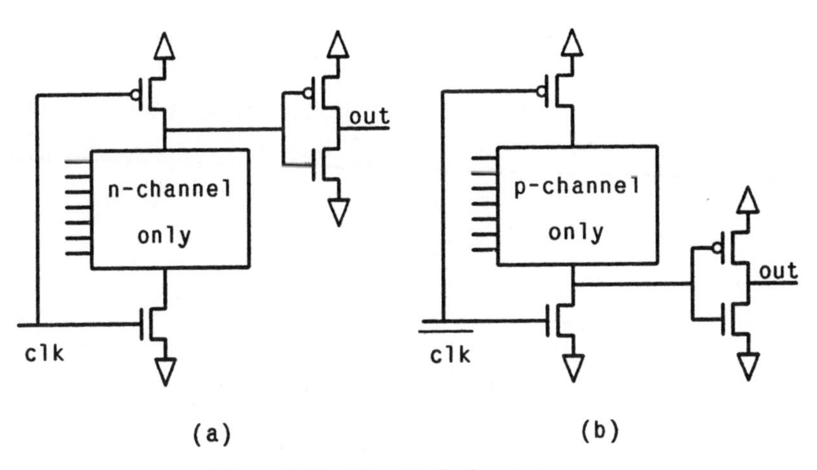

Figure 4-21: Domino logic gates.

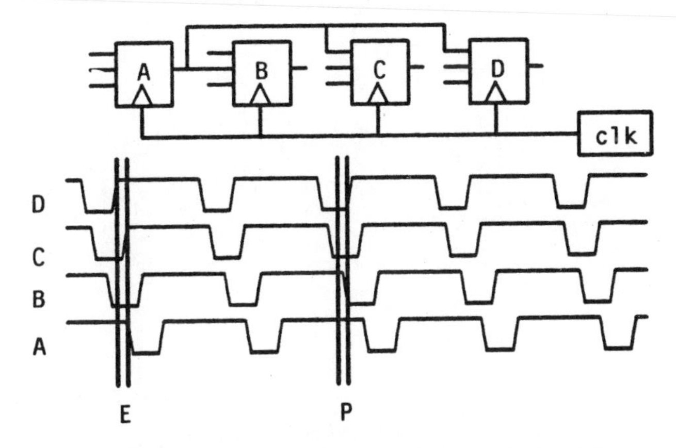

Figure 4-22: An extreme example of incorrect layout which might generate incorrect precharging/evaluation.

precharge, the output of the dynamic gate rises — that is, the output of the static inverter goes down. Because the inverter is connected only to n-channel transistors, correct precharging of all the gates can theoretically be achieved. In practice, careful layout is always necessary because large skews between the clock and the signal lines can still cause incorrect precharging and/or evaluation. An extreme case is shown in Fig. 4-22. A, B, C, and D are domino gates; the output of gate A goes to B, C, and D. We assume that the output signal suffers no delay, while the clock signal common to all gates is significantly skewed. This situation can actually occur if the clock line is laid out in polysilicon and the signal lines run in metal, or if the clock line is much longer than the path between A and D.

The clock signal to the four gates in Fig. 4-22 shows that gate D starts evaluating while gate A is still evaluating the previous input set (time interval E). If the output of gate A is high, the dynamic section of gate D might discharge to V_{ss}, depending on the other inputs. The behavior of gate D in Fig. 4-23(a) will still be correct, while gate D in Fig. 4-23(b) will discharge incorrectly.

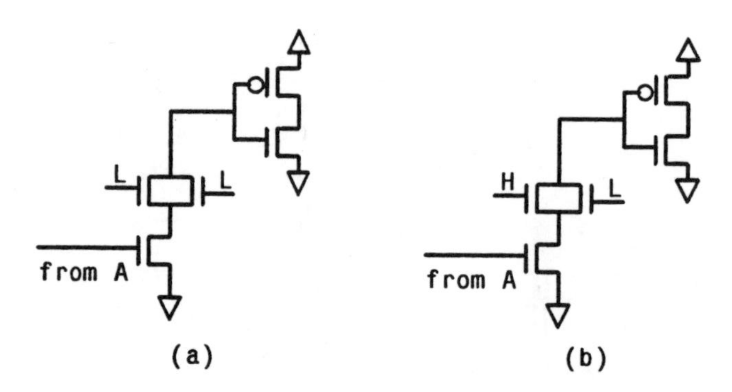

Figure 4-23: Gate D in Fig. 4-22. The gate is not affected by delayed precharging of gate A (a); the gate is affected by delayed precharging of gate A (b).

When gate A finally is precharged, it is too late to recover. Incorrect precharge can occur during the interval P. While gate D is precharging, gate A is still evaluating. Although there is no direct path between V_{dd} and V_{ss} in gate D — the evaluation n-channel device is off — redistribution of charge inside the gate due to a high output coming from gate A might lower the output of D.

What is shown above is an extreme case and is easily solvable by simply laying out the clock line from left to right and using the same material for both signals and clock — this would "align" the signals with the clock. The purpose of the example was to point out that races can still occur and, in more complex situations, the clock layout has to be planned carefully. Finally, this problem is not at all peculiar to domino logic, as we have already seen in the previous sections.

The advantage of domino logic is that only one clock is necessary and therefore an accurate planning of the clock line layout is less difficult. The static inverter provides signal buffering. Slow rising signals, which can also affect the behavior of the circuit, are less likely to occur.

Because of the output inverter, domino logic is a non-inverting logic. Therefore, gates such as an XOR are not possible. Since the logic is fully compatible with static logic, this does not represent a major problem for a large class of applications — the necessary inverting gates can be implemented using any static logic. For low frequency applications, an extra p-channel device connected in parallel with the pull-up p-channel transistor can be used with its gate tied to V_{ss} (for an n-channel device based gate). This increases the power dissipation because a direct path is created between V_{dd} and V_{ss} if the inverter output goes down in the evaluation phase. Finally, the layout of a domino gate is shown in Plate IX.

4.2.5. NORA Logic

NORA (NO RAce) logic [3] is not strictly a logic gate design style, but a comprehensive methodology to implement dynamic logic which is extremely robust to clock skews. It would be more appropriate to talk about "NORA logic blocks," rather than NORA gates.

A NORA logic block is shown in Fig. 4-24 (the two dashed transistors are optional devices for low-frequency operation). It consists of two P-E gates connected *locally*, two inverters for connections to other gates of the same kind, and an output stage which is the typical C^2MOS inverter. Two clocks are necessary. It is worth pointing out that this logic has been considered for use in highly pipelined circuits, and combinational circuits might be more effectively implemented with some of the other logics — both static and dynamic — that have been dealt with in previous sections. Moreover, the number of inputs should be high enough to compensate for the extra circuitry necessary to implement a NORA block — that is, the clocked inverter and the two optional static inverters.

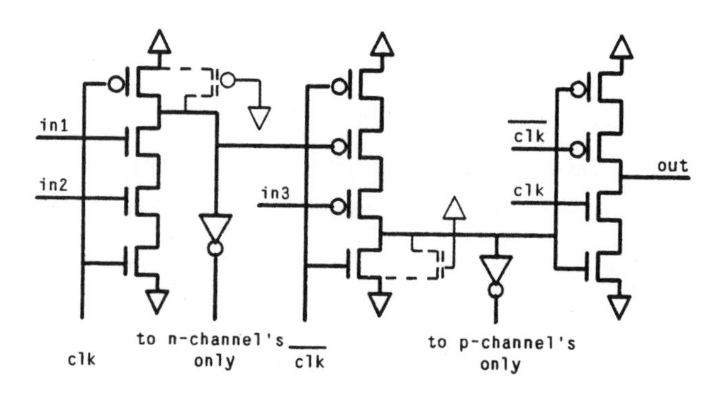

Figure 4-24: NORA building-block. Dashed devices are for low-frequency operation.

Pipelining of blocks, such as the one shown in Fig. 4-24, is achieved by alternating blocks with clk-$\overline{\texttt{clk}}$ clocks (as in Fig. 4-24) with blocks with $\overline{\texttt{clk}}$-clk clocks (as in Fig. 4-24 with clk and $\overline{\texttt{clk}}$ exchanged). We will call the first type of building-block "clk-type" (CT) and the second one "$\overline{\texttt{clk}}$-type" ($\overline{\text{CT}}$). Therefore, CT and $\overline{\text{CT}}$ blocks can be cascaded, and while one is evaluating, the other one is precharging. The clocked inverter acts as a latch, storing the information produced during the evaluation phase until the next block is ready to process it.

Let us now analyze the behavior of the logic block. When `clk` is low and $\overline{\texttt{clk}}$ is high, both the first and second stage are precharging. The output of the second stage is low and the clocked inverter is in the high impedance state. When `clk` is high and $\overline{\texttt{clk}}$ is low, both stages are evaluating and the clocked inverter will provide the output of the second stage negated. When the two clocks switch again, the clocked inverter switches off, *storing the information into the parasitic capacitance* formed by its output capacitance, the stray capacitance of the interconnection, and the gate capacitance(s) of the input transistor(s) of other blocks.

The block shown in Fig. 4-24 features no internal race, because the inputs are allowed only one transition. This is also accomplished by connecting different gates (e.g., n-channel transistor based gates connected to p-channel transistor based gates) directly, while gates of the same kind (e.g., p-channel transistor based gates with p-channel transistor based gates) are connected through an inverter, in domino-like fashion.

A NORA block holds the information produced during the evaluation phase even in the presence of significant skews between the two clock phases. This is possible because the clocked inverter-latch is used. If the last gate before the inverter changes its status during evaluation — for instance, if the output of the gate goes up, as in Fig. 4-24 — the actual output of the clocked inverter is produced by *only one of the two clocks, not by both.* In our case, the output of the inverter will go down when `clk` goes up, regardless of the timing relationship that `clk` has with $\overline{\texttt{clk}}$. Even if there were an overlap between the two phases, the output would be glitch-free and stable. If the output of the last stage confirms the value of precharging (that is, if it is low in the circuit shown in Fig. 4-24), the actual output of the clocked stage is driven by $\overline{\texttt{clk}}$. Again, the output will be stable and glitch-free, regardless of the timing relationship between $\overline{\texttt{clk}}$ and `clk`.

Problems can arise when clock fall-time and rise-time are long: the output clocked inverter can corrupt the stored information because of charge redistribution inside its nodes [3, 8]. A way to avoid this phenomenon is to use a clock pair — and its complement — instead of one clock — and its complement [8]. Fig. 4-25 shows a gate of this type; four clocks are necessary.

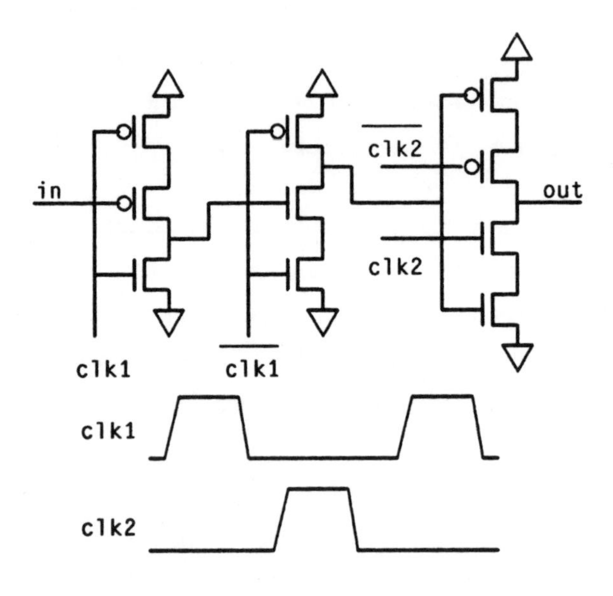

Figure 4-25: Four clocks can be used to avoid corruption of the value stored at the output of the clocked inverter due to slow rising/falling clocks.

Finally, just as in all other dynamic gates we have considered up to now, internal charge sharing is not eliminated. As was pointed out earlier, NORA logic is oriented to highly pipelined circuits and cannot be considered a general-purpose approach. Moreover, its cost in terms of extra gates is justifiable only when the logic functions implemented are sufficiently complex, otherwise the overhead of the clocked inverter becomes a dominant factor as far as transistor count is concerned.

4.3. Charge Sharing

Dynamic logic works because the parasitic capacitors associated with output nodes can store information. Such capacitors can be gate capacitances, interconnect capacitances, and so on.

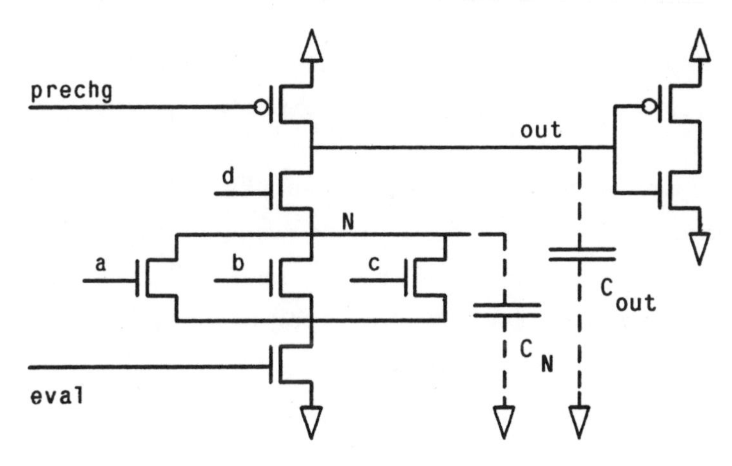

Figure 4-26: Capacitances responsible for charge sharing.

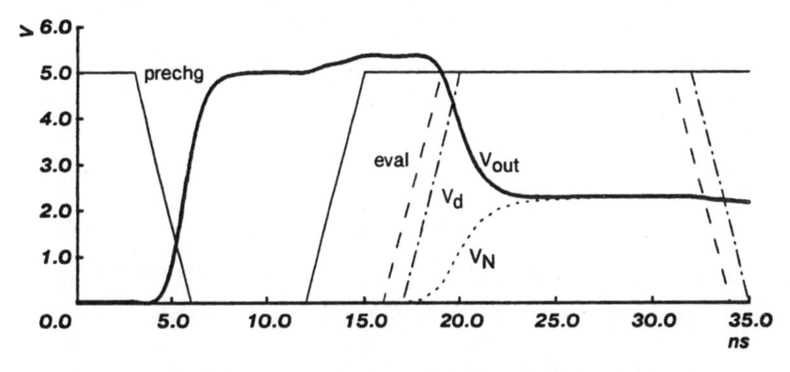

Figure 4-27: Waveforms showing the effect of charge sharing for the circuit in Fig. 4-26.

Dynamic logic is based also on the storing-sensing concept. After the information has been stored into a node capacitance, its status is sensed through a different node.

If the capacitor is charged we will have a "1" logic level, if the capacitor is discharged we will have a "0" logic level (positive logic is assumed). Note that the sensing node will have *its own capacitance.* This observation leads us directly to the phenomenon of charge sharing. In other words, a correct sensing operation can be done *only when the value of the sensing capacitor is much smaller than the value of the capacitor which stores the information.* If this is not the case, the sensing operation will be destructive, because the charge will redistribute itself between the two capacitors.

This is explained better by the example shown in Fig. 4-26. C_{out} is the capacitance associated with the storage node. C_{out} is the sum of the drain capacitance of the precharge p-channel transistor, the input capacitance of the inverter driven by the node, and the drain capacitance of the n-channel transistor whose input is d. Fig. 4-27 shows the effect of charge sharing:

1. Precharge takes place: the node out is correctly pulled-up and reaches V_{dd} at 7ns.

2. The precharge signal goes up and reaches V_{dd} at 15ns. Because the evaluation signal is still low, the four n-channel devices are isolated from both V_{dd} and V_{ss}; the output node is floating at V_{dd}.

3. At about 16ns the evaluation signal starts to rise and reaches V_{dd} at 18ns. The input signals to the nodes a, b, and c — not shown in the figure — stay low. The input signal to node d (V_d) starts to rise and reaches V_{dd} at 20ns. The corresponding n-channel device enters conduction.

4. The node out is coupled, through the n-channel transistor whose input is d, to the capacitance associated with node N. If this capacitance is large enough — the capacitance of node N is the contribution of the drain capacitance of three transistors — the charge stored in C_{out} redistributes between C_{out} and C_N, incorrectly lowering the voltage at the output node.

Eventually, the voltage drops below $V_{dd}/2$ and switches the inverter.

There are no simple solutions to avoid this effect in the circuit shown in Fig. 4-26. To avoid charge sharing effects the designer can follow three approaches:

- Rewrite the boolean equations to eliminate the three-input OR.

- Make sure that V_d can make a high-going transition only during the precharge phase and never during the evaluation phase.

- Provide "shunt" transistors to discharge nodes likely to cause charge sharing. An example of this technique is presented in Fig. 6-8.

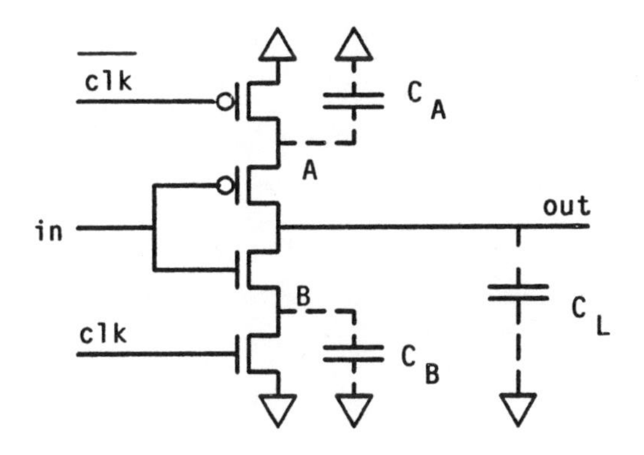

Figure 4-28: Charge sharing from a different interconnection scheme in C^2MOS logic.

Another interesting example of charge sharing is a C^2MOS gate with inverted interconnections, as shown in Fig. 4-28 together with the two junction capacitances that are responsible for the effect. Let us suppose that `clk` is high and `in` is low: the output of the gate is high. When `clk` goes down, the gate output is disconnected and stays high. Now, with the clock still low, the input goes up. The n-channel transistor whose gate is driven by the input signal conducts, and, if the junction capacitance C_B is large enough, the charge stored in C_L is redistributed between C_B and C_L. The voltage at the output might go below $V_{dd}/2$ and create a spurious glitch or even a wrong logic value in connected gates. The same situation occurs when both `in` and `clk` are high. In this case the output is low. If the input makes an low-going transition after `clk` goes down, the p-channel whose gate is connected to the input starts to conduct and the C_A junction capacitor will pull up the output, which may go above $V_{dd}/2$. C^2MOS allows the input to change only when `clk` is high. Note that a clocked inverter connected as shown in Fig. 4-20 features no charge sharing.

A logic that has been claimed to be free of charge sharing is shown in Fig. 4-29 [7]. It uses four different clocks and aims to separate the evaluation phase from the holding phase, when the output is fed to the next gate. When `clk1` goes low, the node X is precharged high. Note that the transistor n_A is conducting. Although there might be charge redistribution because both transistors are conducting, there is no corruption at node X because it is strongly pulled up to V_{dd}. When `clk1` goes up, the transistor n_A is still conducting and allows evaluation of the logic function. Then, `clk2` goes down, and the node X is physically isolated (holding phase). Because all the inputs are kept stable during precharging — these inputs come from other gates that are "holding" their output — charge sharing during precharging is eliminated. However, charge sharing during evaluation — that is, charge sharing caused by logic functions, as in Fig. 4-26 — is not eliminated.

4.4. Bootstrap Logic

The major advantage of bootstrap logic is the ability to speed up a circuit with a small increase in power dissipation. On the other hand, bootstrap logic is not suitable for a bulk CMOS process. Therefore, bootstrap logic is presented only for the sake of completeness, but we will also point out when its use is possible. Fig. 4-30 shows a simple bootstrap OR gate: C_a and C_b are the bootstrap capacitors, an essential

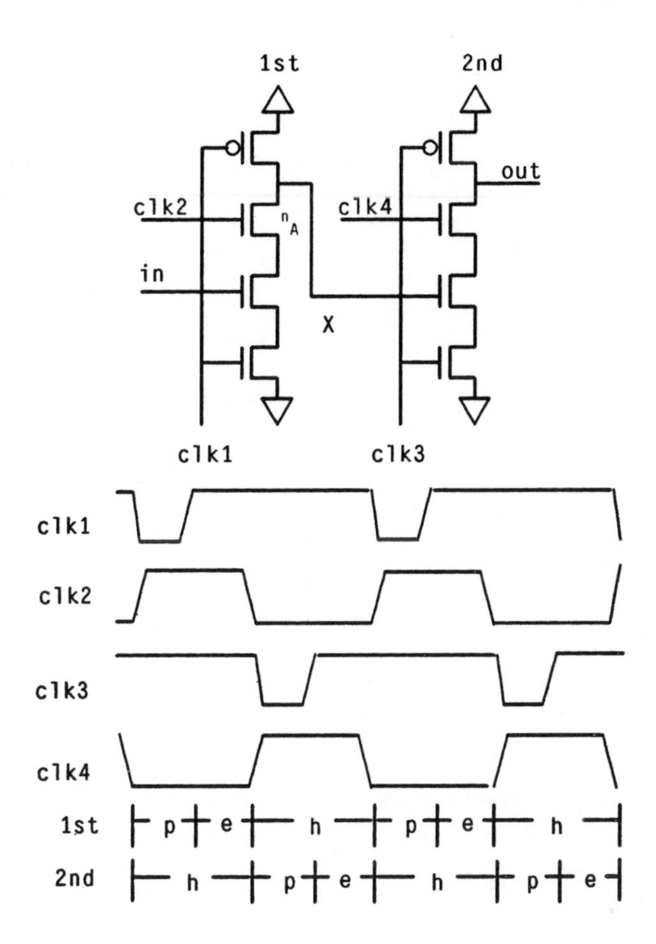

Figure 4-29: A logic gate with no charge sharing: four clocks are used to separate the evaluation phase from the holding phase, when the output is fed to the next gate.

component of this logic.

When `prechg` goes high (`eval` is low), the two pass-transistors conduct, and the inputs a and b are transferred to the nodes A and B. At the same time, the precharge signal drives the drain and source of the leftmost transistor to V_{ss} together with the

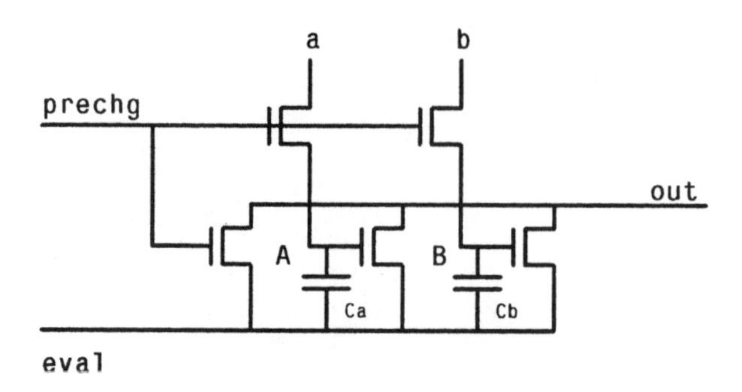

Figure 4-30: Bootstrap OR gate.

drain and source of the other two transistors. The output is therefore precharged low. The load that each input signal coming from the pass-transistor sees is the gate capacitance in parallel with the bootstrap capacitor. Let us assume now that at least one of the inputs is high, say a. The capacitor C_a and the gate capacitance at node A are charged by the signal. If b is low, C_b and the gate capacitance at node B are not charged.

When prechg goes low and eval goes high, the node A is disconnected from the source of the pass-transistor, which is turned off by the precharge signal going low. Not only will the node A be driven high by the high-going transition of the evaluation signal, but the capacitor C_a will "pump" the stored charge into it. As a matter of fact, the voltage at node A will go well above V_{dd}. This creates an overdrive situation for the transistor because its gate voltage can be as high as 7V, making the transistor enter conduction more quickly. The evaluation signal will pass through the transistor and appear at the output.

This circuit clearly demonstrates the effect of bootstrapping, although it cannot be considered a straightforward approach to making an OR gate. In fact, the effect of bootstrapping is decreased, because the n-channel device used as a pass transistor degrades the signal. There is a voltage drop when the n-channel device passes a high signal, which does not allow full charging of the bootstrap capacitor — unless both pass-transistors themselves are overdriven by the precharge signal.

The circuit shown in Fig. 4-30 cannot actually be implemented without some precautions. First, the precharge signal must overdrive, that is to say, it should reach 7V. Although the gate capacitance itself could be used as a bootstrap capacitor, current technologies feature small gate capacitances, and an extra transistor must be used — as a sort of "integrated capacitor" — to provide sufficient bootstrap action. Therefore, the device count increases.

It was mentioned at the beginning of this section that bootstrap logic is not recommended in a bulk CMOS process. The reason for this is very simple: bootstrapped gates are based on the concept of overdriving nodes and therefore some of these nodes reach voltage levels well above V_{dd} or below V_{ss}. As we saw in Chapter 3, this is the principal cause of latchup. Moreover, bootstrap logic requires extreme care for small feature sizes. The overdrive can cause catastrophic failures even in nMOS fabrication processes because it increases the likelyhood of hot-electron effects. Very few examples of bootstrapped circuits in bulk CMOS can be found in the literature. One of these is a clock generator for a 1-Mbit dynamic RAM chip [15]. However, great care has been taken to isolate the clock circuitry from the rest of the chip. The design of the clock generator uses one kind of device only (i.e., n-channel transistor).

Bootstrapped gates have been used extensively in nMOS technology, especially for on-chip clock generators. A circuit of this kind is described in [2]; it can drive pulses with 1ns rise and fall-time into a 50pF load. A comprehensive use of bootstrapping techniques in nMOS design is reported in [6, 13].

4.5. Logic Design at the System Level

After reviewing the most important static and dynamic logics, the problem of which logic to use is finally addressed. The problem does not have a unique solution and depends on the particular circuit being implemented. However, some general guidelines can be offered.

First, dynamic logic is inherently faster than static logic, *when the precharge phase is not considered.* If we consider a precharge-high, n-channel device based gate, pulling down the output is faster than pulling down the output of a static gate, because:

1. The gate will start to pull down when the input signal goes above the *device threshold voltage* — 0.8V, for instance. A static gate will pull down only when the input signal goes above the *gate threshold voltage* which is half of (V_{dd} – V_{ss}) for a balanced gate.

2. The output of the dynamic gate has to drive a capacitance which is the sum of all the gate input capacitances of the n-channel (or p-channel) devices connected to it; a static gate sees both p- and n-channel transistor gate capacitances. Moreover, we have ignored stray capacitances. Because dynamic gates occupy less area than static gates do, it is likely that stray capacitances will be larger in static logic design.

However, if a fair comparison is to be attempted, the precharge time of dynamic gates must be factored into the analysis. In this case a comparison becomes more difficult and the analysis has to be done on a case-by-case basis.

As a simple example, let us consider a critical path implemented both in static and dynamic logic and assume that precharging takes 20ns and domino logic is used. If the complete evaluation of the path in domino logic is faster than the corresponding path implemented with static logic and the difference between the two is more than 20ns, it pays to use dynamic logic, because the sum of precharge and evaluation will be smaller than the delay of the static logic path. This, of course, does not consider other parameters, such as the design of a clock generator — if on-chip — the problems associated with clock layout, charge sharing, etc. Overall costs should take this into account, and a complete analysis should include the cost in terms of design time of a faster logic. Although the number of devices decreases in dynamic logic, the typical device per man/day figure is lower in dynamic logic design.

Speed and design time are not the only parameters to be considered. Dynamic logic requires less area than complementary static logic, and this leads to smaller circuits and higher yield. The final decision will be based on a global evaluation of all the above parameters, and it is likely that only part of the circuit will be implemented in dynamic or static logic. As a matter of fact, dynamic and static logics can both exist on the same chip.

Another important consideration relates to the integration of the chip in a system. The "speed" of the chip should be evaluated at the system level: it does not make much sense to optimize I/O and internal timing when the chip will behave poorly within a system. This issue is even more complex than the ones dealt with earlier, because an in-system evaluation before the actual implementation is impossible for many chips. This is especially true for general-purpose components, such as microprocessors, where the external requirements might be unspecified throughout most of the design period. However, special-purpose components can have clearer requirements. In this case both the internal timing and I/O interface can influence the choice between static logic and dynamic logic or the choice of a particular logic within a family.

There is no reason to state that one logic is always better. The designer should carefully analyze the internal and external constraints and choose according to the various points of merit we have presented.

References

[1] Chatterjee, P.
Device Design Issues for Deep Submicron VLSI.
In *Proc. of the 1985 International Symposium on VLSI Technology, Systems and Applications - Taiwan*, pages 221-226. May, 1985.

[2] Cook, P.W. *et al.*.
1μm MOSFET VLSI Technology: Part III - Logic Circuit Design Methodology and Applications.
IEEE Journal of Solid-State Circuits SC-14(2):255-267, April, 1979.

[3] Goncalves, N.F. and H.J. De Man.
NORA: A Racefree Dynamic CMOS Technique for Pipelined Logic Structures.
IEEE Journal of Solid-State Circuits SC-18(3):261-266, June, 1983.

[4] Heller, L.G. and J.W. Davis.
Cascode Voltage Switch Logic.
In *ISSCC Dig. Tech. Pap.*, pages 16-17. 1984.

[5] Krambeck, R.H., C.M. Lee and H.S. Law.
High-speed Compact Circuits with CMOS.
IEEE Journal Solid State Circuits SC-17(3):614-619, June, 1982.

[6] Lutz, C., *et al.*.
Design of the Mosaic Element.
In *Proc. of 1984 Conference on Advanced Research in VLSI*, pages 1-10. M.I.T., January, 1984.

[7] Meyer, J.E.
MOS Models and Circuit Simulation.
RCA Review 32:42-63, March, 1971.

[8] Murray, A.F. and P.B. Denyer.
A CMOS Design Strategy for Bit-Serial Signal Processing.
IEEE Journal of Solid-State Circuits SC-20(3):746-753, June, 1985.

[9] Pfennings, L.C.M.G. *et al.*.
Differential Split-Level CMOS Logic for Subnanosecond Speeds.
IEEE Journal of Solid-State Circuits SC-20(5):1050-1055, October, 1985.

[10] Pfiester, J.R., J.D. Shott and J.D. Meindl.
Performance Limits of CMOS ULSI.
IEEE Journal of Solid-State Circuits SC-20(1):253-263, February, 1985.

[11] Radhakrishnan, D., S.R. Whitaker and G.K. Maki.
Formal Design Procedures for Pass Transistor Switching Circuits.
IEEE Journal of Solid-State Circuits 20(2):531-536, April, 1985.

[12] Sakamoto, H. and L. Forbes.
Grounded Load Complementary FET Circuits: Sceptre Analysis.
IEEE Journal of Solid-State Circuits :282-284, August, 1973.

[13] Seitz, C.L. *et al.*.
Hot-Clock nMOS.
In *Proc. of 1985 Chapel Hill Conference on VLSI*, pages 1-17. 1985.

[14] Suzuki, Y., K. Odagawa and T. Abe.
Clocked CMOS Calculator Circuitry.
IEEE Journal of Solid-State Circuits SC-8(6):462-469, December, 1973.

[15] Taylor, R.T. and M.G. Johnson.
A 1-Mbit CMOS Dynamic RAM with a Divided Bitline Matrix Architecture.
IEEE Journal of Solid-State Circuits SC-20(5):894-902, October, 1985.

[16] Wu, M.Y., W. Shu and S.P. Chan.
A Unified Theory for MOS Circuit Design - Switching Network Logic.
Int. J. Electronics 58(1):1-33, January, 1985.

Chapter 5
Circuit Design

Circuit design requires a global optimization of various parameters, given the specifications of the chip. The most important parameters that the designer has to consider are *speed*, *area*, *noise margin*, and *power dissipation*. The last parameter is somewhat less important than the others in CMOS technology, but with the frequency of circuit operation steadily increasing this no longer holds true. Moreover, as we shall see in Chapter 7, power dissipation considerations become extremely important when we consider the input/output section of the chip.

No general theory is presently available for circuit performance optimization, by which we mean the minimization of an objective function which includes all four parameters listed above. Lacking a general theory, even sophisticated CAD tools cannot perform global optimization and aim to optimize a power/speed figure, for instance. The major problem arises when an "area" function has to be factored in. So many parameters are involved (design rules, logic design methodology, etc.), that achieving a general-purpose area-cost function becomes a very difficult, if not impossible, task. This is not true when some constraints are introduced, as with gate-array or standard-cell design. In this case it is possible to come up with an area function which well represents the *actual* area [12, 8], because strong constraints, such as maximum number of inputs to each gate, can be imposed.

This chapter gives the reader the information necessary to *locally* optimize the design of a circuit; this information can also be used to characterize gate-array cells or macro cells. Special emphasis is given to the determination of parasitic parameters, such as resistance, capacitance, and inductance of interconnections, because it is generally agreed that their role is becoming increasingly important with the shrinkage of feature size. In fact, it is the delay of the interconnections which limits the overall speed of circuits.

Once the parasitic parameters have been determined, it is possible to apply design methodologies to minimize the delay through a chain of gates. We will present some results for delay minimization calculation, using an inverter chain as a case-study. We will also show how these results can be extended to multi-input gates. First, an optimal pull-up|pull-down ratio is derived. Second, the inverter chain delay is computed, both with and without stray capacitance. This methodology fails to produce accurate results when very wide channel devices are used. In this case it becomes necessary to consider second-order effects, in particular the RC constant of the gate. This topic will be dealt with in Chapter 7, where a more rigorous treatment of inverter chain delay is presented. Finally, some considerations on static and dynamic logic sizing will be presented.

5.1. Resistance, Capacitance, and Inductance

This section introduces the fundamental formulae to compute the resistance, capacitance, and inductance of various elements in a circuit, namely metal and polysilicon wires, contacts, and diffusion regions. As has already been pointed out, these parameters played an important role in earlier fabrication processes, but the gate delay was the limiting factor in circuit speed. With today's processes, line widths have made interconnection delay the limiting factor, because most parasitic parameters increase for smaller feature size. Fig. 5-1 shows the RC constant for various interconnect materials, namely polysilicon, silicide ($TaSi_2$), and aluminum.

For line widths (or feature sizes) below 1.5μm, the RC aluminum constant per centimeter becomes larger than the RC constant of the MOS capacitor (t_d), and, for submicron values, a few hundred microns of aluminum produce a higher RC constant than a MOS transistor.

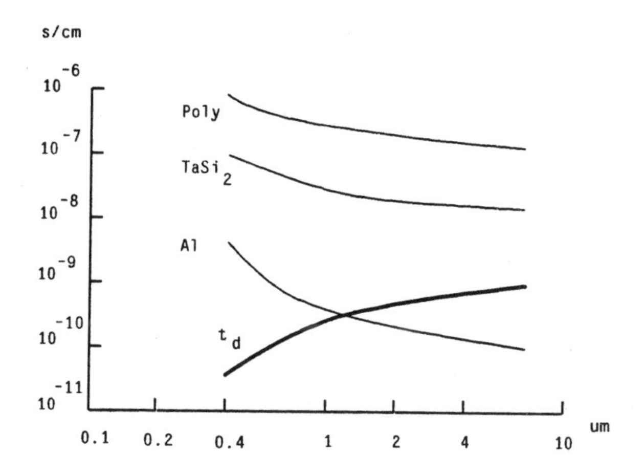

Figure 5-1: RC constant per unit length vs. line width for various conductors and for a minimum feature size MOS capacitor (t_d). *After [17].*

When the propagation delay of the interconnection is comparable to the switching device fall-time or rise-time, the interconnection has to be dealt with as a transmission line and proper impedance-matching procedures carried out. The typical propagation time for an interconnection is given in [10]:

$$\tau_{pd} \approx 33\sqrt{\varepsilon_r}\ \text{ps/cm},$$

where ε_r is the dielectric constant of the material. If we assume that the material is silicon, we have:

$$\tau_{pd} \approx 110\ \text{ps/cm}.$$

A rule of thumb to decide whether the interconnect should be treated like a transmission line is the following:

$$L_c = \frac{1}{3}\,\frac{min[\tau_r,\tau_f]}{\tau_{pd}} ,$$

where L_c is the *critical path length* and τ_f and τ_r are the gate fall-time and rise-time, respectively. If the interconnection path is longer than L_c, transmission line design methodologies must be used. In the case of silicon and with a gate rise-time of 200ps, we have a critical path length of 0.6cm, that is, 6000μm. It is clear at this point that more and more care has to be applied to the layout of interconnections and the use of material with very low dielectric constant becomes necessary.

5.1.1. Interconnect Resistance

An approximate formula for the resistance of a conductor is:

$$R = \rho\left(\frac{L}{TW}\right), \tag{5-1}$$

where L is the conductor length, T is its thickness and W its width. ρ is the resistivity of the material; typical values for ρ are:

$$\rho(\text{polysilicon}) = 3\text{x}10^{-3}\Omega\text{cm} ;$$

$$\rho(\text{aluminum}) = 2.5\text{x}10^{-6}\Omega\text{cm} ;$$

$$\rho(\text{silicide - TaSi}_2) = 125\text{x}10^{-6}\Omega\text{cm} .$$

The interconnection resistance is usually expressed in $\Omega/\square$, where the symbol $\square$ ("square") indicates a geometrical square of conductor. An interconnection 2μm wide and 100μm long will be 50$\square$. If the resistance is 25$\Omega/\square$, the total resistance of the interconnection is 1250Ω.

5.1.2. Interconnect Capacitance

Let us compute the capacitance of a conductor with the geometric parameters shown in Fig. 5-2. L is the wire length, W its width, the thickness of the wire is T, and H is the oxide height.

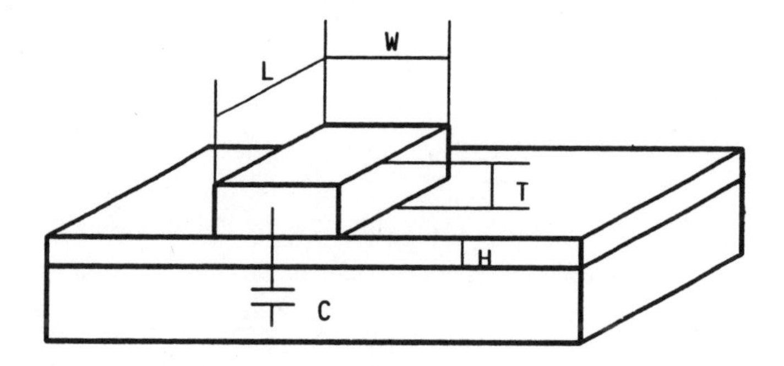

Figure 5-2: Wire capacitance: parameters.

The capacitance per unit length of a single, isolated line can be computed by using the following empirical formula [14]:

$$C = \varepsilon_0 \varepsilon_{ox} \left[1.15\left(\frac{W}{H}\right) + 2.8\left(\frac{T}{H}\right)^{0.222} \right] . \tag{5-2}$$

Eq. (5-2) is the sum of two terms, one relative to the lower and upper surfaces of the conductor, and the second representing the side-wall contribution (fringe effect). If a finite line is considered, Eq. (5-2) has to be modified to take into account the influence that the four corners of a finite wire introduce in the overall capacitance. We have:

$$C = \varepsilon_0 \varepsilon_{ox} \left[1.15\left(\frac{LW}{H}\right) + 1.4(2W + 2L)\left(\frac{T}{H}\right)^{0.222} + \right.$$

$$\left. + 4.12H\left(\frac{T}{H}\right)^{0.728} \right] ; \tag{5-3}$$

that is, an area contribution (first term), a perimeter contribution (second term) and a corner contribution (third term). If L $\gg$ W we obtain:

$$C = \varepsilon_o \varepsilon_{ox} \left[1.15\left(\frac{L}{H}\right) + 2.8L\left(\frac{T}{H}\right)^{0.222} + 4.12H\left(\frac{T}{H}\right)^{0.728} \right]. \tag{5-4}$$

As an example, suppose W = 3μm, L = 400μm, T = 5000Å and H = 1000Å. Using Eq. (5-3) we obtain:

$$C \approx 0.5\text{pF}.$$

A problem which is common to many circuits is the layout of long parallel lines to form buses. When parallel lines are considered, the capacitance of each conductor does not follow Eq. (5-2) because of the presence of interelectrode (coupling) capacitance, as Fig. 5-3 shows. The capacitance of each conductor is the contribution of its own intrinsic capacitance plus the coupling capacitance with the other conductor. We can write:

$$C_a = C_{a(intrinsic)} + C_{ab(coupling)}, \tag{5-5}$$

where $C_{a(intrinsic)}$ follows Eq. (5-2), and

$$C_{ab(coupling)} = \varepsilon_o \varepsilon_{ox} \left[0.03\left(\frac{W}{H}\right) + 0.83\left(\frac{T}{H}\right) + - 0.07\left(\frac{T}{H}\right)^{0.222} \right] \left(\frac{S}{H}\right)^{-1.34}. \tag{5-6}$$

S is the spacing between the two parallel conductors. The relative error of Eq. (5-5) is less than 10% with the constraints:

$$0.3 < \frac{W}{H} < 10, \quad 0.3 < \frac{T}{H} < 10, \quad \text{and} \quad 0.5 < \frac{S}{H} < 10.$$

However, Eqs. (5-2) through (5-6) are valid if the medium above the conductor is

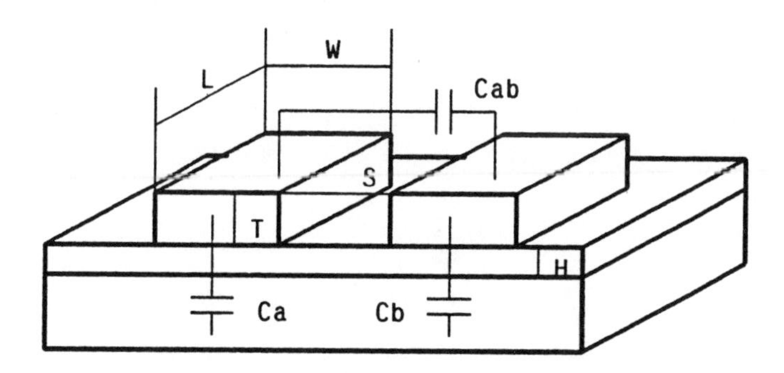

Figure 5-3: Interelectrode (coupling) capacitance.

air. More often, metal lines are covered by an overlay of different material. When this is true, Eqs. (5-2) through (5-4) may still be considered acceptable (with an error higher than 10%), but Eq. (5-6) underestimates the value of the coupling capacitance. To make the analysis simpler, we will assume that the interconnection is deposited over the field oxide (SiO_2) and the overlay can be either air or "overglass" (SiN). In reality, a metal conductor would be deposited over an insulator such as P-glass (phosphorous-doped silicon dioxide) to avoid short circuit with other interconnection layers (e.g., polysilicon). However, this would make the analysis of the capacitance much more complex, because the dielectric below the interconnection would be composite — a layer of silicon-dioxide with P-glass above it. Our model considers the interconnect to be part of a sandwich structure of silicon dioxide (below) and SiN (above). If double metal structures were to be considered, the first metal would be deposited between silicon dioxide and P-glass — or other insulating material — and the second metal layer deposited between P-glass and SiN.

We can achieve higher accuracy by using the basic theory of coupled microstrip structures [5, 11], because the interconnection can be considered a microstrip. We will now present a set of equations that depends on the material both *under* and *above* the conductors. The formulae will be shown for the conductor pair. Each conductor, therefore, has its own intrinsic capacitance and a coupling capacitance. The basic model is a coupled microstrip structure consisting of two parallel, symmetric lines that support only two modes of transmission, that is, even and odd modes (see, [4]); Fig. 5-4 shows the parasitic capacitances in both transmission modes.

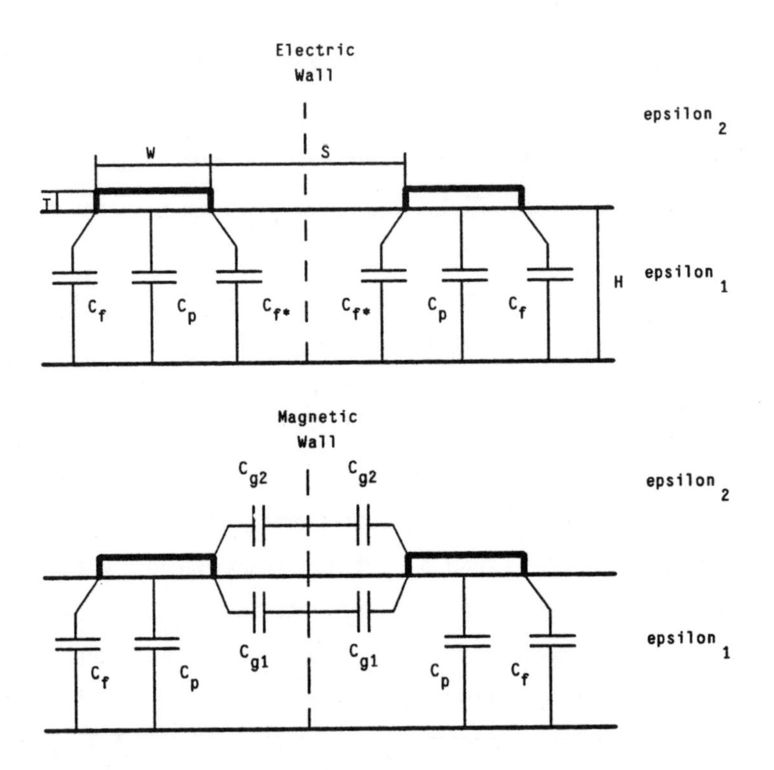

Figure 5-4: Two-conductor structure considered as coupled microstrips.

The intrinsic capacitance of each conductor can be expressed as the average between the capacitance C_{ev} associated with the even mode and the capacitance C_{od} associated with the odd mode, that is:

$$C_{intrinsic} = \frac{1}{2}(C_{ev} + C_{od}),$$

while the coupling capacitance is:

$$C_{coupling} = \frac{1}{2}(C_{od} - C_{ev}).$$

Moreover, we have:

$$C_{ev} = C_p + C_f + C_{f*}, \tag{5-7}$$

$$C_{od} = C_p + C_f + C_{g1} + C_{g2} + C_{gt}. \tag{5-8}$$

The physical meaning of the above parameters is the following:

- C_p is the parallel plate capacitance between the wire and the substrate;

- C_f and C_{f*} are capacitances from fringe effects;

- C_{g1}, C_{g2}, and C_{gt} are from mutual coupling between the two conductors.

Finally, we have C_M, which is the intrinsic microstrip capacitance:

$$C_M = \varepsilon_{eff}[Z + 1.393 + 0.66\ln(Z + 1.444)], \text{ for } W > 2H\ ; \tag{5-9}$$

$$C_M = 2\pi\varepsilon_{eff}\left[\ln\left(\frac{8}{Z} + \frac{Z}{4}\right)\right]^{-1}, \text{ for } W \leq 2H\ ; \tag{5-10}$$

$$Z = \frac{W}{H} + \left[\frac{1.25T}{\pi H}\right]\left[1 + \ln\frac{2H}{T}\right]; \tag{5-11}$$

$$\varepsilon_{eff} = \frac{1}{2}(\varepsilon_1 + \varepsilon_2) + \frac{1}{2}(\varepsilon_1 + \varepsilon_2)\sqrt{\frac{W}{W + 12H}} +$$

$$+ \frac{1}{4.6\sqrt{W/H}}(\varepsilon_1 - \varepsilon_2)(T/H)\ . \tag{5-12}$$

Note that ε_1 and ε_2 are not the relative permittivity values but the actual ones. We have the following equations (see Fig. 5-4):

$$C_p = \varepsilon_1 \frac{W}{H} ; \tag{5-13}$$

$$C_f = \frac{1}{2}(C_M - C_p) ; \tag{5-14}$$

$$C_{f*} = \frac{C_f}{1 + H/S} \sqrt{\frac{\varepsilon_1}{\varepsilon_{eff}}} ; \tag{5-15}$$

$$C_{g1} = \frac{\varepsilon_1}{\pi} \ln \coth\left[\frac{\pi S}{4H}\right] +$$

$$+ 0.65\, C_f \left[\frac{0.02}{S/H} \sqrt{\frac{\varepsilon_1}{\varepsilon_o}} + 1 - \left(\frac{\varepsilon_o}{\varepsilon_1}\right)^2\right] ; \tag{5-16}$$

$$C_{g2} = \frac{\varepsilon_2}{\pi} \left[0.693 + \ln\left(\frac{1 + X'}{1 - X'}\right)\right], 0 \le X^2 \le 0.5 ; \tag{5-17}$$

$$C_{g2} = \frac{\pi \varepsilon_2}{ln(1 + \sqrt{X}) - ln(1 - \sqrt{X})}, 0.5 \le X^2 \le 1 ; \tag{5-18}$$

$$X = \frac{S}{S + 2W} ; \qquad X' = \sqrt{1 - X^2} ;$$

$$C_{gt} = 2\varepsilon_2 \frac{T}{S} . \tag{5-19}$$

The above formulae compute the capacitance of both a coupled microstrip structure — when the dielectric above the structure is air — and a coupled stripline-like structure — when the dielectric above the structure is not air. An example of capacitance computation is presented in Appendix B, where the approximate method presented at the beginning of this section is compared with the more accurate one presented above. The error in the first method is within 10%, as pointed out before. It is the designer's responsibility to decide which of the two methods should be used. The second approach is recommended when the interconnect is on a critical path and is driven at very high frequencies.

5.1.3. Interconnect Inductance

The inductance in air of a coupled microstrip (stripline) structure can be expressed by the following formulae:

$$L_{intrinsic} = \frac{\mu_o \varepsilon_o}{2} \left[\frac{1}{C^a_{od}} + \frac{1}{C^a_{ev}} \right] \text{H/cm} ; \tag{5-20}$$

$$L_{coupling} = \frac{\mu_o \varepsilon_o}{2} \left[\frac{1}{C^a_{ev}} - \frac{1}{C^a_{od}} \right] \text{H/cm} . \tag{5-21}$$

μ_o is the permeability and C^a_{ev}, C^a_{od} are the capacitances *in air* for the even and odd modes. If the medium is not air, and there is a dielectric-air interface, the dielectric constant of the material is introduced in the equations. Typical values for conductor inductance are in the nH/cm range. Line inductance is critical in power and ground lines, where a large current circulates and significant inductive coupling noise can be generated.

5.1.4. Interconnect Discontinuities

Discontinuities in the layout of interconnections, such as the ones shown in Fig. 5-5, introduce further parasitic inductances and capacitances [6]. However, these parasitic parameters can be ignored when the frequency of operation is below 1GHz.

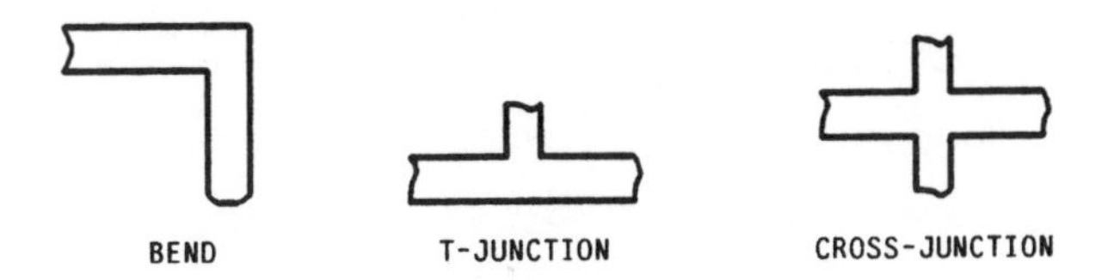

Figure 5-5: Interconnect discontinuities and associated parameters.

5.1.5. Coupling Parameters and Interconnect Delay

Let us consider two inverters which drive two parallel 0.5cm metal conductors. The parasitic parameters for the interconnections are:

$$C_{intrinsic} = 1.756pF\,;$$

$$C_{coupling} = 0.114pF\,;$$

$$L_{intrinsic} = 5.03nH\,;$$

$$L_{coupling} = 4.4nH\,.$$

$C_{intrinsic}$ and $C_{coupling}$ are computed in Section B.2, and $L_{intrinsic}$, $L_{coupling}$ are computed from Eqs. (5-20) and (5-21), where ε_o has been substituted by $\varepsilon_o\varepsilon_{ox}$. In fact, we assume that the conductors run inside SiO_2. The circuit is shown in Fig. 5-6(b), and a simple 1-stage ladder is used to model the interconnection. The size of the devices is in microns.

Fig. 5-7 shows an input waveform and three output waveforms V_{o1}. Waveform 1 refers to the isolated connection shown in Fig. 5-6(a), where only intrinsic capacitance is between the interconnection and V_{ss}. Both waveform 2 and 3 refer to the circuit of Fig. 5-6(b). Waveform 2 has been obtained by driving both inverters with the same signal in Fig. 5-7 (bold). Waveform 3 has been obtained by driving the first inverter with the same signal, while the second inverter is driven by an opposite signal. The influence of the interconnection coupling is evident, even though the parameters used (feature size, oxide thickness, line width, etc.) come from a fairly conventional technology. Finally, curves 1 and 2 are completely overlapped. When both inverters are driven by the same signal, the voltage difference across the inductor-capacitor coupling structure is always zero — or very close to zero. Therefore, conductor coupling is drastically reduced.

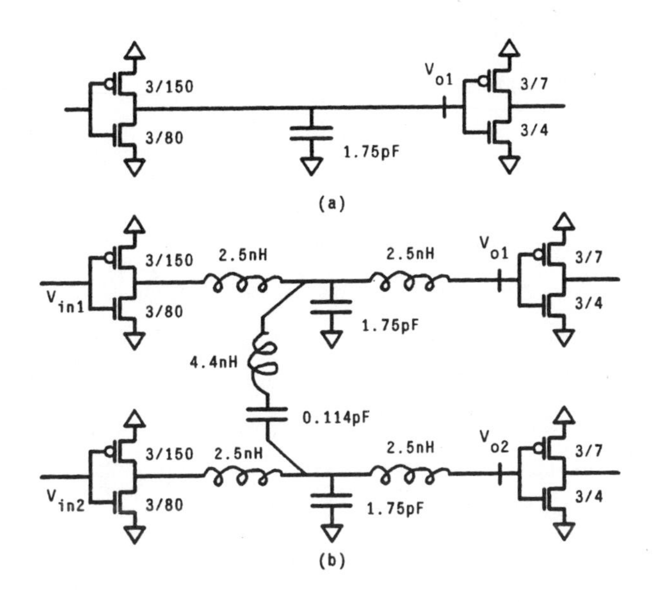

Figure 5-6: Two inverters driving long, parallel lines. Intrinsic and coupling parasitic parameters.

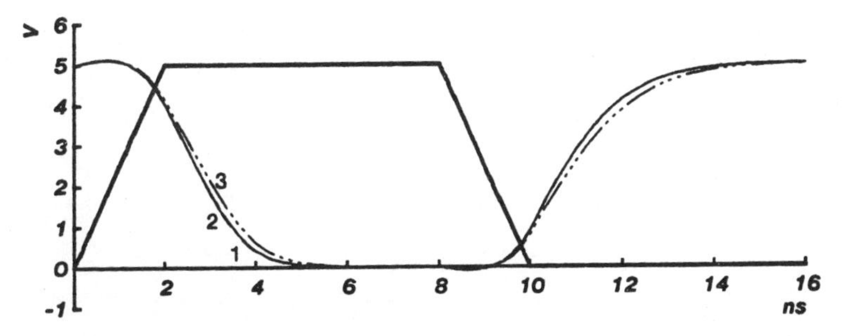

Figure 5-7: Input and output waveforms for the circuit in Fig. 5-6.

Note also that the intrinsic inductance of the interconnection has far less importance than the capacitance because of the voltage-controlled nature of the MOS transistors. However, when large currents are involved, as in the case of power lines, the presence of large inductances does create noise problems.

5.1.6. Diffusion Resistance

The resistance of a diffused region can be expressed as:

$$R_s = \rho_d \frac{1}{(x_j - x)} , \tag{5-22}$$

where ρ_d is the average resistivity of the diffusion region, x_j is the junction depth and x is a vertical coordinate which measures the distance from the surface: when $x = 0$ we measure the surface resistance. The average resistivity ρ_d depends on the surface concentration of the diffusion, the substrate doping concentration and the diffusion profile of the junction, such as Gaussian, erfc, etc. Tabulation of the diffusion resistance for various diffusion profiles and substrate concentrations can be found in so-called "Irvin curves" [7, 1].

For instance, for substrate concentration of $10^{16}cm^{-3}$, surface concentration of $10^{19}at/cm^3$, and $x_j = 1\mu m$ we have the following, for a Gaussian diffusion:

$$\rho_d\,(\text{p-type}) \approx 0.02\Omega cm \rightarrow R_s \approx 200\Omega/\Box ;$$

$$\rho_d\,(\text{n-type}) \approx 0.012\Omega cm \rightarrow R_s \approx 125\Omega/\Box .$$

Typical diffusion resistances in today's processes are around $80\Omega/\Box$. However, new fabrication technologies for sub-micron processes have shown diffusion resistances of the order of $8\Omega/\Box$ [2]. This reduction compares very favorably with lightly doped diffusion techniques, which are currently considered viable candidates for sub-micron processes. Finally, note that the n-type diffusion has higher conductivity than the p-type diffusion and, if diffusion must be used as an interconnect — something which is never recommended, anyway — it is advisable to use n-type diffusion.

5.1.7. Contact Resistance

Contact resistance has commonly been neglected because of its very small influence on overall performance (except the metal-drain or metal-source contact in very wide transistors: see Appendix A). Again, with smaller feature sizes this no longer holds true, and contact resistance must be considered. Contact resistance is usually given in terms of Ω/contact. This is a "general purpose" parameter, as we shall see below, but it is sufficient in most cases. Contact resistance depends on the material forming the contact, such as aluminum, polysilicon, etc., and — in the case of metal-diffusion contacts — the doping concentration of the diffused layer. The *specific contact resistance* ρ_c is an important parameter for computing the contact resistance. For low doping concentrations (10^{17}cm^{-3} or less) ρ_c can be expressed as [18]:

$$\rho_c = \frac{k}{qA^*T} \exp\left[\frac{q\phi_S}{kT}\right] \ (\Omega\text{-cm}^2), \quad (5\text{-}23)$$

where A^* is the Richardson's constant and ϕ_S is the Schottky barrier height. We have:

$$A^* = \frac{4\pi q m^* k^2}{h^3} .$$

h is the Planck's constant and m^* is the effective mass of the charge carrier.

For higher doping concentrations, ρ_c is approximately proportional to $\exp\{\phi_S/(N_{d/a}{}^{1/2})\}$ [18]. Typical theoretical values for ρ_c in this case are $10^{-7}\Omega\text{cm}^2$. In practice, the actual resistance is much higher because of a high-resistivity interfacial layer which is produced during contact fabrication. This layer results from the formation of a composite material between, for instance, aluminum and diffusion. Therefore, typical values are of the order of 10^{-5}-$10^{-6}\Omega\text{cm}^2$. Some ϕ_S values are tabulated below.

Material	ϕ_S (V) for n-type Si	ϕ_S (V) for p-type Si
Aluminum (Al)	0.72	0.58
Molybdenum (Mo)	0.68	0.42
Titanium (Ti)	0.0.50	0.61
Tungsten (W)	0.67	0.45

The *contact resistance* R_c is defined as:

$$R_c = \frac{\rho_c}{A} ,$$

where A is the contact area. If we assume $\rho_c \approx 10^{-6}\Omega cm^2$ and a $4\mu m^2$ contact, we have:

$$\rho_c \approx 25\Omega .$$

Smaller contacts or higher impurities will increase the resistance, and it is not uncommon to have 100Ω/contact [19]. New fabrication processes have been proposed to reduce the contact resistance for micron and sub-micron technologies [3]. In fact, it is assumed that actual values for ρ_c of about $1.8x10^{-7}\Omega cm^2$ are necessary for $0.5x0.5\mu m^2$ contacts. Smaller contacts will require ρ_c of the order of $10^{-8}\Omega cm^2$.

R_c assumes that the current flowing through the contact is distributed uniformly all over the contact area. In reality, current is crowding at the leading edge of the contact — that is, the contact side encountered first by the current flow. This means that the potential drop across the interfacial layer is *not* constant [13] — constant potential drop is assumed in the computation of ρ_c and R_c. When diffused resistors have to be designed, a better characterization of the total resistance is given by R_f rather than R_c. R_f is the 'front" resistance of the contact, which is defined as:

$$R_f = \frac{\sqrt{R_s \rho_c}}{W_c} \coth\left[\left(\frac{R_s}{\rho_c} \right)^{1/2} L_c \right],$$

where R_s, W_c and L_c are the diffusion resistance, and the contact width and length, respectively. The total resistance of the diffused resistor is:

$$R_{dr} = R_s N_{sq} + 2R_f, \tag{5-24}$$

where N_{sq} is the number of squares, and two contacts are included. Note that R_f is necessary because R_c sometimes underestimates the contact resistance.

5.2. Modeling Long Interconnects

Once the parameters of the interconnect have been computed by using the formulae presented in the previous sections, the interconnect has to be modeled by an equivalent circuit which can approximate its characteristics (load, delay, etc.). Because of its very low resistance, a single capacitor can suffice for metal interconnections — inductances can be added if large currents flow through the interconnection, as is the case of power lines. This is not true when material with higher resistance is considered — polysilicon is a typical example. In this case, a model with distributed parameters has to be used, because a simple resistor-capacitor model would be highly inaccurate (relative error of 30% or more).

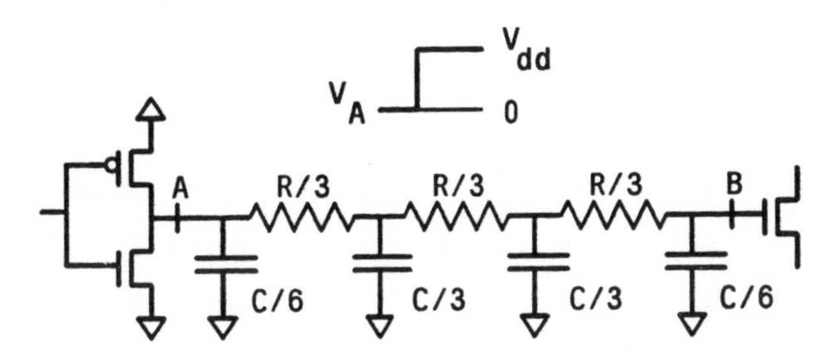

Figure 5-8: π-ladder equivalent circuit for long interconnections.

Among the many equivalent circuits, a 3-element π-ladder circuit provides an accurate model of a high-resistivity interconnection [15]. In fact, increasing the number of stages does not significantly improve the accuracy of the model. The π-ladder circuit is shown in Fig. 5-8. This model introduces an error of about 3%. The circuit can be used in computer-aided simulations of digital circuits. A simple formula for the delay introduced by an interconnection, according to [15], is:

$$\tau_d = 1.02CR + 2.21(c_t r_t + c_t R + r_t C), \tag{5-25}$$

where τ_d is the time interval between a zero voltage at point A and $0.9V_{dd}$ at point B when a step input voltage is applied. The other parameters are:

- C: total interconnection capacitance, which can be computed using one of the formulae presented in Section 5.1.3.

- R: total interconnection resistance, computed with Eqs. (5-1), (5-23) (or approximation) and (5-22).

- c_t: load capacitance, which includes the gate capacitance of the transistor(s) driven.

- r_t: equivalent resistance of driving transistor.

$$r_t = (\text{max drain conductance})^{-1} = \frac{L}{W\mu_{n/p} C_o V_{dd}}$$

Let us consider an inverter that drives a polysilicon line 1000μm long, 2μm wide, 0.2μm thick, and connected to another inverter. The dimensions of the devices are shown in Fig. 5-9 and are in microns. Moreover, we have: t_{ox}=250Å and μ_n=400cm^2/Vs, μ_p=200cm^2/Vs. Finally, the field oxide thickness is 0.5μm.

We have:

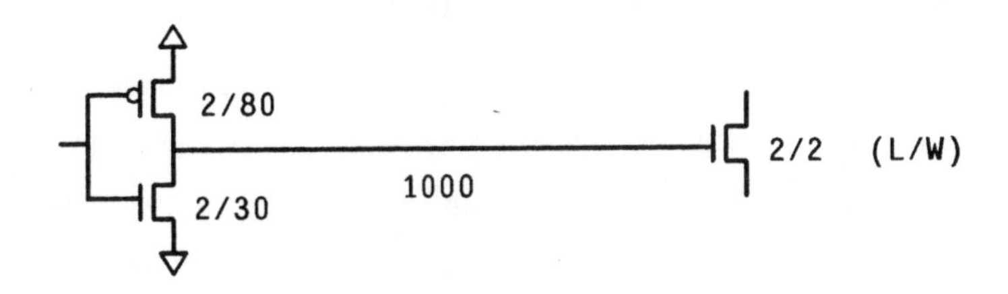

Figure 5-9: Example of delay computation.

$$c_t \approx 5.53\text{x}10^{-3}\text{pF} ;$$

$$r_t \approx 181\Omega ;$$

$$R \approx 75\text{K}\Omega ;$$

$$C \approx 0.51\text{pF} .$$

From Eq. (5-25) we have:

$$\tau_d = 40.1\text{ns} .$$

Note that for very large values of C and R, the delay is mainly dictated by the RC constant of the interconnect, and wider driving gates *cannot decrease the delay*, as Eq. (5-25) shows. The minimum τ_d is equal to 1.02CR. If we assume that $TaSi_2$ is used instead of polysilicon, all the parameters remain unchanged with the exception of R. We have:

$$R = 3.1\text{K}\Omega ,$$

which results in a much shorter delay:

$$\tau_d = 1.85\text{ns} .$$

5.3. The Concept of Equivalent Gate Load

Using physical units during the design of a circuit can be cumbersome and can easily lead to mistakes. One way of making the computation simpler is to introduce a new unit, called "equivalent gate load." This approach consists of expressing all the capacitances, whether they are stray or gate capacitances, in terms of multiples of the gate capacitance of a minimum-size transistor. This method is better explained by the case-study shown in Fig. 5-10. The first step is to determine the gate capacitance of a minimum size transistor; this can be accomplished by using Eq. (2-10). This will be the unit used throughout the entire computation. Then, all the parasitic capacitances must be computed so the equivalent load can be computed. Note that the underlying assumption is that a single capacitor will be a sufficiently accurate model for the interconnect capacitance. Therefore, the equivalent gate load is correct only when very low resistivity materials are used as interconnect. If long polysilicon lines are present, a single capacitor model does not suffice, as we saw in the previous section, and the equivalent gate model leads to serious inaccuracies.

Let us now apply this methodology to the example shown in Fig. 5-10. An inverter is connected to a larger inverter and we want to compute the overall load that the first inverter sees. Let us suppose that the minimum feature size is $2\mu m$, that is, the smallest transistor is $2\mu m$ x $2\mu m$. The load that the first inverter sees is basically the sum of the following capacitances:

1. *Gate capacitance* of the second inverter.

2. *Stray capacitance* of the interconnection between the output of the first inverter and the input to the second one.

3. *Intrinsic output capacitance* of the first inverter, which is basically the sum of the drain-substrate junction capacitances of the two devices.

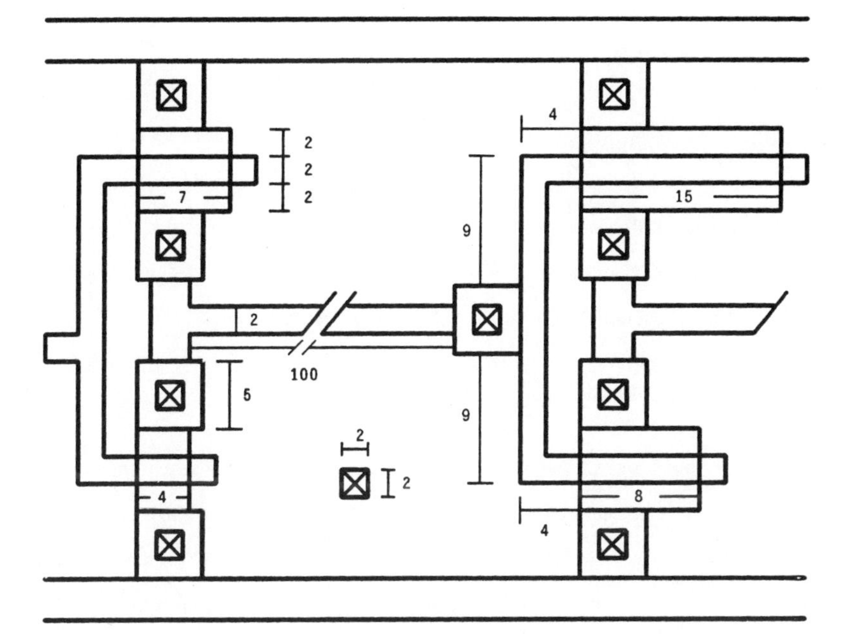

Figure 5-10: Layout example used to compute the different contributions to the first inverter's output load.

We have ignored poly-metal overlap capacitance of the contact. Let us suppose we have the following fabrication parameters:

t_{ox} = 250Å = 0.025μm ;

H (field oxide thickness) = 0.5μm ;

T (interconnect thickness) = 0.2μm ;

$\mu_n = 400\text{cm}^2/\text{Vs} \quad \mu_p = 200\text{cm}^2/\text{Vs}$.

From Eq. (2-10) we have that the gate capacitance of a minimum-size transistor equals 5.52x10^{-3}pF. The capacitance per unit length of both the metal conductor and the polysilicon conductor can be derived from Eq. (5-2):

$$\text{capacitance per unit length} = 2.38 \text{ pF/cm} .$$

The metal interconnect is 100μm long; its capacitance is:

$$C_{metal} \approx 0.024 \text{ pF} .$$

Finally, we have the two branches of polysilicon interconnect after the metal-poly contact; their total length is 26μm, which adds another 0.006pF. The total capacitance introduced by the interconnect is therefore:

$$C_{interconnect} = 0.024\text{pF} + 0.006\text{pF} = 0.030 \approx 5.5\, C\square ;$$

that is, the interconnect has the same gate capacitance of 5.5 minimum-size gates. The symbol C□ indicates the "gate capacitance of a minimum-size transistor." For the input capacitance of the second inverter we consider the worst case, that is, both gate capacitances are in parallel and equal to AC_{ox}. The n-channel device of the second inverter has capacitance 4 C□, while the p-channel device has capacitance 7C□. The total is therefore:

$$C_{inv} = 11\, C\square.$$

The last capacitance, that is, the drain-to-substrate capacitance, is not fixed but depends on the size of the n-channel and p-channel devices of the first inverter, which we still have to calculate. Some approximation is necessary, and, given the (small) dimensions of the transistors, it will not significantly influence the results. However, when much wider transistors are used, two or more iterations might be necessary. First, a reasonable guess of the drain areas and perimeters is made and the capacitance calculated. Then, when the sizes of the driving transistors have been determined, the drain areas and perimeters are computed again, the overall load updated and the size of the driving transistors compared to the new output load. If the size is consistent with the delay minimization technique adopted, the iterative procedure stops here. If the increased size of the driving transistors has significantly changed the junction capacitance, another iteration is necessary.

Given all the necessary parameters (such as metallurgical junction depth, substrate doping concentration, etc.), we can compute the drain-to-substrate junction capacitance by following the approach presented in Chapter 2. Let us suppose that each drain-to-substrate junction capacitance is 5×10^{-3}pF. The contribution of the two capacitances in parallel will therefore be:

$$C_{output} = 10 \times 10^{-3} \text{pF} \approx 2\, C\square \,.$$

The total load that the first inverter sees is therefore:

$$C_{total} = C_{interconnect} + C_{inv} + C_{output} = 5.5 + 11 + 2 = 18.5\, C\square \,.$$

The fan-out is 18.5 minimum-size transistors. We can now apply a scaling methodology to determine the size of the first inverter. We will present one such methodology in the next section. Note that considering only the input capacitance of the second inverter would create an error of about 40%; moreover, note that the interconnect *is not extremely long*. Finally, most of the interconnect is metal, and this allows us to apply the equivalent gate load methodology. Such simple approach can lead to large errors if the interconnect is mostly polysilicon. As was pointed out earlier, polysilicon requires much more careful and time-consuming design — mostly computer-assisted — to derive a full characterization of the load capacitance.

5.4. Delay Minimization

Minimizing the delay in a critical path is a problem of global optimization which becomes very difficult when the path consists of multiple-input gates, because the delay also depends on the input pattern. This problem of global optimization will be dealt with in Chapter 7 for the simple case of an inverter chain. The necessity of global optimization can be explained better with an example. Let us assume that we want to minimize the delay of an inverter chain given the value of the output load and the size of the first gate. Increasing the width of the devices in the second stage decreases the delay of the second gate, but, at the same time, this increases the output capacitance seen by the first gate. We decrease the delay in the second stage and we increase the delay in the first stage: is the net result a shorter delay? To find this out, global optimization becomes necessary.

Sizing of gates involves two basic steps: computing the optimum pull-up|pull-down ratio for each gate and computing the optimum sizing ratio between consecutive gates. These two problems are somewhat interdependent, which is why global optimization is necessary; however, they can be treated separately — in a first-order approximation. First, we will determine the optimal pull-up|pull-down ratio for a single gate. Second, we will determine the optimal number of stages in an inverter chain and the widths of all the devices. Finally, we will consider stray capacitances and re-evaluate some of the previous results.

5.4.1. Inverter Delay and Sizing

We want to compute the optimal pull-up|pull-down ratio of a CMOS inverter, where "optimal" means the ratio which minimizes the delay regardless of area and power dissipation. Fig. 5-11 shows the problem: given the widths of the two n-channel devices and the mobility μ_n and μ_p of the transistors, and assuming that

$$|V_{Tn}| = |V_{Tp}|,$$

we want to find the pull-up|pull-down ratio z which minimizes the delay through the first inverter. All four transistors have the same channel length. If the n-channel devices are W_{n1} and W_{n2} microns wide, respectively, the p-channel devices will be zW_{n1} and zW_{n2} microns wide.

We have:

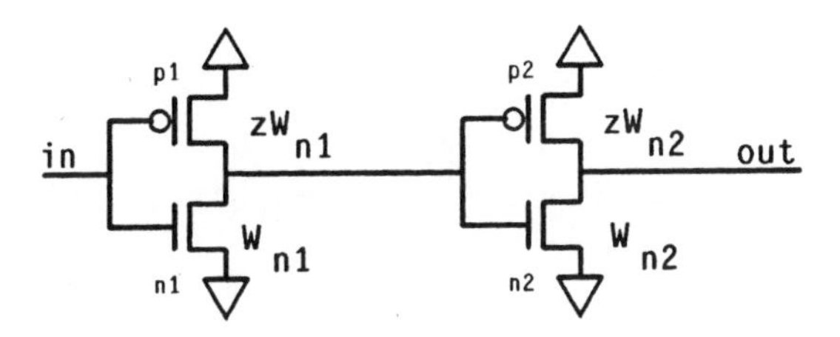

Figure 5-11: Inverter pair.

$\Delta\downarrow$: delay when the input signal performs a low-going transition;

$\Delta\uparrow$: delay when the input signal performs a high-going transition.

$\Delta\uparrow$ and $\Delta\downarrow$ can be expressed as:

$$\Delta\uparrow \propto \frac{zW_{n1} + W_{n2} + zW_{n2}}{\mu_n W_{n1}} + \frac{W_{n2}}{z\mu_p W_{n2}} ;$$

$$\Delta\downarrow \propto \frac{W_{n1} + zW_{n2} + W_{n2}}{z\mu_p W_{n1}} + \frac{zW_{n2}}{\mu_n W_{n2}} .$$

The equations can be explained as follows. Let us consider $\Delta\uparrow$, for instance:

- Numerator of the first term: the delay is directly proportional to the load, which is proportional to $(W_{n2} + zW_{n2})$ and directly proportional to zW_{n1}, because p1 must be turned off. Denominator of the first term: the delay is inversely proportional to the mobility of the n-channel transistor n1 and to its width.

- Numerator of the second term: when the input makes a high-going transition, the output of the first stage will make a low-going transition. For this reason, p2 will turn on while n2 will turn off. The larger W_{n2} is, the longer the delay. Denominator of second term: the larger p2 and the higher its mobility, the shorter the delay.

The delay we want to minimize is the *average delay*, which is half the sum of the rise-time and fall-time. We have to find the value ξ of z such that:

$$\min[\Delta] = \min\left[\frac{1}{2}(\Delta\uparrow + \Delta\downarrow)\right] = \min[f(z)] = f(\xi)\,.$$

We have:

$$\frac{df}{dz} = \frac{1}{\mu_n} + \frac{W_{n2}}{2\mu_n W_{n1}} - \frac{W_{n2}}{2z^2\mu_p W_{n2}} - \frac{1}{2z^2\mu_p} - \frac{W_{n2}}{2z^2\mu_p W_{n1}} = 0\,;$$

$$z^2 = \frac{\mu_n}{\mu_p}\,\frac{2W_{n1}W_{n2} + W_{n2}^2}{W_{n2}(2W_{n1} + W_{n2})}\,; \tag{5-26}$$

$$\text{if } \mu_r = \frac{\mu_n}{\mu_p} \rightarrow \xi = \sqrt{\mu_r}\,. \tag{5-27}$$

That is, the p-channel transistor width should be the square root of the mobility ratio times the width of the n-channel. This result agrees with the result presented in [9], which was obtained through a more rigorous treatment. Actually, this result holds true only when the threshold voltages of the two transistors are equal. If the two threshold voltages differ significantly, Eq. (5-27) becomes more complex: see [9] for details. Note also that the result shown in Eq. (5-27) is independent of the relative size of W_{n1} and W_{n2}. If $W_{n2} = 3W_{n1}$, for instance, ξ would still be as shown in Eq. (5-27).

Another case is when the output of the second inverter is connected to a capacitive load C. By using the same approach, we can derive the following equation:

$$z^2 = \frac{\mu_n}{\mu_p}\,(W_{n2} + C + W_{n2}^2)\,\frac{W_{n1}}{W_{n2}(2W_{n1} + W_{n2})}\,,$$

and if we assume that C, W_{n2}, and W_{n1} are scaled up equally, that is:

$$W_{n1} = W;$$

$$W_{n2} = tW;$$

$$C = tW_{n2} = t^2W.$$

We have:

$$\xi = \sqrt{\mu_r}\sqrt{\frac{1+2t}{2+t}}\ .$$

For $t = 3$, $z \approx 1.18\sqrt{\mu_r}$. Once again, this result shows the importance of global optimization. If we change the assumption about the output load, as we did in the second case, we obtain a different value for ξ.

The above examples dealt with the very simple case of an inverter pair. Normally, logic circuits consist of multiple-input gates connected to form very complex interconnection graphs, not just a simple chain. The question at this point is: can we use the results presented above in more general and complex cases? The answer is yes, provided that the designer understands the *limitations* of such an approach. First, the "best" pull-up|pull-down ratio is not a critical value. Ratios that differ slightly from the exact square root of the mobility ratio will not increase the delay dramatically. If we are to design a multiple-input gate, we can still apply the "square-root rule" without worrying too much about possible speed degradation. Moreover, the variations in process parameters are usually such that exact scaling does not fundamentally influence overall chip performance. Some of these considerations do not apply to the design of the I/O section of a chip, where power dissipation, noise and speed degradation become critical parameters, especially with small line widths.

5.4.2. Inverter Chain Sizing

In this section we present a simple approach to determine the number of inverters and their size when a load M has to be driven. In particular, we have two constraints:

1. The number of inverters is either odd or even, because we want an

inverting (odd) or non-inverting (even) chain.

2. Maximization of speed, *regardless of area and power dissipation.*

The problem is shown in Fig. 5-12 for a three-inverter chain. The values M and m are known. We assume that the input and output capacitances of a gate are proportional to the device width. The numbers above the inverters show the driving capability of each gate, which is again directly proportional to the gate width. First, we derive an expression for the *relative delay* Δ_r as a function of x and y. Then, we determine the values of x and y that minimize Δ_r.

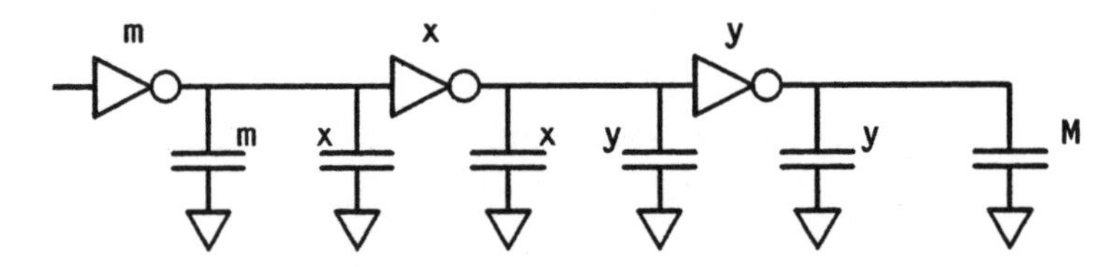

Figure 5-12: Inverter chain sizing with parametrized loads.

We have:

$$\Delta_r \propto \frac{1}{m}(m + x) + \frac{1}{x}(x + y) + \frac{1}{y}(y + M) =$$

$$= 3 + \frac{x}{m} + \frac{y}{x} + \frac{M}{y} = f(x,y)\,.$$

We now must find the minimum of f(x,y). We have:

$$\frac{\partial f(x,y)}{\partial x} = \frac{1}{m} - \left[\frac{y}{x}\right]^2;$$

$$\frac{\partial f(x,y)}{\partial y} = \frac{1}{x} - \left[\frac{M}{y}\right]^2.$$

We find that the minimum of Δ_r is reached when:

$$y = m\left[\frac{M}{m}\right]^{2/3};$$

$$x = m\left[\frac{M}{m}\right]^{1/3}$$

When x and y satisfy the above equations, Δ_r is equal to:

$$\Delta_r = 3 + 3\left[\frac{M}{m}\right]^{1/3}.$$

The result can be generalized by saying that, given a chain of n inverters that drive a load M with the first inverter having size m, the minimum Δ_r is reached when the size of the inverters is:

$$m\left[\frac{M}{m}\right]^{1/n}, m\left[\frac{M}{m}\right]^{2/n}, \ldots, m\left[\frac{M}{m}\right]^{\frac{n-1}{n}}.$$

These results are summarized in the set of figures presented in Appendix C. As we noted in the previous section, these results can be extended to more complex gates, such as multi-input gates. This information can be used to size the devices roughly. More accurate sizing can follow when more information — on area, power dissipation, etc. — is gathered from this first phase of the design. A common problem is that gate size depends on the load, which in turn depends on the fan-out and the physical length of interconnect; this may depend on the area occupied by the gates, which depends on their size. This loop can be broken by applying a multi-pass approach. First, a rough estimate of the size of all the gates is computed by using the simplified equations presented in previous sections. Then, further iterations are carried out using the same equations. Finally, critical sections of the circuit, e.g. output pads or bus drivers, require more accurate approaches, such as the ones presented in Chapter 7.

5.4.3. Inverter Chain Sizing with Stray Capacitance

In the previous section the inverter chain has been optimized without considering the effect that stray capacitance can have on the delay. In this section we introduce stray capacitance and re-evaluate the results obtained in Section 5.4.2. The problem is shown in Fig. 5-13 for a two-inverter chain.

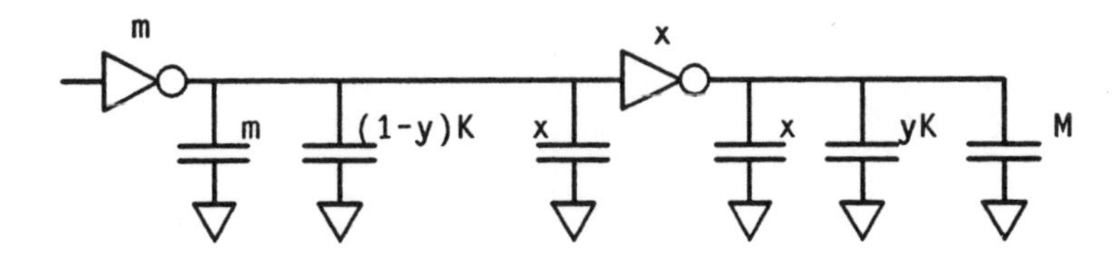

Figure 5-13: Inverter chain with stray capacitances.

Both a load M and a stray capacitance K are present. We want to determine how to subdivide the capacitance K between the two stages of the inverter chain to minimize the delay. In a typical example, a signal has to go from a gate to an output driver which features large input capacitance. Gate and output driver are connected through a long interconnect, which introduces some stray capacitance. We want to find out whether it is more effective to scale up all the gates and consider the net load capacitance as the sum of input capacitance of the output driver and the stray capacitance or — in the physical sense — to distribute the gate chain and subdivide the total stray capacitance among all the chain stages. The problem shown in this section deals with a two-inverter chain, but the results hold true for any inverter chain length and can be extended to multi-input gate chains without affecting overall results. We subdivide the stray capacitance into two parts, namely $(1-y)K$ and yK. We have the following constraints:

$$0 \le y \le 1; \; m < M; \; x > m$$

We can write the relative delay as:

$$\Delta_r = \frac{1}{m}\left[m + (1 - y)K + x\right] + \frac{1}{x}(x + yK + M) = f(x,y) .$$

We now proceed as in Section 5.4.2, that is we minimize Δ_r:

$$\frac{\partial f(x,y)}{\partial x} = \frac{1}{m} - \frac{1}{x}(yK + M) = 0 ;$$

$$\frac{\partial f(x,y)}{\partial y} = \frac{K}{x} - \frac{K}{m} = 0 .$$

There is no solution for $0 \leq y \leq 1$ and the minimum in the acceptable range of y is for $y = 1$. This confirms the intuitive notion that, when a large load has to be driven, it is more effective to scale up the driving capability of the chain and then to drive all the capacitances, both stray and gate input capacitances, rather than to distribute the stray capacitance among the chain stages.

5.5. Transistor Sizing in Static Logic

The methodology used to determine the optimal size of the devices in an inverter or an inverter chain can be used for multi-input gates. Let us assume we have a complementary three-input NAND gate, as shown in Fig. 5-14.

Let us also assume, for the sake of simplicity, that all six transistors have the same channel length, and that the mobility of the n-channel devices is twice the mobility of the p-channel devices. What is the optimal width of the devices which minimizes the delay for all possible input combinations? To start, we assume that all the input transitions are equally probable. Let us first consider the case where both n1, and n2 and n3 are conducting; the output of the gate is low. Then, one of the three inputs goes down. The fastest output rise-time will be achieved when input a goes down, while the slowest output rise-time takes place when input c goes down. When input b goes down the rise-time has a value between the two previous cases. The closer the

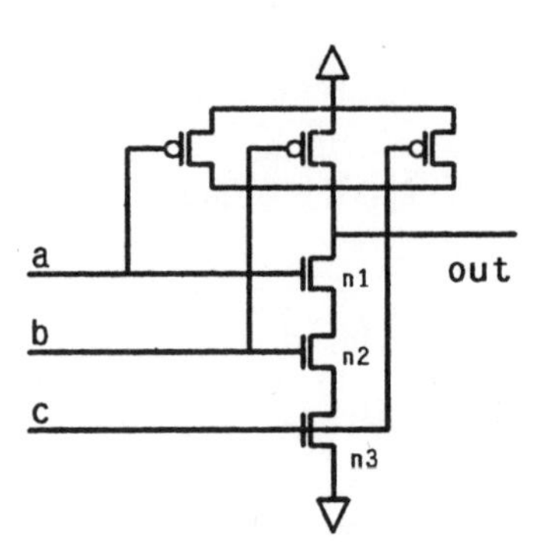

Figure 5-14: Three-input NAND gate.

device which turns off is to the output stage, the faster the output rise-time is. We should scale the channel widths so that n3 is the widest and n1 is the narrowest.

Now we consider the opposite case, that is two inputs are high and the third one performs a high-going transition: the output is pulled down. We obtain the complementary behavior: a high-going transition of input c results in the shortest output fall-time, while if a rises this produces the slowest output fall-time. To balance the dynamic behavior of the gate, n1 should have the widest channel width, followed by n2 and n3. This requirement conflicts with the previous case, which required exactly the opposite scaling.

For equally probable input signals, the conclusion is that the sizing of the gate can be carried out without regard to the input pattern. First, we shall determine the width of a *single* n-channel device pulling down the load, as if the gate were an inverter. Let us assume that this width is 20μm. Because we actually have three transistors in series, we have to multiply the width of each n-channel device in the NAND gate by three to have the same output fall-time — in a first-order approximation. Body effects and drain capacitance increase the actual fall-time, but we shall ignore them in our simple model. Therefore, each n-channel device is 60μm wide.

The width of the p-channel transistor is determined as follows. The worst case is when one input falls, while the other two remain high: one p-channel device is pulling up. If we consider the worst case, each p-channel should have a width equal to the square root of the mobility ratio times the pull-down *equivalent width*, which in this case is one third of the channel width of each transistor, i.e., 20μm. Given a mobility ratio of two, each p-channel device should be 28μm wide. To compensate for the second order effects mentioned above, the p-channel width should be increased by a factor of 20%, i.e., to about 33μm. However, this increases the input gate capacitance and requires a much wider gate to drive the three input NAND. Another possibility is to assume that, on the average, two p-channel transistors always pull-up; this requires each p-channel transistor to be 16μm wide. The long rise-time when one p-channel device is turned on is partially compensated for by a smaller input gate capacitance, that is, the preceding gates can be smaller and occupy less area. The above approach can be applied to any complementary gate, regardless of its complexity. Finally, the designer can decide that a longer fall-time can be traded for narrower n-channel devices and, therefore, narrower p-channel devices. Smaller input capacitance will partially compensate for the performance degradation.

5.6. Transistor Sizing in Dynamic Logic

Dynamic logic sizing differs from static logic sizing because other considerations, such as charge sharing, have to be taken into account. Most, if not all, dynamic logics behave poorly if slow precharging takes place. The sizing of the p-channel pull-up device — we assume an n-channel based dynamic gate throughout the section — should therefore allow short rise-time. Long rise-time in P-E or domino gates results in having the two clocked transistors both conducting for a long period of time. This can lead to excessive power dissipation or, even worse, charge sharing or an incorrect output.

The two gates shown in Fig. 5-15 illustrate the problem. The gate in Fig. 5-15(a)

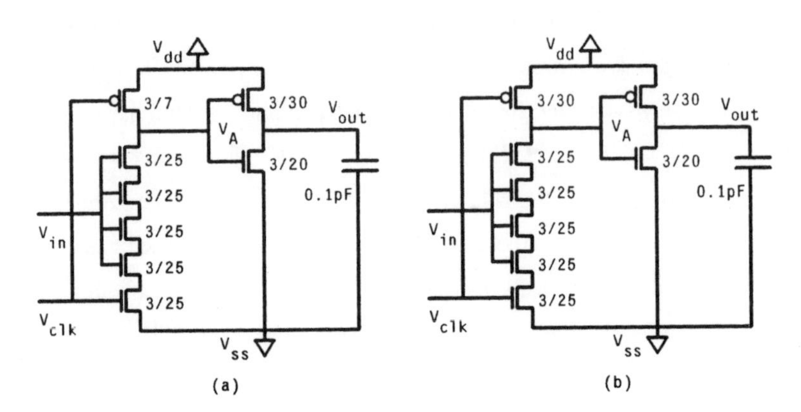

Figure 5-15: Two four-input AND domino gates: gate (a) has a narrow p-channel device, while gate (b) has a much wider precharge transistor. All dimensions are in microns.

has a narrow p-channel device. To compute its width, the width of an n-channel device with a beta equal to the beta of the five series n-channel devices is multiplied by 1.4. That is, it has been assumed that the pull-down section of five n-channel transistors can be replaced by one, 3μm x 5μm n-channel transistor. Fig. 5-16 shows input and output waveforms for the gates, and we can note that the rise-time of V_A is very long.

If we now consider the gate shown in Fig. 5-15(b) and its associated waveforms in Fig. 5-17, the rise-time of V_A is significantly shorter. This affects V_{out}, which reaches V_{ss} much earlier than in Fig. 5-16. Note that V_{out} does not have a shorter fall-time, because the rate of change in the output of a static gate is largely independent of the shape of the input: V_{out} in Fig. 5-17 simply starts falling earlier, because V_A reaches the inverter threshold earlier.

The gate in Fig. 5-15(b) has a much shorter rise-time than the gate in Fig. 5-15(a). This protects against charge sharing during precharge and can also reduce power dissipation. As a rule of thumb, the p-channel device is sized in such a way that its

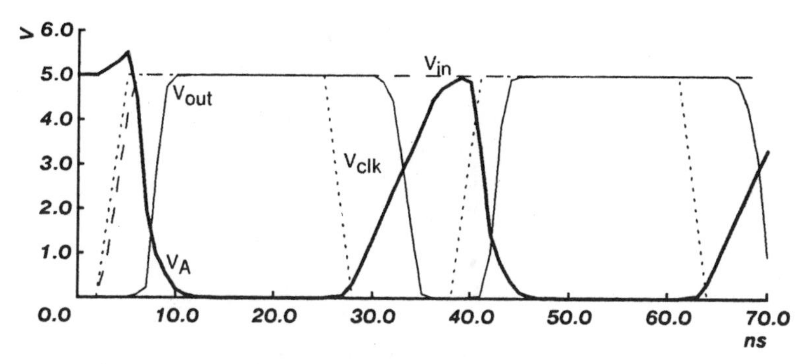

Figure 5-16: Waveforms for the gate shown in Fig. 5-15(a)

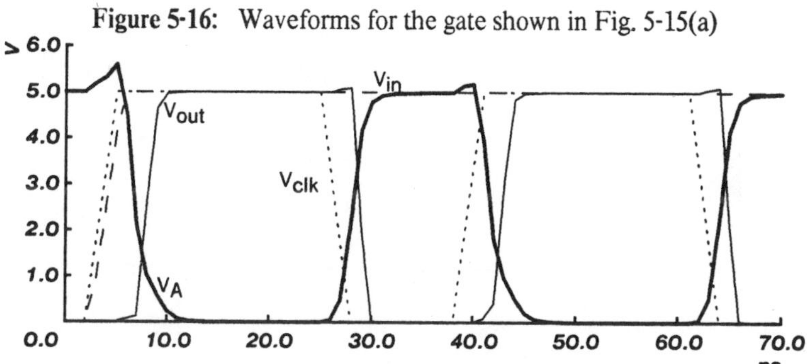

Figure 5-17: Waveforms for the gate shown in Fig. 5-15(b)

width is equal to the average width of the n-channel devices in the evaluation network multiplied by the square root of the mobility ratio. It is necessary to say "average," because gates are usually much more complex than a simple four-input NAND, and, therefore, the n-channel device widths can be very different from one another.

With regard to sizing of dynamic gates, particularly domino gates, an important result is reported in [16], which shows how scaling of n-channel devices (in an n-channel device based gate) can speed up the evaluation phase. Scaling of a gate is shown in Fig. 5-18. Here, the n-channel devices have been scaled *although the total channel area is the same as that shown in Fig. 5-15(a) and (b).*

Fig. 5-19 shows the rise-time and fall-time of two circuits: the bold curves refer to

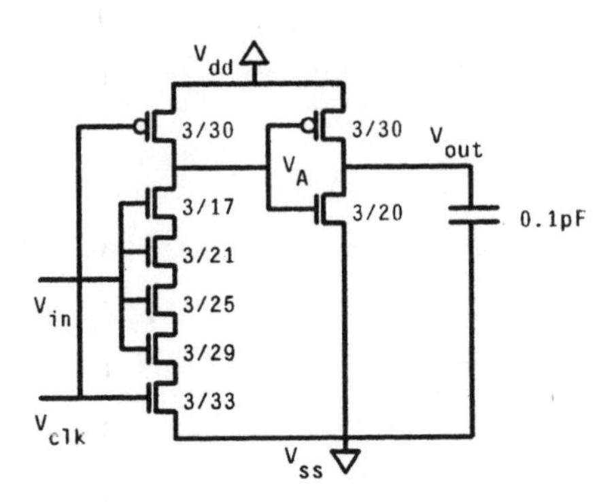

Figure 5-18: Domino gate with scaled n-channel devices.

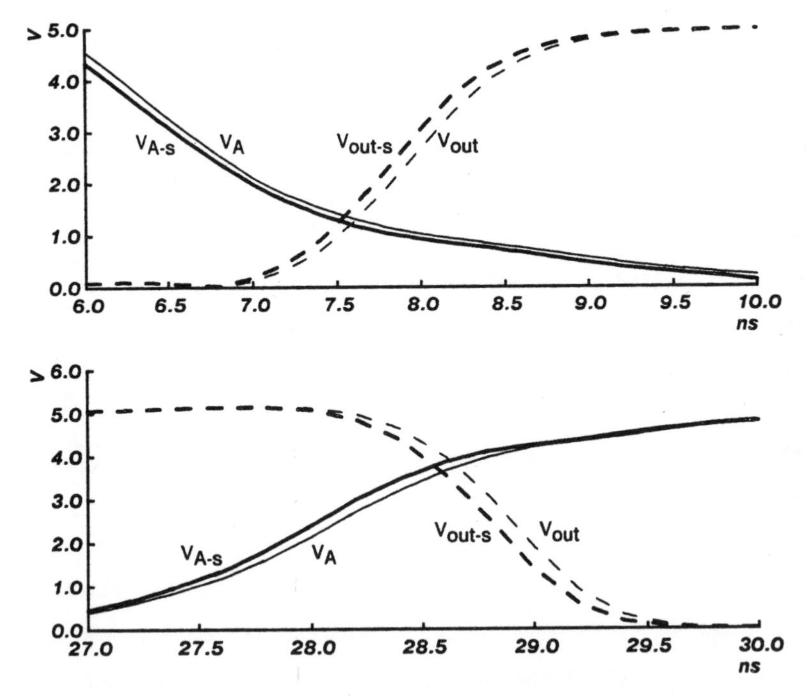

Figure 5-19: Waveforms for the two domino gates of Fig. 5-18 (bold) and Fig. 5-15(b). V_{in} and V_{clk} and other parameters are the same as in Fig. 5-15 through 5-17.

the same gate shown in Fig. 5-15(b) but with the n-channel devices scaled (from top to bottom) as: 17μm, 21μm, 25μm, 29μm, 33μm. The other curves refer to the gate shown in Fig. 5-18(b).

The effect of scaling can be explained as follows. Let us consider a NAND dynamic gate and let n1 be the n-channel device whose drain terminal is the gate output; from top to bottom we will have devices n2, n3, n4, etc. in series. If we decrease the width of n1, we obtain two opposite effects:

1. The parasitic capacitances of n1 decrease, and this speeds up the gate.

2. The narrower channel has increased resistance, and discharging the output node takes longer.

The first effect increases with the increased length of the n-channel transistor chain, because the parasitic capacitance of n1 discharges through the chain and the time constant is proportional to the overall resistance of the pull-down chain. On the other hand, the second effect is *minimized* in very long chains, because the increased resistance of one device does not significantly change the total resistance of the chain, *if the chain is sufficiently long.*

Because the positive effect — that is, decreased parasitic capacitance — is maximized in long chains, while the negative effect — that is, increased device resistance — is minimized in long chains, we can expect that a narrower device n1 will decrease the overall delay. This argument can be applied to the other elements of the chain. However, the closer we proceed towards V_{ss}, the higher the total parasitic capacitance to be discharged becomes. That is why the transistors are *scaled*, rather than shrunk by a constant amount. The last transistor, i.e., the one whose source is connected to V_{ss}, has to discharge the parasitic capacitances of all the devices preceding it, and, therefore, its width has to be the widest in the chain. Scaling of about 30% has been shown to produce optimal results [16]. Note that excessive scaling slows down the gate, because the resistance of n1 does not allow fast discharging of the output node to V_{ss}. This means that the effect of scaling should be checked through circuit simulation; too aggressive scaling produces the opposite result and, in fact, slows down the gate. Finally, it has been shown in [16] that the best results can be achieved if the total area of the devices in the pull-down section is maintained when scaling is applied. This technique can be used in most dynamic gates, because the above results are largely independent of the logic used.

References

[1] Botchek, C.M.
VLSI - Basic MOS Engineering.
Pacific Technical Group, Inc. - Publication Division, Saratoga, CA, 1983.

[2] Butler, A.L. and D.J. Foster.
The Formation of Shallow Low-Resistance Source-Drain Regions for VLSI CMOS Technology.
IEEE Journal of Solid-State Circuits SC-20(1):70-75, February, 1985.

[3] Chen, J.Y.-T. and D.B. Rensch.
The Use of Refractory Metal and Electron-Beam Sintering to Reduce Contact Resistance for VLSI.
IEEE Trans. on Electron Devices ED-30(11):1542-1550, November, 1983.

[4] Davis, W.A.
Microwave Semiconductor Circuit Design.
Van Nostrand Reinhold Co., 1984.

[5] Gupta, K.C., R. Garg and I.J. Bahl.
Microstrip Lines and Slotlines.
Artech, 1979.

[6] Gupta, K.C., R. Garg and R. Chadha.
Computer-Aided Design of Microwave Circuits.
Artech House Inc., Dedham, MA, 1981.

[7] Irvin, J.C.
Resistivity of Bulk Silicon and of Diffused Layers in Silicon
Bell System Technical Journal XLI(2):387-410, March, 1962.

[8] Kang, S.M.
A Design of CMOS Polycells for LSI Circuits.
IEEE Trans. on Circuits and Systems CAS-28(8):838-843, August, 1981.

[9] Kanuma, A.
CMOS Circuit Optimization.
Solid-State Electronics 26(1):47-58, January, 1983.

[10] Lewis, E.T.
High-Density High-Impedance Hybrid Circuit Technology for Gigahertz Logic.
IEEE Trans. CHMT CHMT-2(4):441-450, December, 1979.

[11] Lewis, E.T.
An Analysis of Interconnect Line Capacitance and Coupling for VLSI Circuits.
Solid-State Electronics 27(8/9):741-749, August/September, 1984.

[12] Pretorius, J.A., A.S. Shubat and A.T. Salama.
Analysis and Design Optimization of Domino CMOS Logic with Application to Standard Cells.
IEEE Journal of Solid-State Circuits SC-20(2):523-530, April, 1985.

[13] Proctor, S.J., L.W. Linholm and J.A. Mazer.
Direct Measurements of Interfacial Contact Resistance, End Contact Resistance, and Interfacial Contact Layer Uniformity.
IEEE Trans. on Electron Devices ED-30(11):1535-1542, November, 1983.

[14] Sakurai, T. and K. Tamaru.
Simple Formulas for Two- and Three-Dimensional Capacitances.
IEEE Trans. on Electron Devices ED-30(2):183-185, February, 1983.

[15] Sakurai, T.
Approximation of Wiring Delay in MOSFET LSI.
IEEE Journal of Solid-State Circuits SC-18(4):418-426, August, 1983.

[16] Shoji, M.
FET Scaling in Domino CMOS Gates.
IEEE Journal of Solid-State Circuits SC-20(5):1067-1071, October, 1985.

[17] Sinha, A.K.
Refractory Metal Silicides for VLSI Applications.
Journal Vac. Sci. Technology 19:778, 1981.

[18] Sze, S.M.
Physics of Semiconductor Devices.
John Wiley & Sons, New York, 1969.

[19] Tago, H. *et al.*.
A 6K-Gate CMOS Gate Array.
IEEE Journal of Solid-State Circuits SC-17(5):907-912, October, 1982.

PLATES

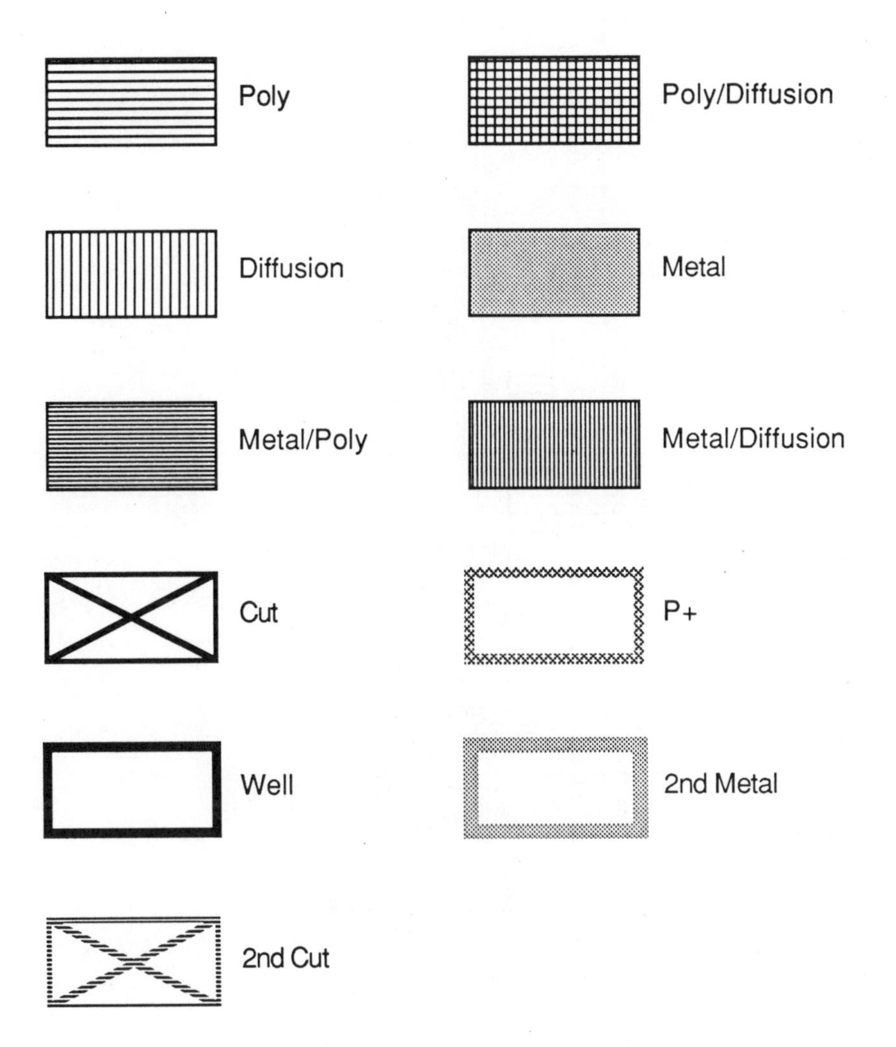

Plate I: Layer patterns.

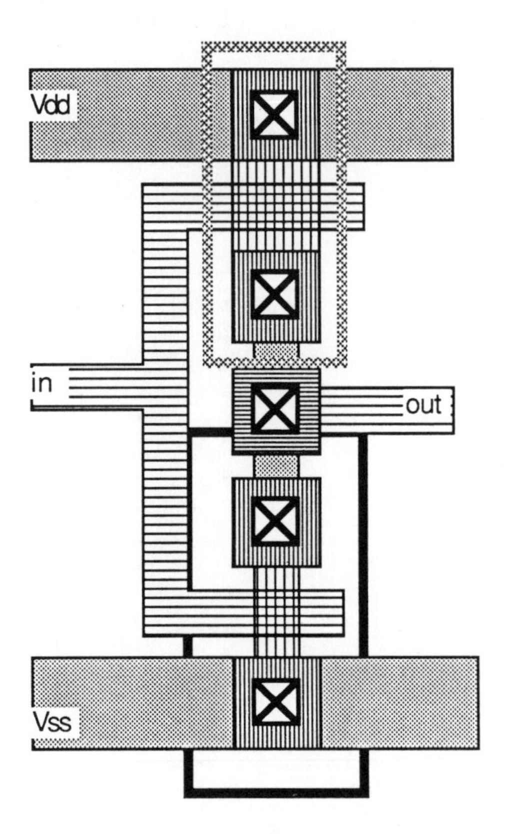

Plate II: Inverter.

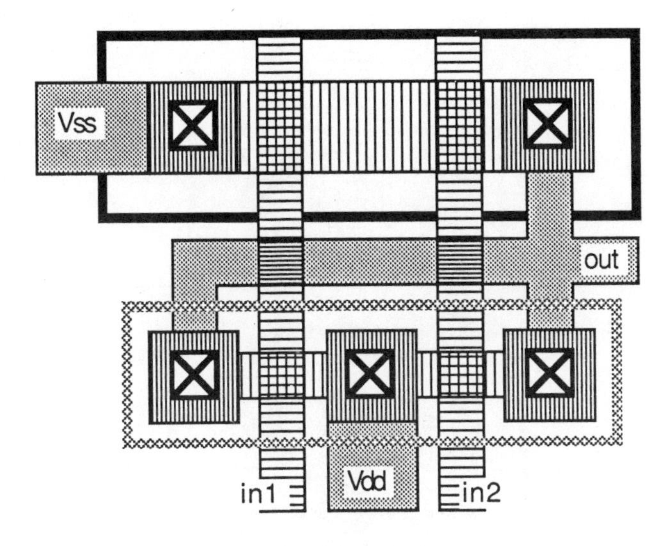

Plate III: NAND gate.

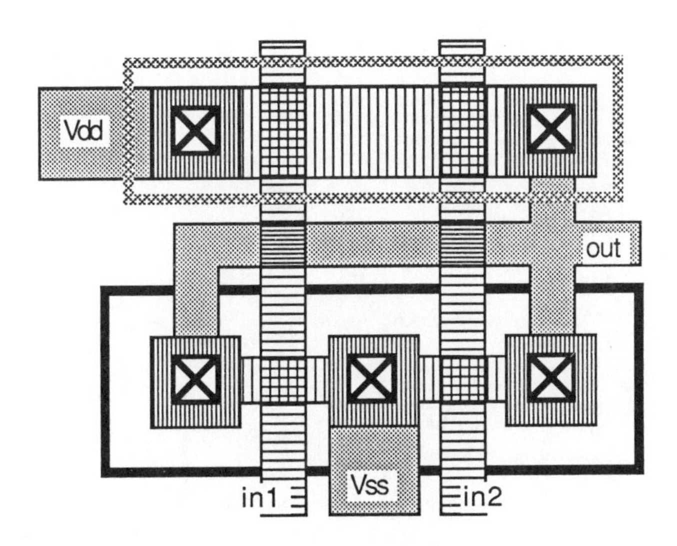

Plate IV: NOR gate.

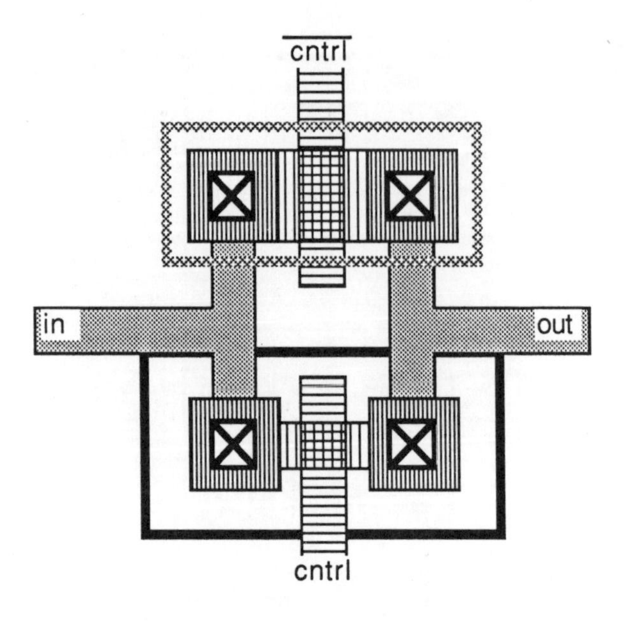

Plate V: Transmission gate.

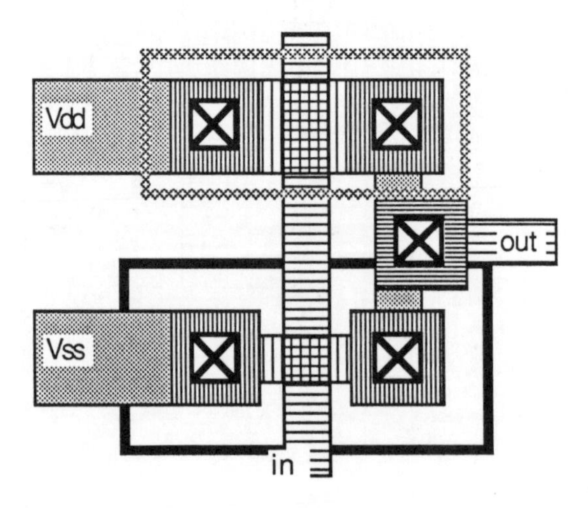

Plate VI: Inverter with different aspect ratio and terminal location.

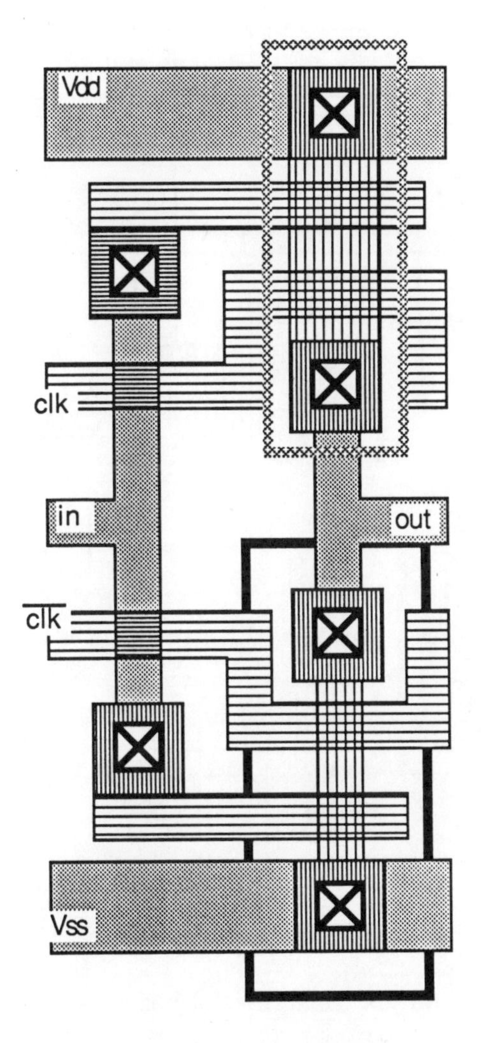

Plate VII: Clocked CMOS gate.

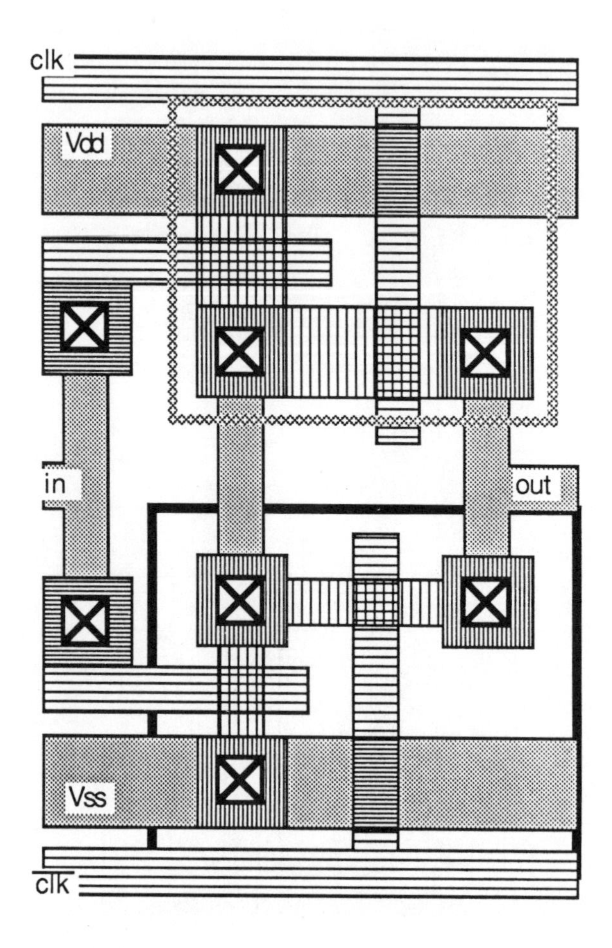

Plate VIII: Shift register building-block.

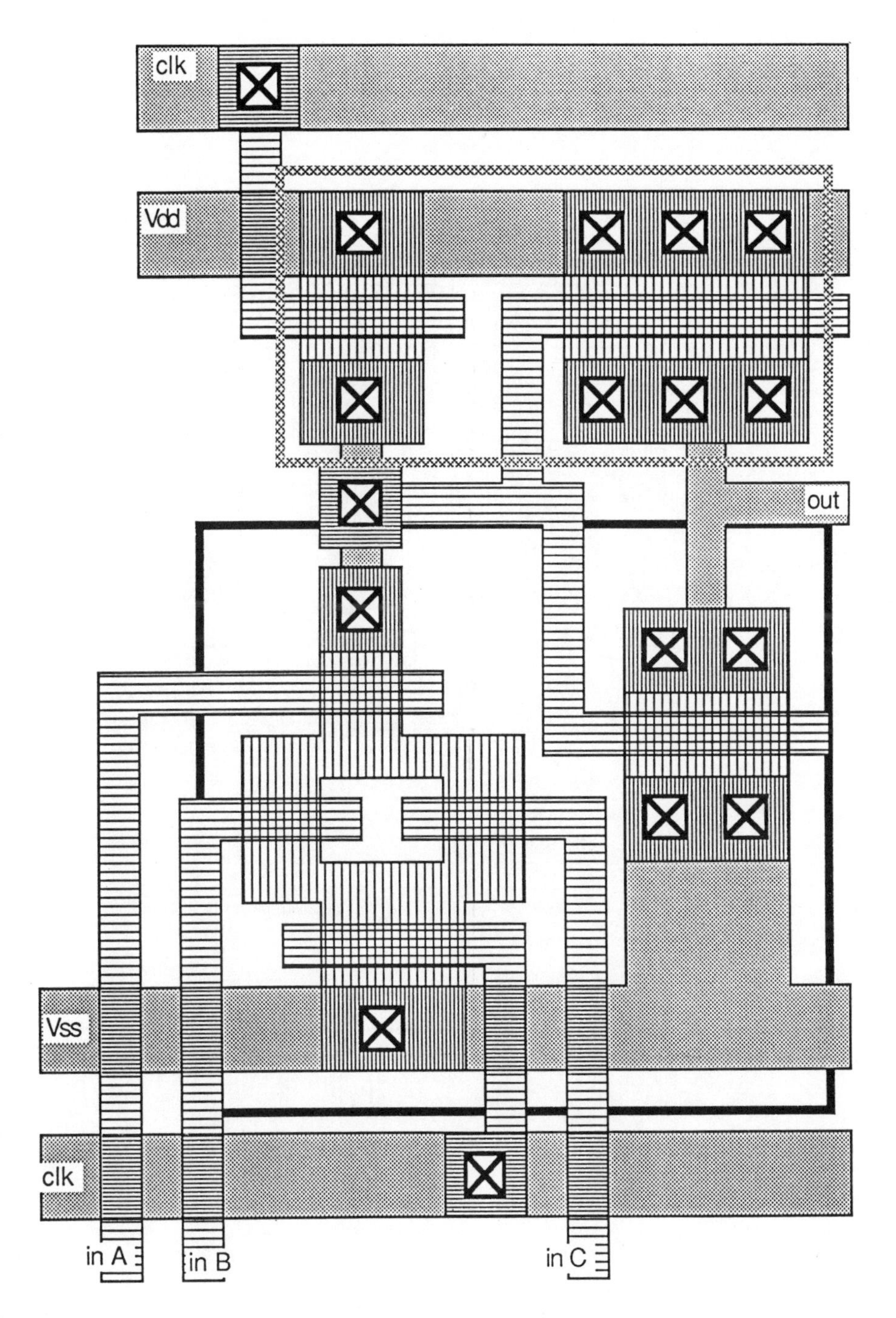

Plate IX: Domino logic gate.

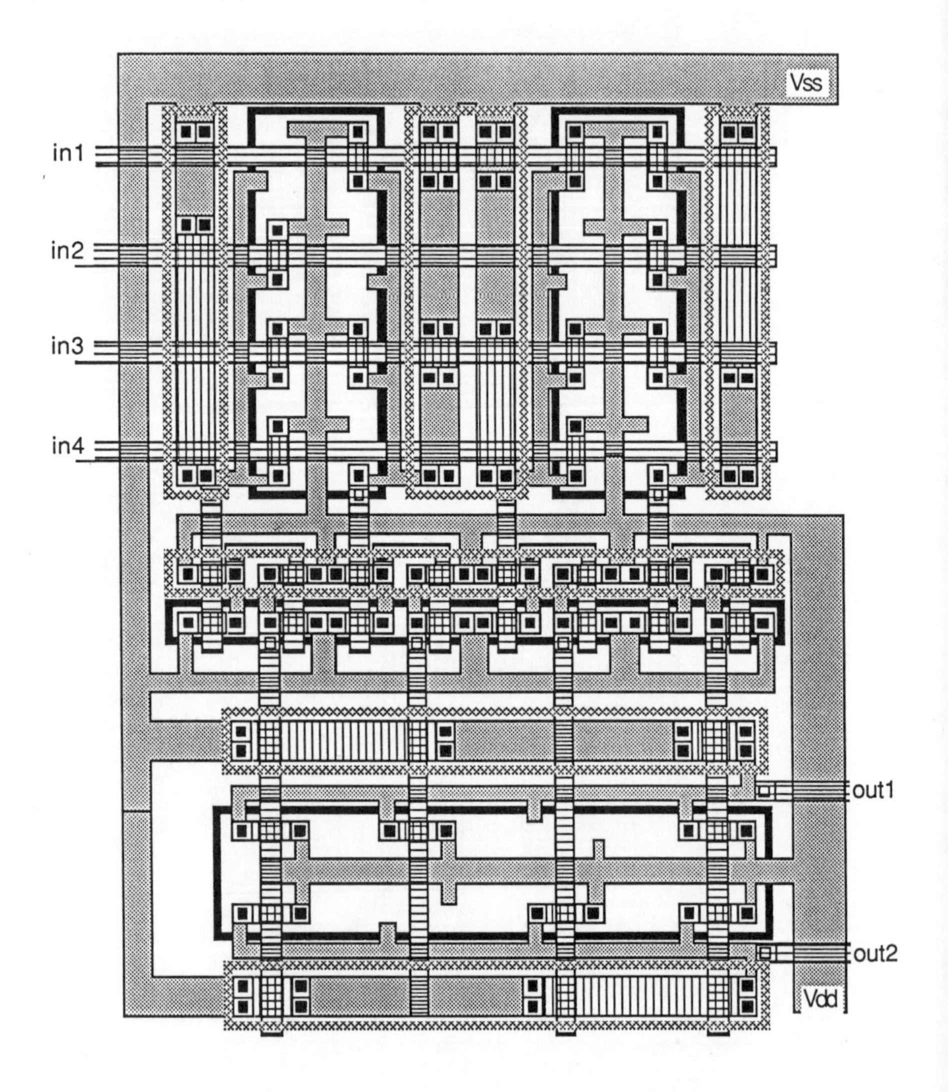

Plate X: CMOS PLA in complementary logic.

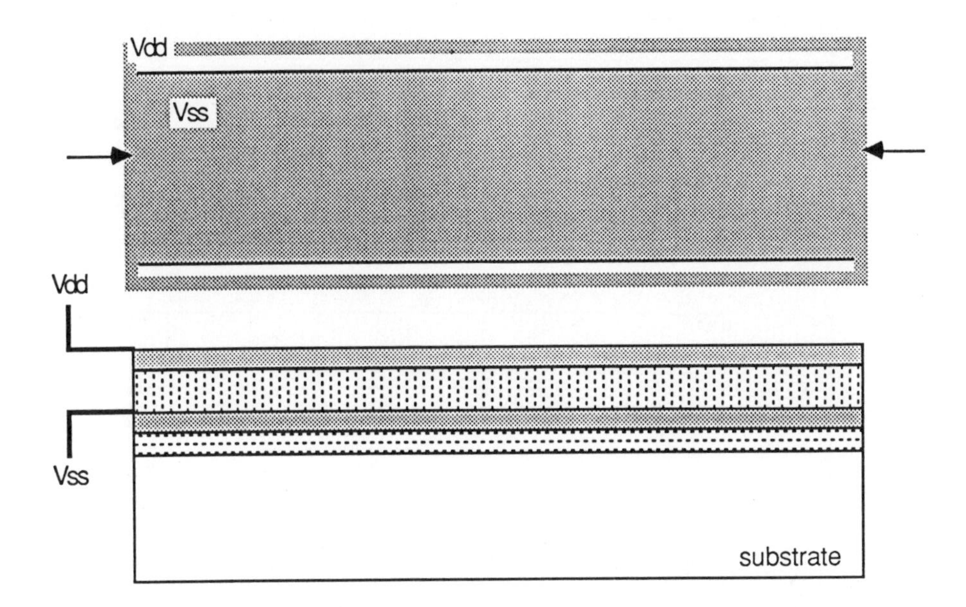

Plate XI: First metal - second metal capacitor.

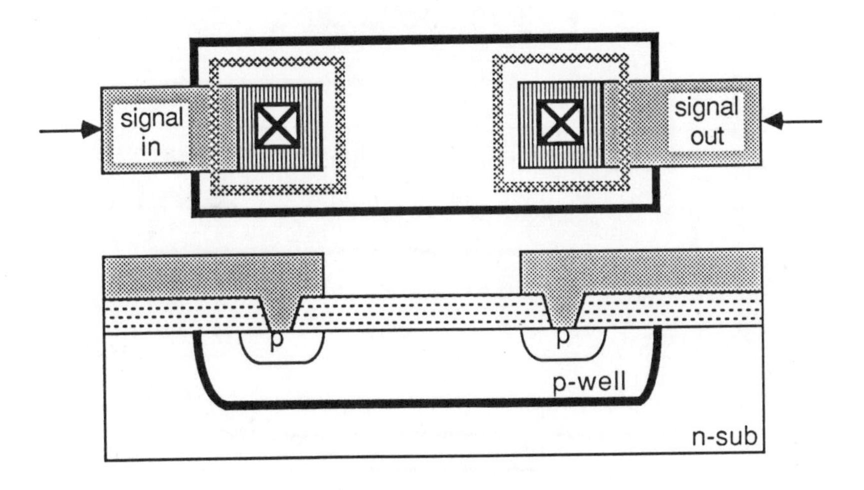

Plate XII: Well resistor.

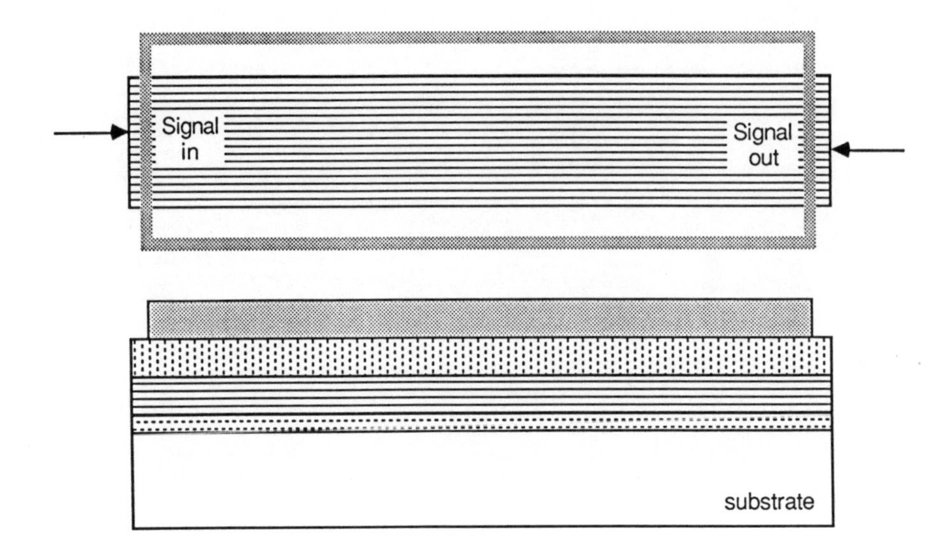

Plate XIII: Polysilicon resistor covered by second metal.

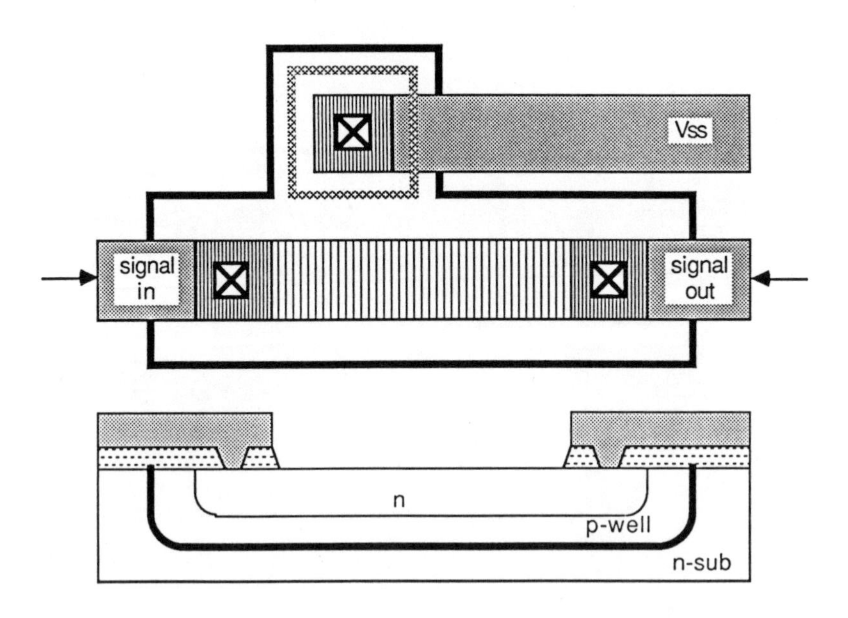

Plate XIV: Diffused diode-resistor structure with substrate contact.

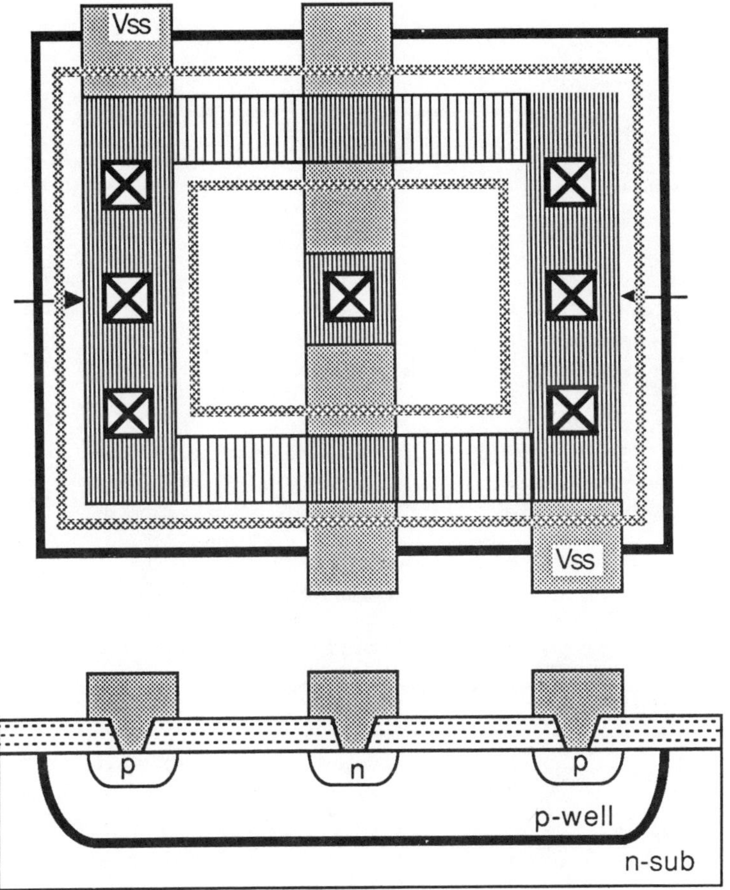

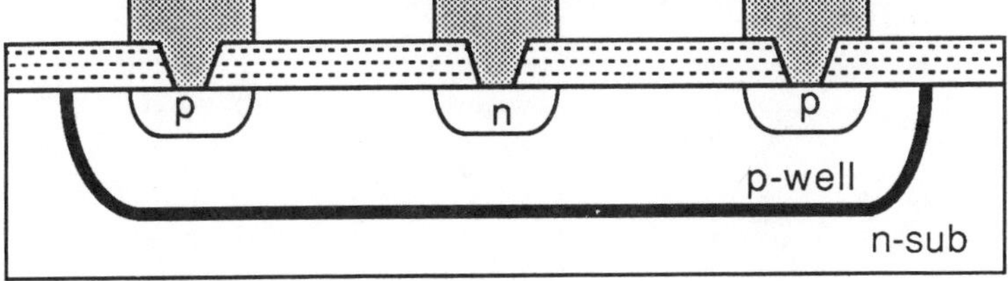

Plate XV: n-p diode with guard-ring.

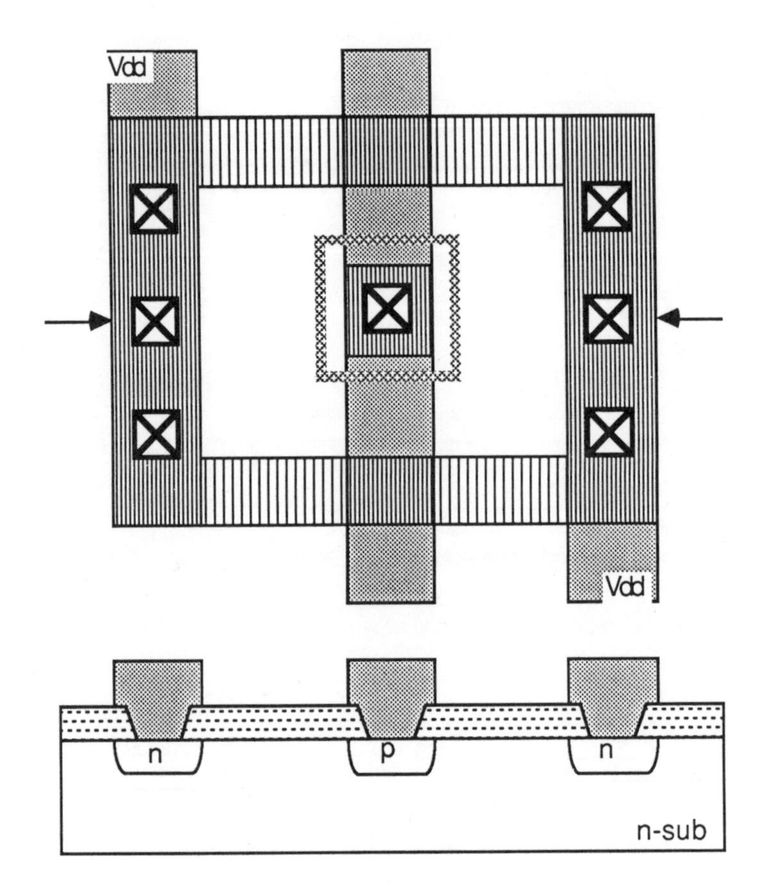

Plate XVI: p-n diode with guard-ring.

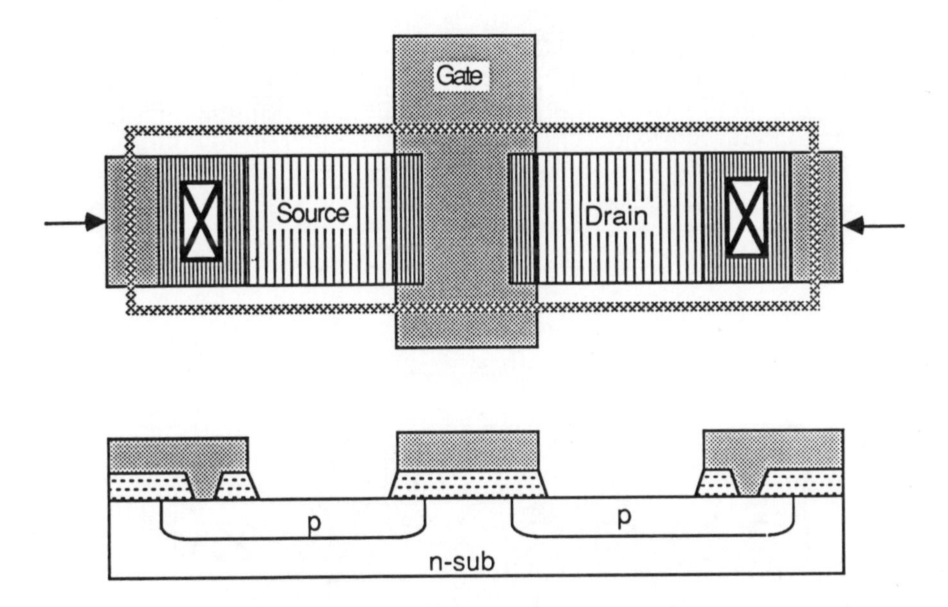

Plate XVII: Thick-oxide gate transistor.

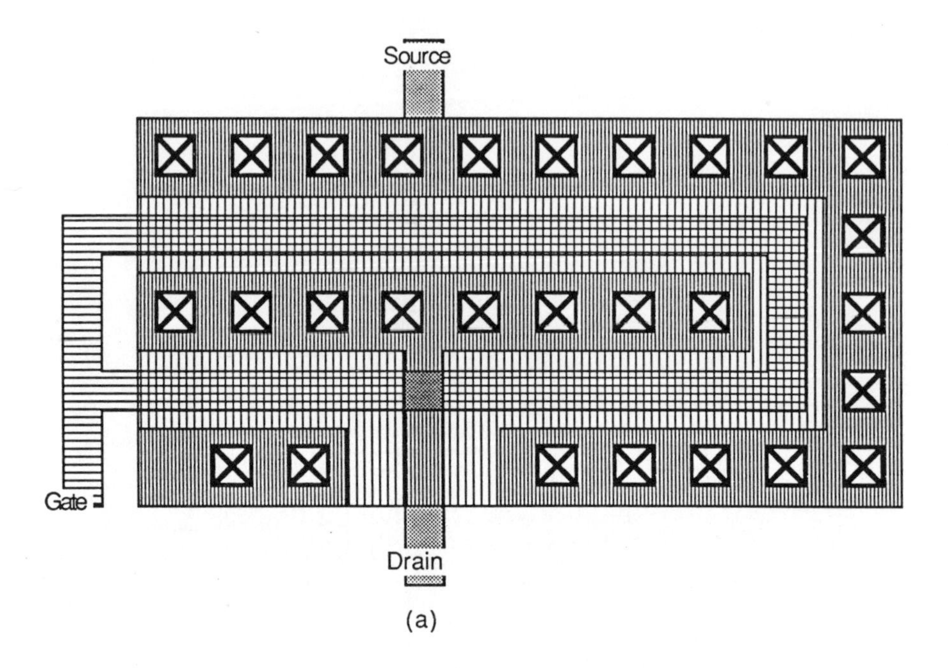

(a)

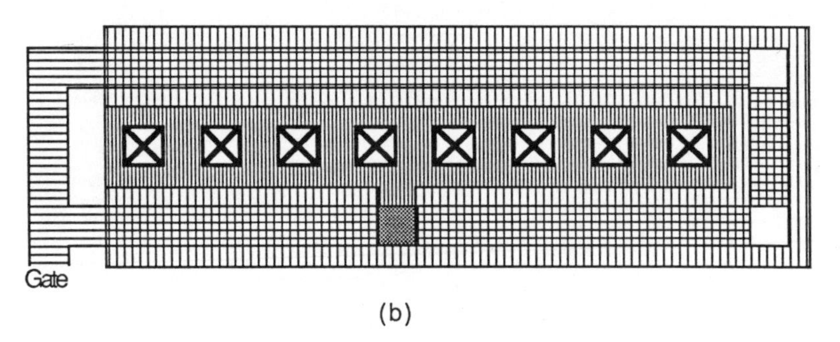

(b)

Plate XVIII: Wide channel device layout (a). The white areas in (b) do not contribute to the channel width.

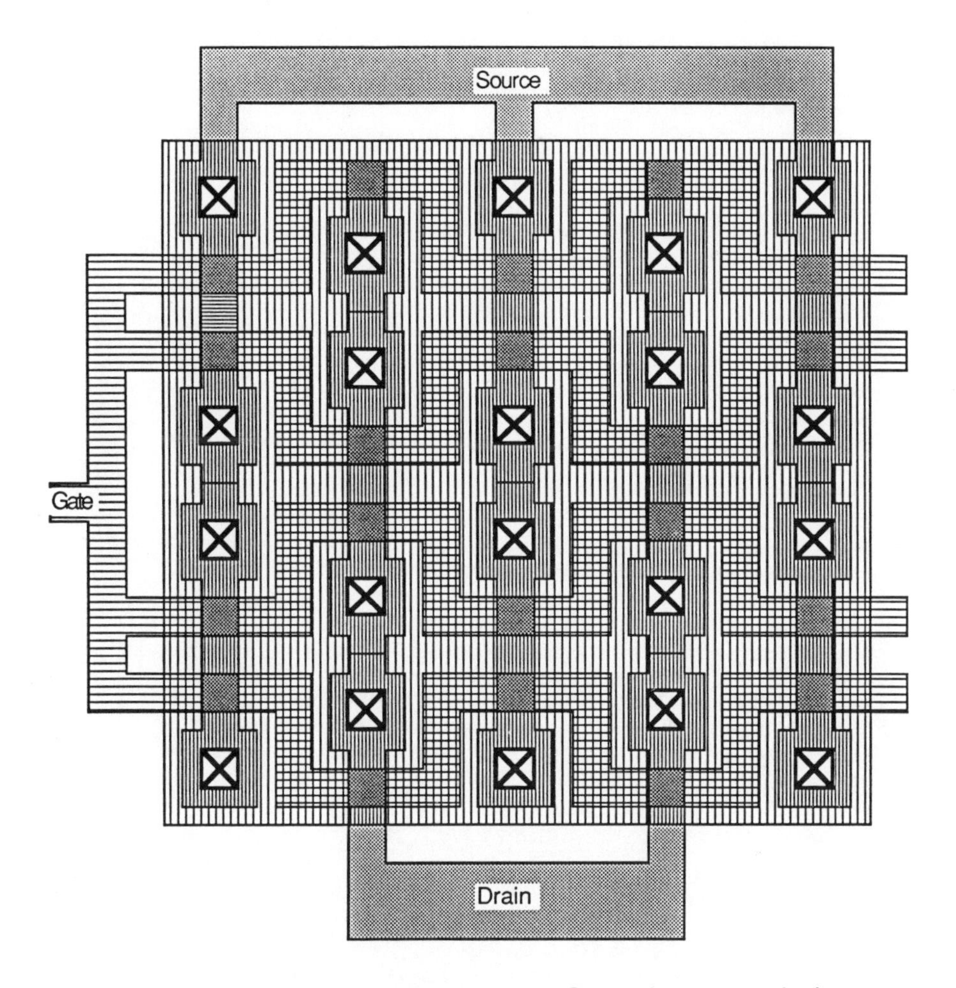

Plate XIX: Very wide channel device layout. Several narrower devices are laid out in parallel. P+ mask is not shown.

Chapter 6

Design of Basic Circuits

The aim of this chapter is not to present a complete, in-depth analysis of every possible building-block that can be found in today's VLSI circuits. Rather, a limited number of circuits is dealt with. This chapter presents circuits of increasing complexity: it starts with flip-flops and latches and ends with multipliers. More complex circuits — such as floating-point or memory management units — have not been considered, because an in-depth treatment — including architectural issues, logic design methodologies, and actual implementation — would have eventually occupied a large portion of the book.

Flip-flops, latches, and shift-registers are dealt with first. These are the simplest storage elements but represent essential building-blocks in more complex structures, such as register files, FIFO's, etc. Then, PLA's are dealt with. PLA design is far from being trivial, especially when dynamic PLA's are implemented. Charge sharing and noise play an important — and annoying — role in dynamic PLA's, and special care must be taken to design robust and reliable circuits.

Random-access memories are then considered. Three basic sub-units are discussed: memory cell, row/column address decoder and sense amplifier. Special emphasis is placed on sense amplifiers, because of the lack of basic literature on the subject. Note

---

Plates referred to in this chapter are found in the PLATES section located between pages 172 and 173.

that decoders and sense amplifiers are used in other memory technologies as well (e.g., ROM's, PROM's, EPROM's).

Finally, two arithmetic building-blocks are considered: integer adder and multiplier. In both cases a design example of a parallel unit is presented. After introducing the various families of integer adders, the implementation of a carry look-ahead adder is discussed in detail. Bit-serial arithmetic is not discussed, because parallel units are still preferred for high-performance components and bit-serial units are used mainly in specialized and dedicated architectures — such as cellular machines. The integer multiplier is based on an improved version of the modified Booth algorithm and allows a reduction in area, as well as in number of devices.

6.1. Storage Elements

A simple 1-bit shift register is shown in Fig. 6-1(a). It is controlled by a two-phase, non-overlapping clock. The information is stored in the capacitance at node B when `clk1` rises and is presented at the output when `clk2` goes high. The same circuit, implemented with C^2MOS logic, is shown in Fig. 6-1(b).

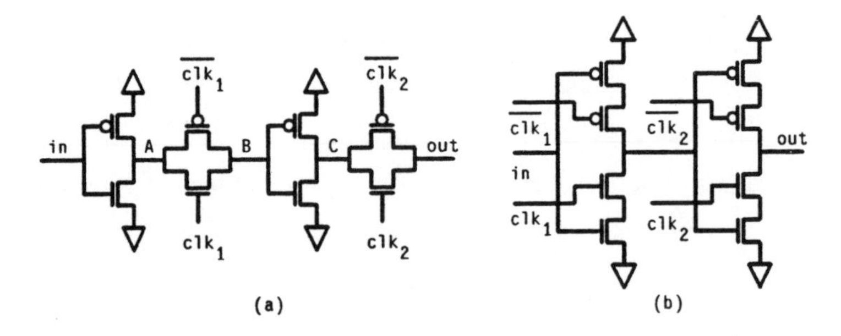

Figure 6-1: 1-bit shift register controlled by a two-phase, non-overlapping clock (a). C^2MOS implementation (b).

The two circuits shown in Fig. 6-1 are an example of dynamic storage element. Static storage elements can be used, as well, and two of them are shown in Fig. 6-2.

Both feature a cross-coupled inverter structure and implement a D-type latch. When `clk` is low, the information is stored in the cross-coupled structure. When `clk` goes high, the latch is updated. Although the circuit in Fig. 6-2(b) seems to have a redundant transmission gate, it is much easier to implement than the circuit in Fig. 6-2(a). Let us consider the circuit in Fig. 6-2(a) in more detail.

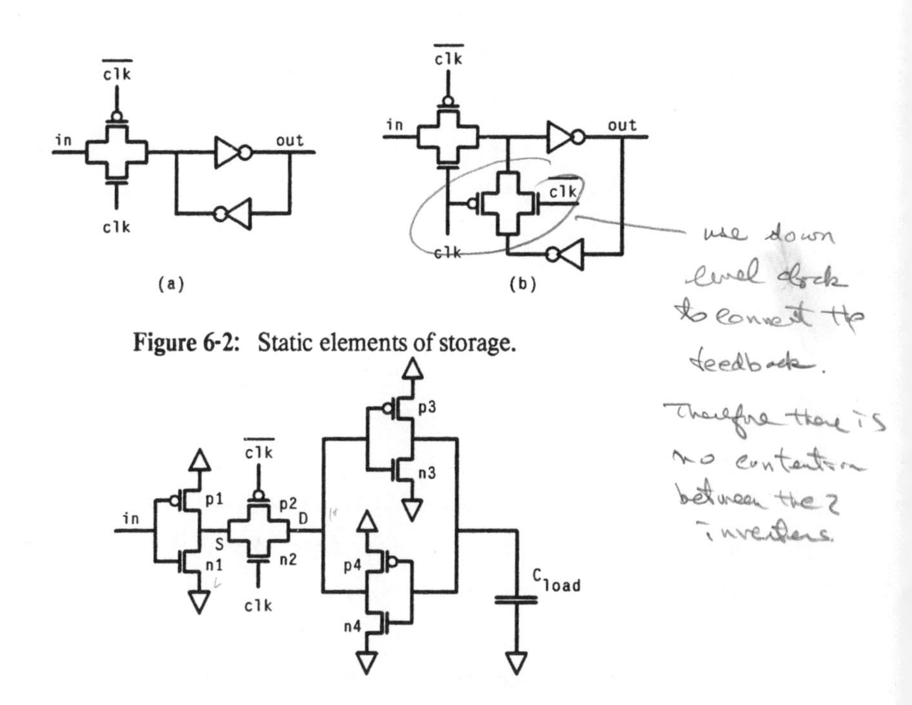

Figure 6-2: Static elements of storage.

Figure 6-3: Detailed representation of the circuit in Fig. 6-2(a).

In Fig. 6-3, the two nodes marked "S" and "D" indicate the source terminal and drain terminal of p2 and n2, respectively. Let us assume that node S is low, node D is high, and `clk` is low. n1, n3, and p4 are conducting, while p1, p2, n3, p3, and n4 are off. When `clk` rises, both p2 and n2 enter conduction, while p4 and n1 *are still on.* Therefore, the circuit is ratioed. In fact, this circuit can work correctly only if the transistors are sized properly. If the devices are not sized properly, the information stored in the latch cannot be changed. For instance, node D stays high even though `clk` goes high and node S is low.

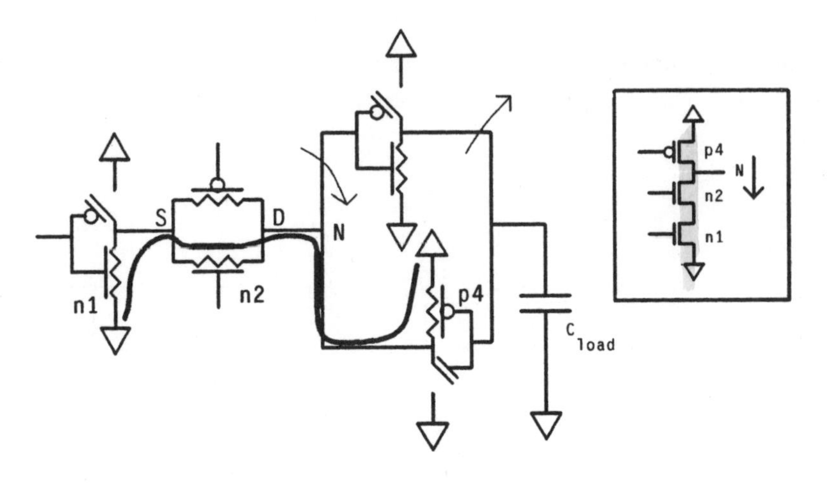

Figure 6-4: Changing the information stored in the latch (see text).

Let us assume that the voltage at node N in Fig. 6-4 is high, `clk` is low, and node S is low. When `clk` rises, both n2 and p2 enter conduction. When this happens, a path from V_{dd} to V_{ss} is formed. Conduction goes mainly through n2, rather than p2, because the drain (node D) is at higher potential than the source (node S). The V_{dd}-V_{ss} path consists of p4, n2, and n1. To pull down node N, it is necessary to have a long p4 transistor. If, for instance, p4 had "normal" size and n2 were a long channel transistor, node N could not pull down and make the top-most inverter switch — and therefore modify the content of the latch.

Let us assume that the voltage at node N is low, `clk` is low, and node S is high. When `clk` rises, both n2 and p2 enter conduction. Once again, a V_{dd}-V_{ss} path is formed and consists of p1, p2, and n4. p2 is conducting more than n2, because the source is at higher potential than the drain. n4 has to be a long channel device to correctly pull up node N. If p2 or p1 are long channel devices, the voltage increase at

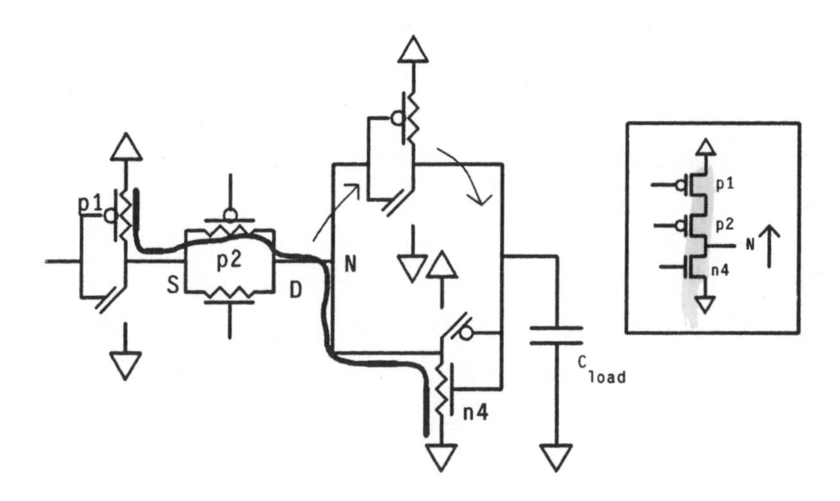

Figure 6-5: Changing the information stored in the latch (see text).

node N will not be sufficient to make the top-most inverter switch, and the content of the latch will stay unchanged.

After examining both cases, we can conclude that both n4 and p4 should be long channel devices, otherwise the circuit cannot work. The other devices can be minimum feature size — with p1 and p3 wider than n1 and n3. However, a more rigorous treatment would involve the analysis of the closed-loop gain of the cross-coupled structure. In particular, p4 can often be minimum size and yet the latch works properly. This can be explained by noting that what actually counts is the *transconductance* of the devices, rather than their size. The transconductance of a saturated MOS device can be defined as:

$$g_m = \frac{W \mu_{n/p} C_{ox}}{L} (V_{gs} - V_{Tn/p}) \qquad (6\text{-}1)$$

As Eq. (6-1) shows, the transconductance depends on the mobility. This helps intuitively to explain why p4 need not be a long channel device, because its mobility is usually from one-half to one-fourth that of the n-channel device. A detailed analysis of the cross-coupled inverters' closed-loop gain is not very useful because it cannot assure correct behavior of the circuit unless all the parasitic elements in the circuit are included. Therefore, a rough sizing, as suggested above, should be carried out, followed by careful circuit simulation.

Finally, the circuit shown in Fig. 6-2(b) has much more reliable behavior. This is due to the extra transmission gate, which usually guarantees correct behavior of the circuit, even when all devices are minimum feature size.

6.2. Full-adder

The full-adder has been chosen as a case-study because it is a simple, but very important building-block in many applications. We will present three different implementations: the first one uses both complementary and transmission gate intensive logic; the second one is an excellent example of transmission gate intensive logic, and the third one uses dynamic logic and allows us to focus once again on the phenomenon of charge sharing.

The first implementation of a one-bit full-adder is shown in Fig. 6-6 [22]. It is a very reliable, fast scheme and requires 34 devices. This full-adder features just two gate delays. If the input signals are not directly available in both polarities, three inverters — that is, 6 more transistors — are necessary.

A transmission gate intensive design of a full-adder is shown in Fig. 6-7 [12]. Neither V_{dd} nor V_{ss} lines are required if input A is available in both polarities. The total number of devices is 16, less than half the number in the previous design. Although this design is slower than that of Fig. 6-6, the fact that neither V_{dd} nor V_{ss} lines are necessary makes it interesting in applications where minimization of area is a critical requirement.

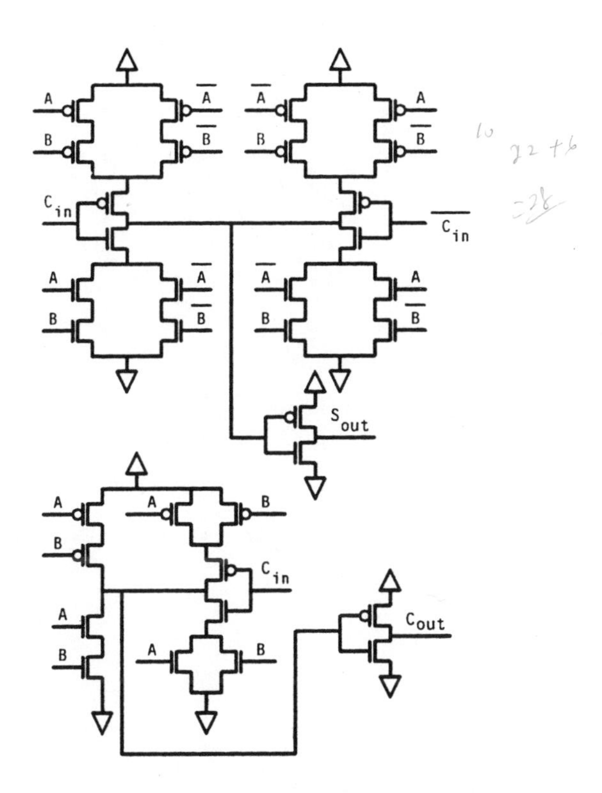

Figure 6-6: Full-adder using both complementary and transmission gate intensive logic.

A full-adder implemented in dynamic logic is shown in Fig. 6-8 [8]. P-E logic is used: both gates are precharged and evaluated in parallel. The first gate is precharged high, while the second gate is precharged low. Transistor n2 is not used to implement the boolean equations of the full-adder, but serves the purpose of protecting the second gate against charge sharing. Let us assume, first, that $\overline{ABC_{in}}$ = 001 and transistor n2 is not present. During evaluation, node 1 is pulled to node 2 — that is, to V_{dd}. If the input sequence $\overline{ABC_{in}}$ = 111 follows, the charge accumulated on node 1 discharges through n1 into node 3 — which is isolated from V_{ss} during evaluation — and the voltage on S_{out} moves from V_{ss} and rises. Transistor n2 makes sure that

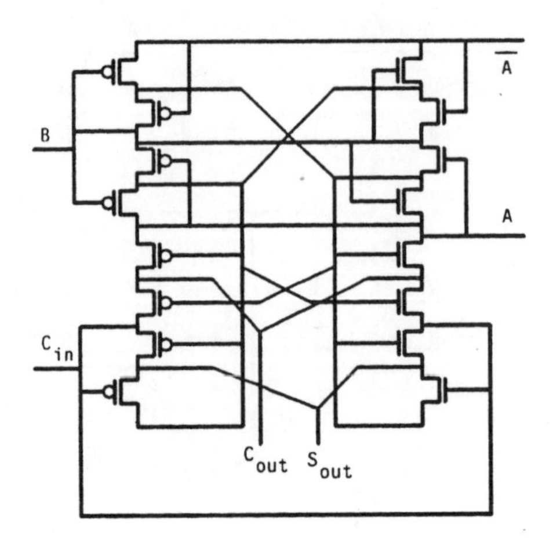

Figure 6-7: Transmission gate intensive full-adder.

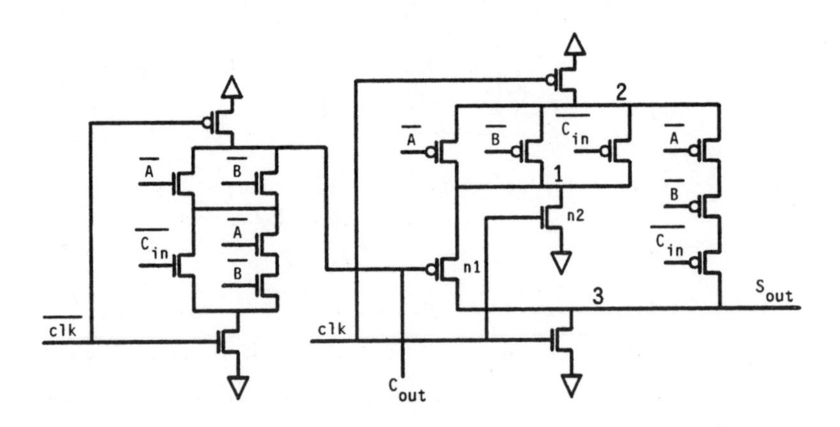

Figure 6-8: Implementation of a full-adder using P-E dynamic logic.

node 1 is pulled down during precharge. This scheme has only 17 transistors. The delay introduced by one-bit full-adders can be as short as 2.0ns, using state-of-the-art 1.2μm fabrication processes.

6.3. Programmable Logic Array

Programmable Logic Arrays (PLA's) are an essential part of many designs and can be used for various purposes, from building-blocks in finite-state machines to complex decoders. CMOS PLA's can be implemented by using most of the logic design styles presented in Chapter 4, and, therefore, both static and dynamic PLA's are used. Which one to choose depends mainly on the overall timing strategy used inside the chip and especially on the logic design style. Clock unavailability is a strong incentive for using static PLA's, because the implementation of a clock generator just to drive the PLA might be too expensive. Other constraints, such as speed and power dissipation, play an important role. A further constraint in choosing the logic design style comes from the necessity to generate a layout which is regular enough to be produced automatically by software tools.

A PLA is composed by two planes, an AND plane and an OR plane, which are cascaded. This AND-OR combination is not always optimal and, as we shall see below, is often changed according to the logic design style used. Static PLA's can be implemented both in nMOS-like logic and in complementary logic. nMOS-like logic is discussed first, because it closely resembles nMOS static logic, and it is assumed that the reader is already familiar with nMOS PLA's. An AND plane is not used in static nMOS PLA's because the size of the pull-up — the n-channel depletion transistor — depends on the number of inputs (the logic is ratioed). Automatic PLA generators would need pitch-matching procedures which complicate the program. The AND-OR configuration is usually changed into an INV-NOR-NOR-INV configuration with input and output inverters. The same argument applies to nMOS-like PLA's in which two NOR planes are used, as shown in Fig. 6-9. nMOS-like PLA's feature the same advantages and disadvantages of the logic used. When the number of inputs is limited, power dissipation and speed are not major constraints, and chip area must be minimized, PLA's can be implemented in nMOS-like logic. Otherwise this approach is not recommended, because of its high static power

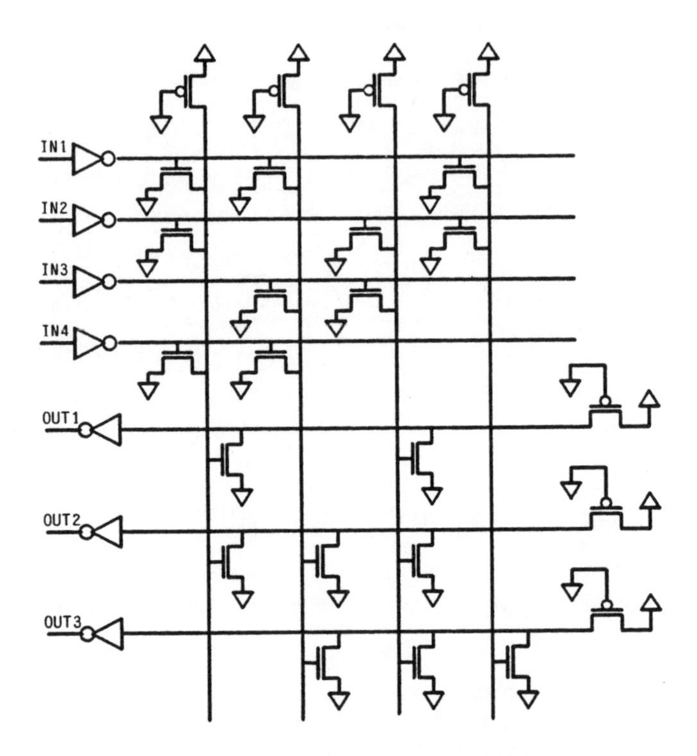

Figure 6-9: PLA implementation in nMOS-like logic.

dissipation.

The INV-NOR-NOR-INV approach is not used in complementary logic PLA's for performance reasons. As already pointed out in Chapter 4, NOR gates should be avoided in complementary CMOS logic, because the p-channel devices are in series. It is therefore recommended to rearrange the planes in such a way that AND structures are used. In complementary PLA's, the AND-OR plane is implemented with a NAND-INV-INV-NAND, as shown in Fig. 6-10, where the bold lines represent metal connections. A layout example of the circuit in Fig. 6-10 is shown in Plate X. This implementation occupies more area, but has no static power dissipation and good speed characteristics. When very large PLA's are used, the increase in area

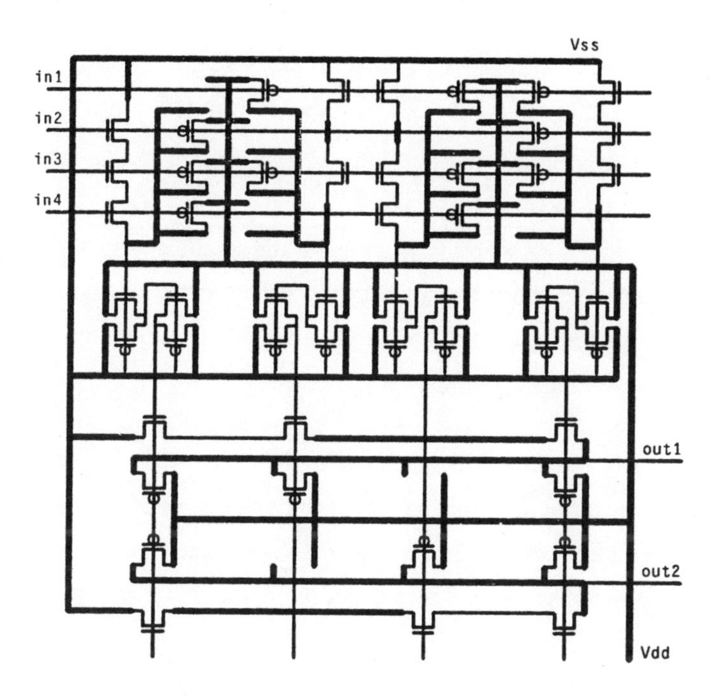

Figure 6-10: PLA implementation in complementary logic: two NAND planes are used.

with complementary implementation can become unacceptable; in this case the only effective solution is a dynamic PLA.

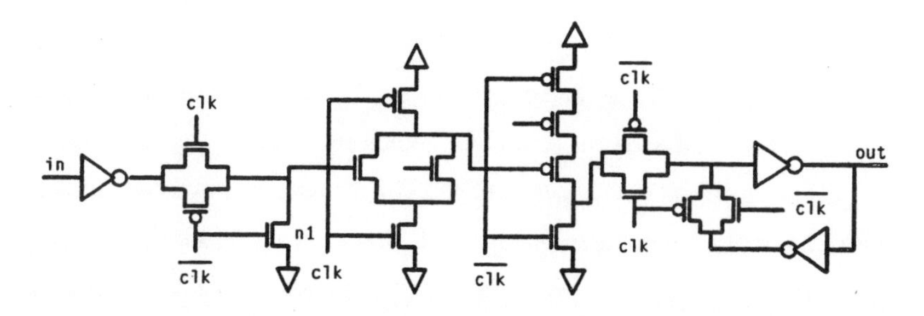

Figure 6-11: A simple dynamic PLA scheme.

There are several implementations of PLA's in dynamic logic. According to Chapter 4, ripple-through PLA's cannot be implemented, because when the first plane is precharging, the second one is unable to do so. A short circuit between V_{dd} and V_{ss} would take place. Therefore, other schemes have to be used. Dynamic PLA's can be implemented in domino logic. Because the logic is non-inverting, AND and OR planes are used. A simple PLA scheme is shown in Fig. 6-11: the PLA operates with `clk` and $\overline{\texttt{clk}}$ and is based on an INV-NOR-NOR-INV combination. This scheme is preferred, because the NOR gate is slightly faster than the corresponding NAND gate in a ratioless logic.

When `clk` is low, both planes are precharging. The n-channel devices in the first plane are kept off by transistor `n1`. This precaution is necessary because dynamic PLA's suffer more extensively from charge sharing than hand laid out gates do. Because regular layout patterns are required, drain and source areas are more difficult to minimize, and larger junction capacitances can be expected. The p-channel devices in the second plane are also off, because their inputs come from precharged-high NOR gates. The output latch is holding the previous value. When `clk` goes high, the input transmission gate conducts, and the first plane evaluates. The signal ripples through the two planes and updates the information stored in the output latch. When `clk` goes down and the input transmission gate goes off, the n-channel devices in the first plane are pulled down, while both planes start precharging again. Finally, note that both input and output stage can be implemented with a C^2MOS inverter.

Large dynamic PLA's suffer from noise problems. Because of their regular structure and automatic layout generation, source terminals can be connected to V_{dd} or V_{ss} through diffusion, rather than metal. Source terminals expected to be at 0V, instead can be at a few hundreds millivolts. Likewise, source terminals expected to be at 5V, can be at 4.5V — or even less. This resistive coupling can create internal glitches and slow down the logic. Let us consider the circuit shown in Fig. 6-12: R is the resistance of a diffusion interconnection to a V_{dd} terminal. The circuit in

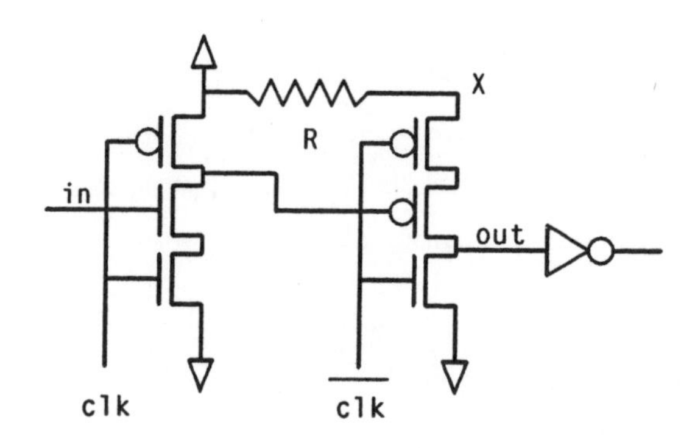

Figure 6-12: Dynamic gate with diffusion interconnect to V_{dd} terminal.

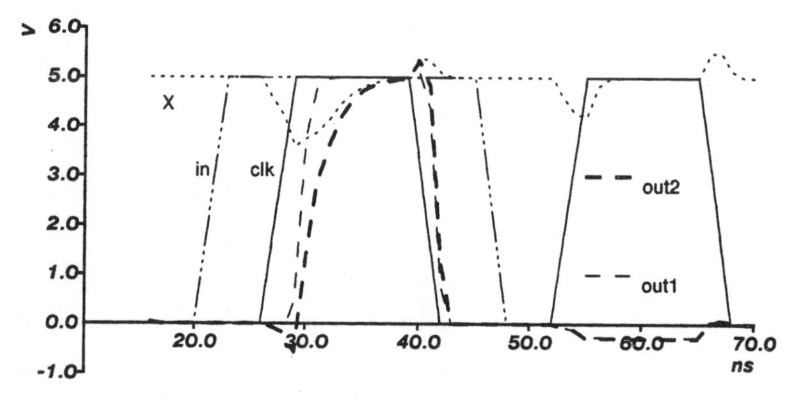

Figure 6-13: Waveforms for the circuit in Fig. 6-12.

Fig. 6-12 has been simulated with R = 1Ω and R = 10KΩ. Note that this last value can be considered appropriate for large PLA's. 200μm of diffusion can have such a resistance. Fig. 6-13 shows the results: `out1` is the output of the circuit with R = 1Ω, while `out2` is the output of the circuit with R = 10KΩ. Note that the voltage drop on node X (R = 10KΩ) is almost 1.8V. Note, also, that `out2` drops to -0.6V, coming very close to forward-biasing the emitter-base junction of the vertical bipolar transistor — with possibility of latchup — and overshoots to 5.4V. Finally, the large resistance slows down the gate.

6.4. Random-access Memory

The block-scheme of a random-access memory (RAM) is shown in Fig. 6-14. The *memory array* is a matrix which contains the memory cells. Each cell stores one bit of information. The *row decoder* selects 1 out of 2^n possible rows of memory cells, while the *column decoder* selects 2^p out of 2^l columns.

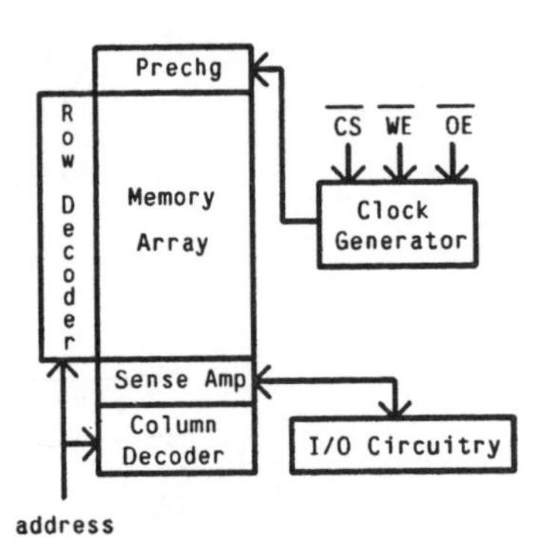

Figure 6-14: Block-scheme of a random-access memory.

Sense amplifiers are used to speed up the access time of the memory. Finally, a *clock generator* circuit generates the internal timing. The clock generator can be absent in some memories, because the off-chip (system) clock is used. An on-chip clock can be triggered by signals such as "chip select" ($\overline{CS}$), "output enable" ($\overline{OE}$), and "write enable" ($\overline{WE}$). The memory clocking scheme is, therefore, independent of the system clock and works asynchronously. Both synchronous and asynchronous memory chips are currently available. Note that the clock generator shown in Fig. 6-14 is connected only to the *precharge logic*. Other designs can have the clock driving either the sense amplifiers, the decoders, the input-output circuits, or any combination thereof, as we shall see in the next pages. CMOS clock generators are usually implemented with a chain of scaled-up inverters [19, 16]. While it was normal to find nMOS bootstrapped clock generators in dynamic RAM's (DRAM's), problems of latchup make this circuit riskier in CMOS, although some examples can be found in the literature (see, for

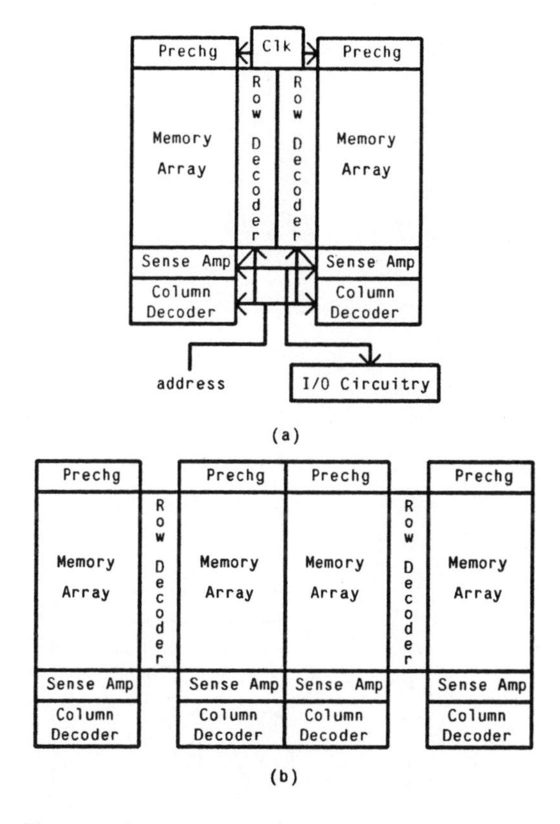

Figure 6-15: Two floor-plans of memory chips.

instance, [21]).

If we were to design a 64Kx8 memory chip and had only one matrix of 64Kx8 cells, the *column decoder* would not be necessary. In reality, a 64Kx8 array of cells would be very tall and narrow. This aspect ratio is not a positive feature because very long lines, usually in polysilicon — or polycide — have to be used to access the uppermost cells, and this slows down the circuit. Moreover, such aspect ratio does not match with the chip cavity, which is approximately square. Therefore, the internal word width is usually much wider than the word width of the memory chip. For instance, four 512x256 arrays of cells are used in 64Kx8 memory chips. Two different floorplans of memory chips are shown in Fig. 6-15. These arrangements aim to reduce the length of data and control lines in the vertical direction. They also distribute the total load to more than one row decoder. Fig. 6-15(b) does not show clock generator, data and control lines, and I/O circuitry. Whether two or four memory arrays should be used depends on the word width of the memory chip, total memory capacity, clock distribution, and layout issues.

6.4.1. Memory Cell

Memory cells can be either static or dynamic. In the first case, information is stored in a static structure such as the six-transistor cell shown in Fig. 6-16, which has two cross-coupled inverters. This is the typical memory cell in static RAM's (SRAM's).

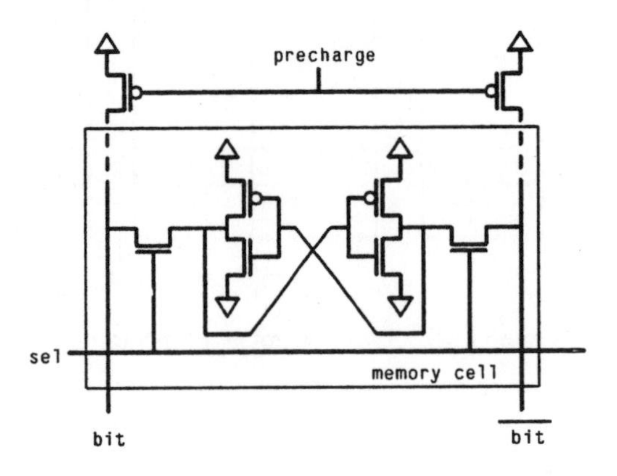

Figure 6-16: Six-transistor static memory cell.

The selection line (`sel`) comes from the row decoder and is shared by all the cells which belong to the same row in the matrix. The cell provides complementary outputs, that is `bit` and $\overline{\texttt{bit}}$. The same two lines are used for reading the content of the memory cell and to write new information into it. A "read" operation consists of two steps. First, the `bit` and $\overline{\texttt{bit}}$ lines are precharged high, while `sel` is low. Then, `sel` goes high and either `bit` or $\overline{\texttt{bit}}$ is pulled down. Depending on which line goes down, we can determine whether a logic "0" or a logic "1" is stored into the cell. In a "write" sequence, the precharge transistors are off, the data is placed on the `bit` line — the complementary value is placed on the $\overline{\texttt{bit}}$ line — while `sel` is low, and, finally, `sel` goes high.

The size of the cell is one of the most critical parameters in the design of a memory, because up to 80% of the total area can be occupied by the memory array. The cell shown in Fig. 6-16 consists of four n-channel and two p-channel enhancement devices. Therefore, some area is wasted because of the spacing necessary between n-type and p-type devices — a more compact layout can be achieved in SOS or SOI memories, because this spacing is not longer required. To decrease the area of the memory cell it is possible to use different pull-up structures, such as n-channel depletion transistors or polysilicon resistors. In these two cases both power dissipation and access time increase, especially when resistive loads are used.

The six-transistor cell is a fairly robust design, even under significant variations of some parameters — such as voltage on the bit lines, transistor size, and so on. Theoretically, charge sharing can take place during "read" operations and even flip the cell. In reality, this is unlikely to happen if the two pull-down n-channel transistors are two or three times wider than the pass transistors. The two p-channel transistors should have minimum size to minimize leakage current, although pulling up the output of one of the inverters during a "write" operation will take longer.

In dynamic memory cells, the information is stored into capacitors rather than in a static structure of cross-coupled inverters. These capacitors need to be refreshed to insure that the information is not lost. Refresh can be a problem when DRAM's are used in purely synchronous architectures, such as systolic arrays [1]. Nonetheless, DRAM's feature much higher storage capacity than SRAM's because the number of transistors in the basic cell can be reduced drastically. Different schemes of dynamic

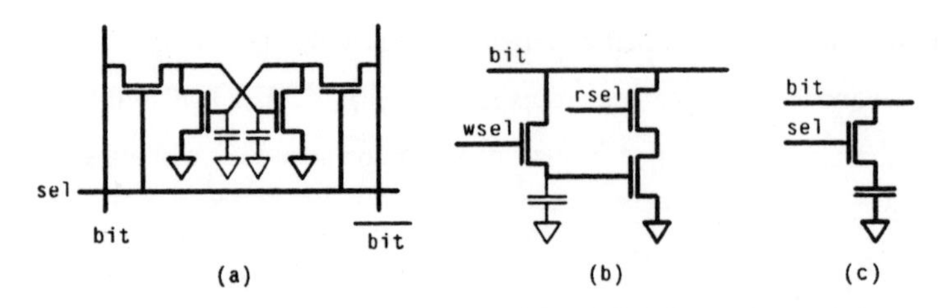

Figure 6-17: Four-, three-, and one-transistor dynamic memory cells.

memory cells are shown in Fig. 6-17: four-, three-, and one-transistor cells are shown.

The dynamic memory cell shown in Fig. 6-17(a) works in a way similar to the six-transistor static cell [23]. Information is stored into the gate capacitance of the cross-coupled devices. The cell is written by placing a value and its complement on the `bit` and $\overline{\texttt{bit}}$ lines, while `sel` is kept low. Then, `sel` goes high and the information is stored in the cell. Fig. 6-17(a) does not show two pull-up transistors that are controlled by the precharge signal, like in Fig. 6-16. When the cell is written, these pull-up transistors are turned off. To read the cell, both pull-up devices are turned on and `sel` goes high. One bit line will go low, while the second one will stay high. Note that this allows the cell to be refreshed just by reading it.

The three-transistor dynamic cell [24] shown in Fig. 6-17(b) works as follows. The charge is stored in the gate capacitance of the pull-down transistor. Like in Fig. 6-17(a), the bit line is connected to a pull-up device not shown in the figure. To write the cell, a value is placed on the bit line while the pull-up transistor of the bit line is off. Then, `wsel` goes high. If the value placed on the bit line is low, the capacitor will discharge — if it was previously charged. Similarly, if the value placed on the bit line is high, the cell capacitor is charged — if it was previously uncharged. To read the cell, the bit line is first pulled up. Then, `rsel` goes high. If the voltage across the capacitor was high (i.e., if the capacitor was charged), the bit line is pulled down. If the voltage across the capacitor is zero (i.e., the capacitor is uncharged) the bit line stays high. Note that this cell, unlike the six-transistor and four-transistor cell, does not have bit lines carrying complementary values. This has some impact on the design of sense amplifiers, as we shall see in Section 6.4.3.

A one-transistor cell is shown in Fig. 6-17(c) [19]. The capacitor is a characteristic of the fabrication process and normally uses SiO_2 as dielectric. To write the cell, a value is placed on the bit line and `sel` goes high. The charge on the bit line is transferred to the storage node and the capacitor is charged, discharged, or left unchanged. To read the cell, the bit line is first pulled up. Then the cell is selected. If the voltage across the capacitor is zero, the bit line is pulled down, otherwise it stays high — and at the same time refreshes the cell capacitor. Both p-channel and n-channel devices can be used in one-transistor cell DRAM's [19, 21].

Much more could be said on the design of memory cells, but this would go beyond the purpose of this book. This is especially true for state-of-the art memory chips, where cell size and characteristics are influenced by the fabrication process parameters to such an extent that memory cell design becomes exclusively an exercise in semiconductor physics, rather than in logic design.

6.4.2. Decoder

Both static and dynamic decoders are currently used in memory chips. Decoders can be single stage circuits or, more often, multiple stage circuits. A simple single stage circuit is shown in Figs. 6-18 and 6-19. The NOR-based decoder of Fig. 6-19 is usually preferred because the n-channel devices are in parallel — rather than in series — as in the NAND-based decoder of Fig. 6-18. The address is provided in both polarities, and each decoder has different inputs. These nMOS-like circuitries suffer from static power dissipation. To eliminate static power dissipation, the same circuits can be implemented in dynamic logic (e.g., in domino logic). Note that NAND-based decoders can be used in high performance memory design when address decoding can be overlapped with bit line precharging [17].

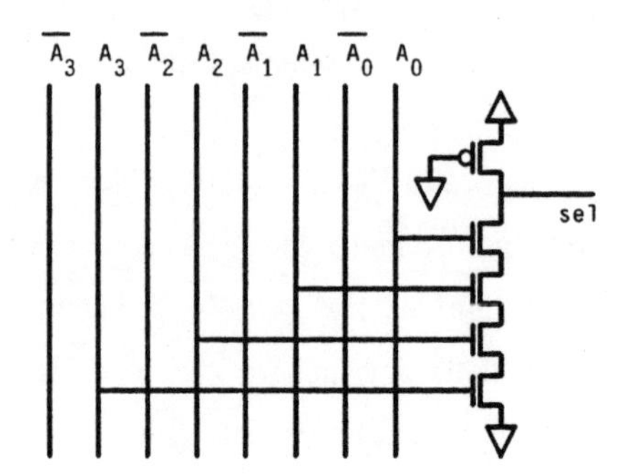

Figure 6-18: Simple nMOS-like, NAND-based row decoder.

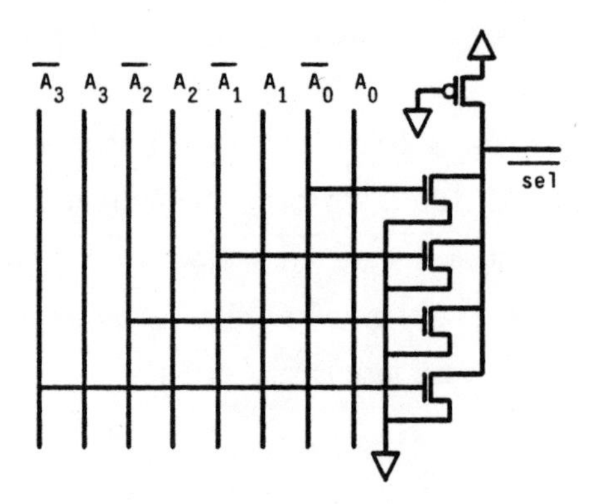

Figure 6-19: Simple nMOS-like, NOR-based row decoder.

The decoders shown in Figs. 6-18 and 6-19 occupy a significant amount of area, and, therefore, smaller circuitries are more often implemented. These schemes are based on multiple-stage decoders. One of these circuits is shown in Fig. 6-20 and uses both complementary and transmission-gate intensive logic [7, 20]. If four of these circuits are used and each one receives a different combination of A_1 and A_2, we obtain eight output lines, of which only one is active. If addresses with more than 3 bits are used, multiple stages of the same circuit shown in Fig. 6-20 can be used. This allows us to put only the last stage in the column(s) containing the row decoder(s), and this simplifies pitch matching with the size of the memory cell.

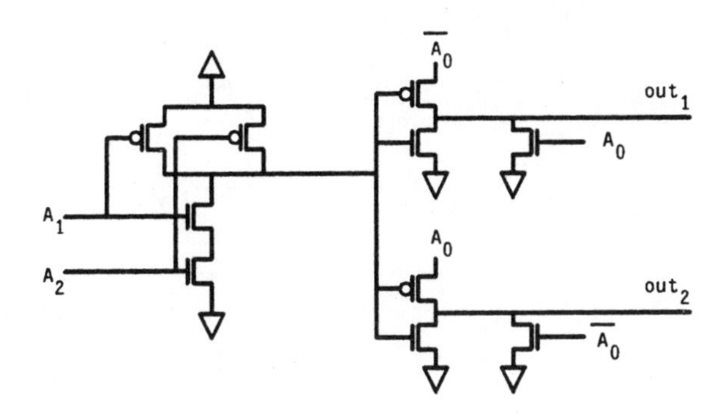

Figure 6-20: Building-block for a multi-stage row decoder.

6.4.3. Sense Amplifier

Before reading the content of a memory cell, both `bit` and $\overline{\texttt{bit}}$ lines are precharged high (assuming a six-transistor memory cell). When the precharge devices are turned off and the cell is selected, one of the two bit lines goes down, while the other stays high. Depending on which one goes down, we can determine whether a logic "1" or a logic "0" is stored into the cell. Because the transistors inside the cell are very small and the capacitance of the `bit` and $\overline{\texttt{bit}}$ lines is large, pulling down one of the two lines takes a significant amount of time. This results in a very slow access time. Sense amplifiers are circuits which speed up the reading operation by sensing and amplifying small voltage differences between the two lines. Sense amplifiers are particularly difficult circuits to design; in fact, most sense amplifiers contain positive feedback loops and exploit — rather than avoid, as in conventional logic design — charge sharing to improve performance.

This section, far from being an in-depth analysis of design methodologies for sense amplifiers, aims to present some basic topologies and to point out the fundamental issues in the design of such circuits.

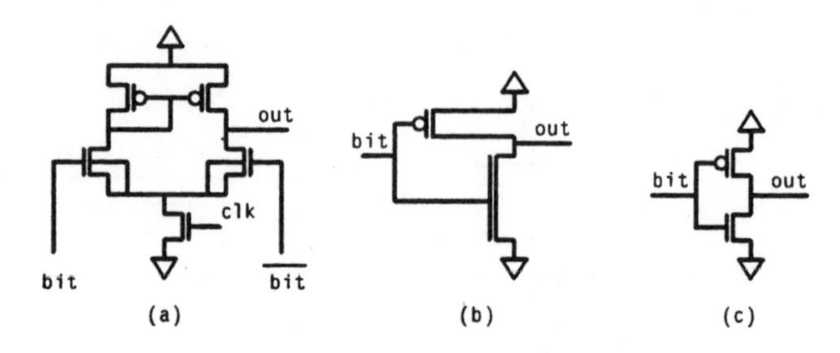

Figure 6-21: Three possible circuits at the end of a bit line. The circuit shown in (a) is the most effective.

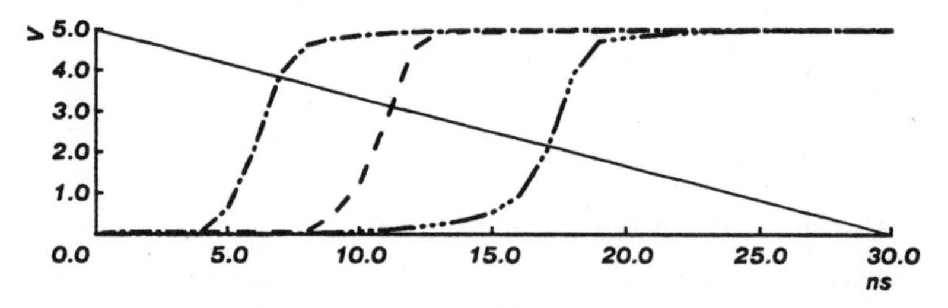

Figure 6-22: Slow voltage drop on a bit line and output voltages of the three circuits shown in Fig. 6-21. The bit line is connected to the input of the circuits shown in Fig. 6-21(b) or (c). See text for the connections to the circuit in Fig. 6-21(a).

To show how effective sense amplifiers can be, let us start with an example. First, the bit line whose voltage drops is connected to an inverter like the one shown in Fig. 6-21(c). This inverter features "normal" ratios with the p-channel device somewhat wider than the n-channel device — both transistors have the same channel length. The rightmost rising curve in Fig. 6-22 shows the dynamic behavior of such a circuit when the input voltage drops slowly. As expected, the output of the inverter does not switch until the input voltage reaches the inverter threshold voltage. One approach for increasing speed involves designing an unbalanced inverter, as depicted in Fig. 6-21(b). Again, the input to this inverter is the bit line whose voltage drops. The inverter consists of a very wide p-channel device and a very long n-channel device. By choosing the widths and lengths appropriately, the inverter threshold voltage can be set close to, say, 4V. The output of the inverter switches to V_{dd} when the input voltage goes below 4V, rather than below 2.5V, as is the case with the balanced inverter of Fig. 6-21(c). Note that, by using this unbalanced inverter, the noise margin worsens, because even a small voltage drop in the bit line can make the inverter switch. Finally, the circuit shown in Fig. 6-21(a) produces the leftmost rising curve in Fig. 6-22. The inputs to this circuit are the `bit` and the $\overline{\texttt{bit}}$ line. This behavior is far better than the behavior of the previous two circuits, and the noise margin is not significantly affected. In fact, the two bit lines are so close to each other that we can assume that noise will equally affect both of them. Because the sense amplifier detects voltage *differences*, rather than absolute values, the noise margin does not worsen as much as it does in the unbalanced inverter. The circuit in Fig. 6-21(a) is now analyzed in more detail.

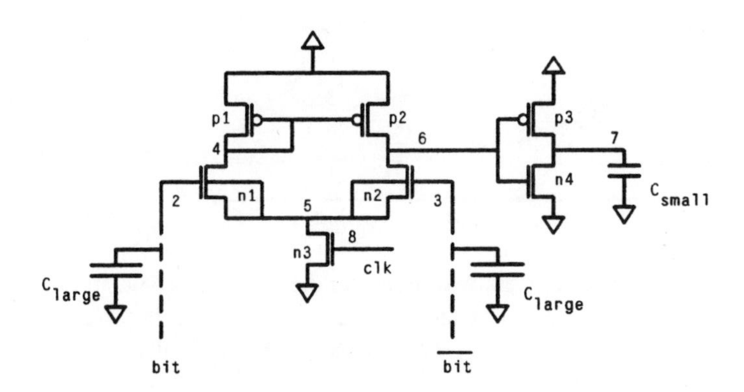

Figure 6-23: A circuit which speeds up the memory read access time.

The circuit shown in Fig. 6-23 is a differential amplifier stage, typically used as a front-end in operational amplifiers. The gates of the two n-channel devices n1 and n2 are connected to the `bit` and $\overline{\texttt{bit}}$ lines. Each bit line is characterized by a large capacitance C_{large}. Transistor n3 is a long channel device which acts as a current source for both branches. The four transistors n1, n2, p1, and p2 are neither long channel devices nor wide channel devices. The n4-p3 inverter is only used to allow us a fair comparison to the two circuits shown in Fig. 6-21(b) and (c), but is not part of the differential amplifier scheme. The small capacitance C_{small} is used simply to terminate the output of the inverter.

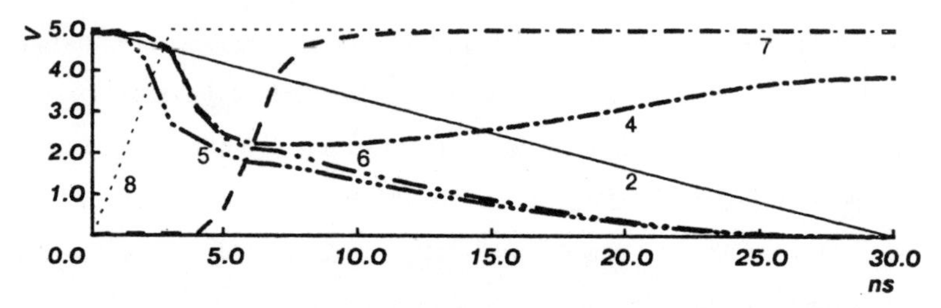

Figure 6-24: Waveforms for the circuit in Fig. 6-23: the voltage at node 2 is falling while the voltage at node 3 (not shown) stays high.

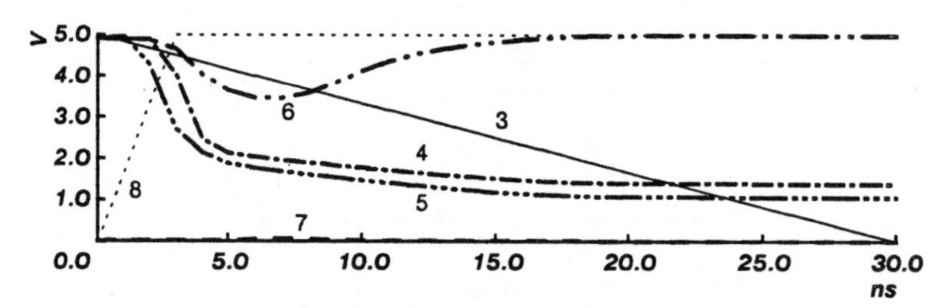

Figure 6-25: Waveforms for the circuit in Fig. 6-23: the voltage at node 3 is falling while the voltage at node 2 (not shown) stays high.

First, the two bit lines are pulled up while `clk` is low, therefore keeping n3 off. Both n1 and n2 conduct, and node 5 is pulled up because of capacitive coupling. Node 6 goes high as well. Therefore, the output of the inverter is low. Then, the voltage on node 2 starts to drop (see Fig. 6-24), and — at the same time — transistor n3 is turned on. The voltage on node 6 suddenly drops, and the output of the inverter goes high.

If the voltage on node 3 — rather than on node 2 — drops, this turns n2 off and does not allow node 6 to discharge to V_{SS}. The output of the inverter will stay low. Therefore, the behavior of this circuit is different depending on whether the voltage on node 2 or node 3 drops — with the other node high. If the voltage on node 2 drops, the output of the inverter performs a high-going transition; if the voltage on node 3 drops, the output of the inverter does not change (i.e., it stays low). This allows us to detect whether a logic "1" or a logic "0" is stored into the memory cell.

Not all fabrication processes allow us to use this circuit, because the two transistors n1 and n2 have their substrate terminal connected to the drain of transistor n3. This is possible using twin-tub [20], SOS [11], or p-well fabrication processes. In an n-well fabrication process, where the n-channel devices are formed over the substrate, rather than over a well, the circuit of Fig. 6-23 cannot be implemented, because all the n-channel devices have their substrate terminals connected to V_{SS}.

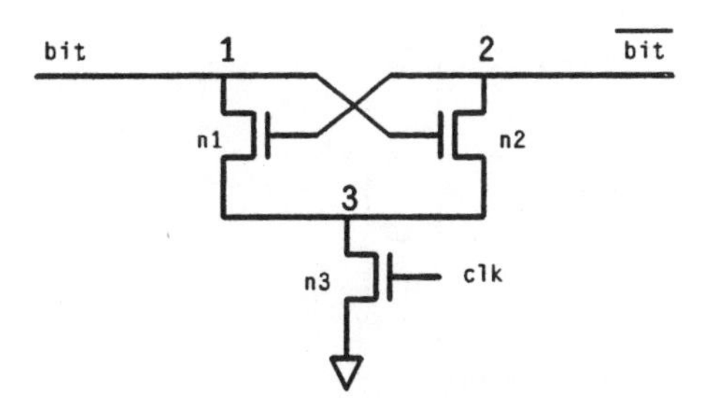

Figure 6-26: Cross-coupled latch.

Another basic topology for sense amplifiers is the cross-coupled latch, shown in Fig. 6-26. Let us assume that both nodes 1 and 2 have been pulled up during precharge and that the voltage on node 2 starts to drop. When n3 is turned on, the voltage at node 1 is higher than the voltage at node 2. Therefore, n2 turns on first and further pulls down the potential at node 2. This makes it more difficult for n1 to conduct. Eventually, n1 turns off completely, node 1 cannot discharge to V_{ss}, and n2 keeps conducting, that is, node 2 goes to V_{ss} through n3.

In reality, this circuit has a much more complex behavior. First, imbalance in the threshold voltages of n1 and n2 can decrease the sensitivity of the circuit[1]. Second, the layout of the circuit has to be geometrically symmetric, because different junction capacitances in the diffused areas can decrease the sensitivity.

One serious problem in sensing bit lines comes from charge sharing between the (large) line capacitance and the (small) memory cell capacitance. Although this is especially critical in dynamic memory cells, charge sharing can destroy the information in static memory cells as well. Even when such a catastrophic occurrence does not take place, charge sharing causes voltage drops in the bit line and decreases the efficiency of the sense amplifier.

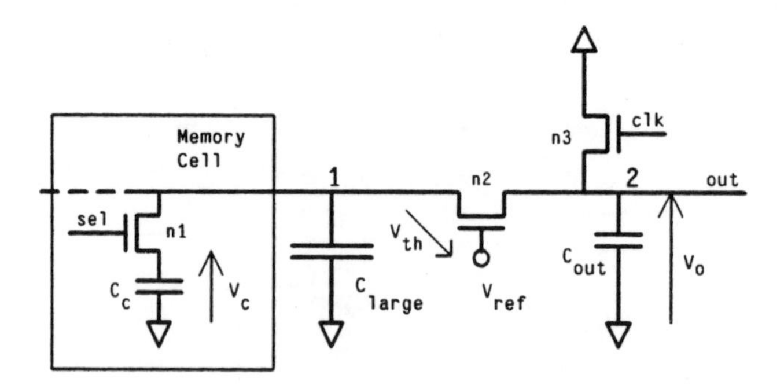

Figure 6-27: Charge transfer based circuit.

One possible solution to the problem above has been presented in the literature

[1]The sensitivity of a sense amplifier is defined as the minimum voltage difference between V_1 and V_2 (see Fig. 6-26) which is detectable by the circuit.

[10] and is now discussed briefly. It consists of a *shield resistor* between the line capacitance and the sense amplifier input node. This resistor can be implemented by an n-channel depletion transistor [24, 23] or by an n-channel enhancement transistor controlled by a reference voltage. We assume that a dynamic memory cell such as the one-transistor cell shown in Fig. 6-17 is being used. Fig. 6-27 shows the memory cell, the bit line capacitance C_{large}, the shield resistor (device n2), a pull-up transistor (n3) and an output capacitance $C_{out} \ll C_{large}$. The bit line is precharged when `clk` goes high and C_{large} is charged. When the potential at node 1 (V_1) reaches $V_{ref} - V_{th}$, n2 turns off. Therefore, the voltage across C_{large} is $V_{ref} - V_{th}$. At the same time, V_o reaches $V_2 > V_{ref}$. Transistor n3 is turned off.

When the cell is selected, there is charge sharing between C_{large} and C_c. Assuming that the cell contains a logic "0", $V_c < V_1$. Because of charge sharing between C_{large} and C_c, V_1 becomes:

$$V_1 = \frac{(V_{ref} - V_{th})C_{large} + V_c C_c}{C_c + C_{large}} .$$

The voltage drop V_{drop1} on node 1 is:

$$V_{drop1} = (V_{ref} - V_{th}) - V_1 ;$$

$$V_{drop1} = \frac{(V_{ref} - V_{th} - V_c)C_c}{C_c + C_{large}} . \qquad (6\text{-}2)$$

This voltage drop is sufficient to make n2 leave the cut-off state and return to saturation. At this point C_{large} and C_{out} are switched in parallel. Because V_o is at higher potential than V_1, some electronic charges flow from C_{large} to C_{out}, and V_o drops (while V_1 increases). The amount of charges flowing to C_{out} is:

$$Q = V_{drop}(C_c + C_{large}) .$$

This mechanism is called "charge transfer" and allows us to drive highly capacitive lines at high speed. When V_1 reaches $V_{ref} - V_{th}$, n2 turns off again. The voltage drop at node 2 is:

$$V_{drop2} = Q/C_{out} = (V_{ref} - V_{th} - V_c)\frac{C_c}{C_{out}}, \qquad (6\text{-}3)$$

and the voltage drop developed at node 1 has now been amplified at node 2. In fact, by comparing Eq. (6-2) with Eq. (6-3) we have:

$$C_c + C_{large} \gg C_{out}.$$

V_{ref} shall be generated on-chip to track process variations. A simple voltage reference generator is shown in Fig. 6-29 and consists of transistors p4, n9, n10, and n8.

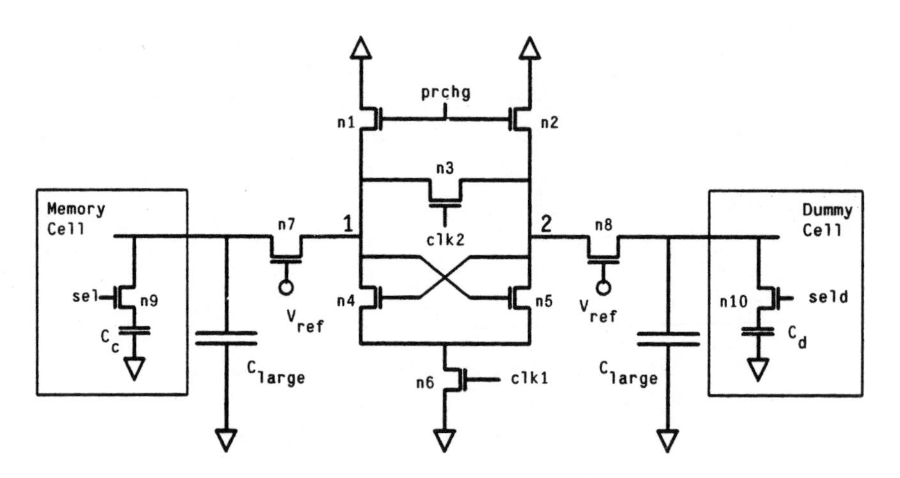

Figure 6-28: Sense amplifier scheme based on cross-coupled devices and charge transfer mechanism.

A sense amplifier based on the previous scheme is shown in Fig. 6-28. One input to the cross-coupled latch is the bit line, while a "dummy bit line" provides the other input. This is a typical arrangement in one-bit memory cells, because the cell does not provide complementary outputs. Let us assume that before precharging both bit lines have a potential between the two logic levels (e.g., $V_{dd}/2$). We will justify this assumption later. When precharge takes place, nodes 1 and 2 are pulled up. Then, n1 and n2 are turned off. Both the memory cell and the dummy cell are selected, and charge transfer takes place. C_{out} in Fig. 6-27 is the drain junction capacitance of transistors n4 and n7 — and of transistors n5 and n8, on the dummy cell side. Its value is slightly larger than that of the memory cell capacitor (e.g., $C_{out} \approx 3C_c$). Assuming that the memory cell contains a logic "0" — that is, the voltage across C_c is zero — the voltage drop on the dummy bit line will be half as much as that on the bit line, because the voltage across C_d is between the two binary levels. Therefore, the potential at node 1 drops by a larger amount and keeps n5 off. The smaller voltage drop on node 2 keeps n4 on, and when `clk1` goes high and n6 starts conducting the potential at node 1 goes to V_{ss}. The voltage drop on node 1 makes n7 leave the saturation region and work in the linear region. The bit line goes to V_{ss}, and a zero voltage level inside the memory cell is restored. When n3 is switched on, the dummy bit line discharges into the bit line. This equalizes the voltage level between the two lines and sets it in between the two binary levels. The same voltage is now across C_d — which is still selected. Finally, the dummy cell is deselected and n3 is turned off. If a logic "1" is stored in the memory cell, node 2 will be at a lower potential than node 1, and n5 conducts while n4 does not. The bit line cannot discharge to V_{ss}, and the information stored into the memory call is maintained. Moreover, the memory cell is also refreshed.

Many memories use a combination of the circuit schemes presented above. A typical approach is to use column sense amplifiers together with main sense amplifiers, as shown in Fig. 6-29. The column sense amplifier provides current amplification on the bit lines and decreases the "read" time. Fig. 6-30 shows the input signals to the circuit of Fig. 6-29. The circuit has already been precharged. When the cell is selected, the voltage on node 2 starts to fall, while node 3 stays high

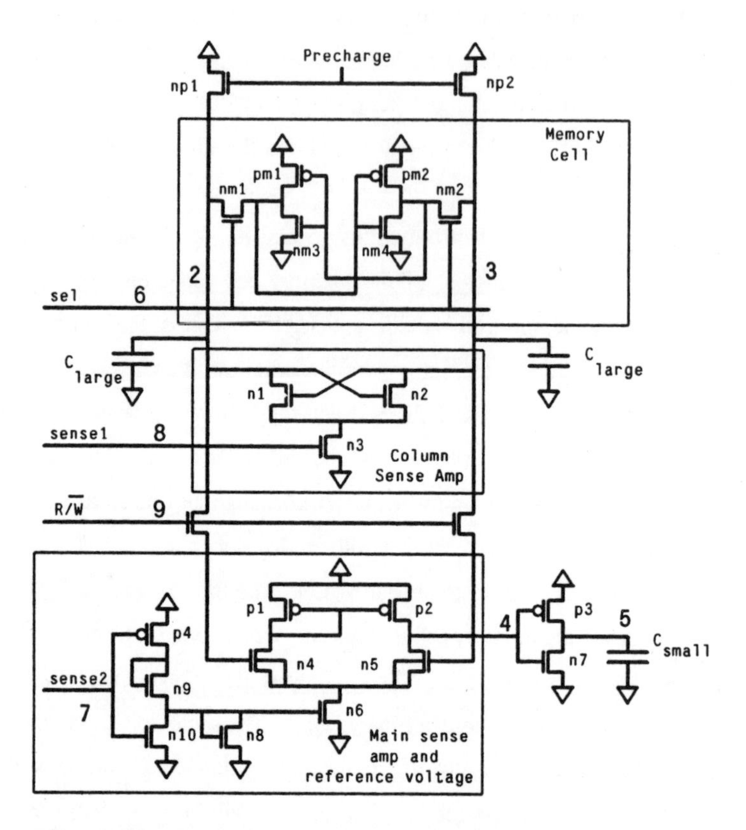

Figure 6-29: Complete circuit diagram: the precharge devices, a six-transistor static memory cell, a column sense amplifier and the main sense amplifier are shown.

(the small drop is due to charge sharing phenomena). Then, the column sense amplifier is selected, together with the Read/$\overline{\text{Write}}$ (R/$\overline{\text{W}}$) circuit and the reference voltage circuit.

Fig. 6-31 shows the voltage waveforms at nodes 2, 3, 4, and 5. The influence of the column sense amplifier is shown in Fig. 6-32. Here we see the same output waveforms for a circuit which is identical to the circuit in Fig. 6-29, except that it has no column sense amplifier. The same waveforms — except for sense1, which is not used — and the same sizes are used for all devices.

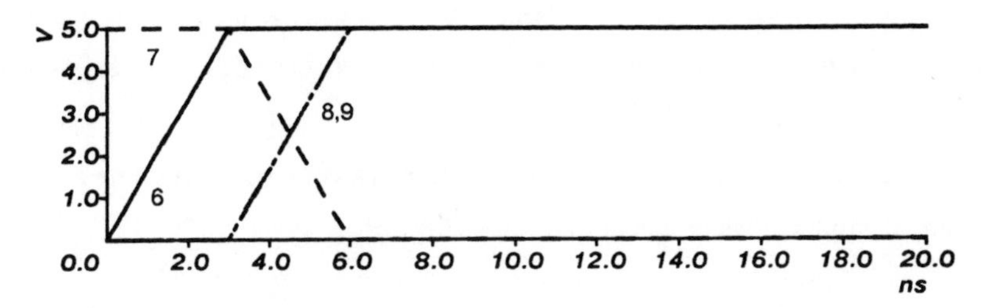

Figure 6-30: Input signals for the circuit shown in Fig. 6-29. The numbers refer to the nodes in the same figure.

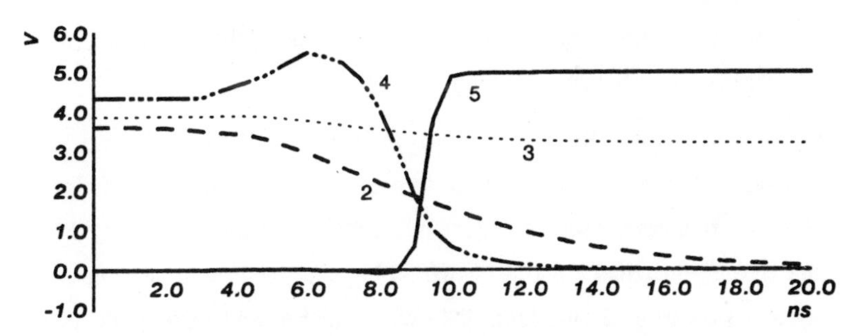

Figure 6-31: Waveforms for the circuit shown in Fig. 6-29.

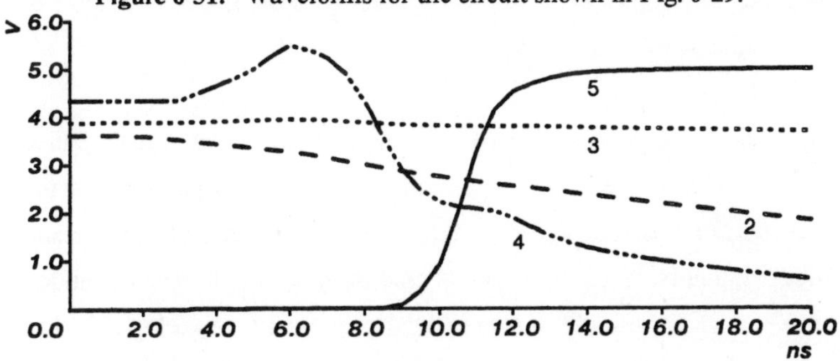

Figure 6-32: Waveforms for the circuit shown in Fig. 6-29, without the column sense amplifier.

Finally, sense amplifiers that use both bipolar and MOS devices have been proposed [15]. These bipolar transistors are not parasitic devices, but are fabricated together with the two enhancement MOS devices, and feature a cut-off frequency of 2GHz (1mA collector current). The front-end of the sense amplifier is a bipolar differential stage, which exploits the low noise characteristics of bipolar devices, thus providing higher sensitivity than that of a CMOS sense amplifier. Once the signal has been amplified by the bipolar front-end, it is sent to a second stage, this time a CMOS differential amplifier, which can output full-swing signals.

6.5. Parallel Adder

Any logic design is constrained by parameters such as speed, circuit density, and power dissipation — and adders are no exception. Among the various schemes that are present in the literature, we can identify three main classes:

- *Ripple-carry adder* (RCA). The carry output of one full-adder is propagated to the carry input of the adjacent full-adder in ripple fashion. The execution time of such a scheme is roughly linear with the data width, that is, a 24-bit adder (that adds two 24-bit data) will be approximately twice as slow as a 12-bit adder. Therefore, an X-bit adder will add two operands in X times the time of a full-adder. The major advantage of such a scheme is limited area.

- *Carry look-ahead adder* (CLA). CLA's generate two signals called "propagate" and "generate." The computation is highly parallel, and the execution time is roughly proportional to the logarithm of the operand width. If a full-adder with delay τ is used inside a carry look-ahead adder, an X-bit adder will add two operands in approximately $\tau\log(X)$. The layout of carry look-ahead is much more complex than the layout of ripple-carry adders and the area increases significantly.

- *Carry-select adder* (CsA)[1]. From the point of view of speed and circuit density, this scheme is between the ripple-carry and the carry look-ahead adder, i.e., it is slower than a CLA, but faster than an RCA. Its area is closer to that of the RCA, and its layout can be considered to be as regular as the layout of RCA's. The basic CsA cell consists of a full-adder that executes *two additions*: one assumes that the carry input is one; the other assumes that the carry input is zero. The actual carry input will select one of the two "sum" outputs with a simple and fast multiplexer.

The use of RCA's should be limited to small adders and to situations in which the area must be minimized. A CsA is much faster, in fact, and the only drawback consists of a small increase in area. As a general rule, CLA's should be used when the operand word length is longer than 16 bits. The design and layout time can be traded for higher speed if enough area is available.

In this section we present one possible implementation of a CLA [4]. This adder is implemented in domino logic and, as has been pointed out, all the carry look-ahead schemes can be rearranged in such a way that only one XOR is needed at the end of the computation. Therefore, only one (static) XOR gate has been used for each bit. The block-scheme of the adder is shown in Fig. 6-33. The first row produces the "propagate" and "generate" terms. The core part of the adder is a tree structure. Three basic cells are needed to implement this scheme: a propagate/generate cell, the internal cell of the tree-structure, and an output XOR. The boolean equations for the propagate/generate cell are:

$$g_i = a_i \wedge b_i \,;$$

$$p_i = a_i \text{ XOR } b_i \,.$$

[1]"CsA" is used, rather than "CSA." In the literature CSA usually refers to "carry-save adders," that are full-adders used in multiplier arrays.

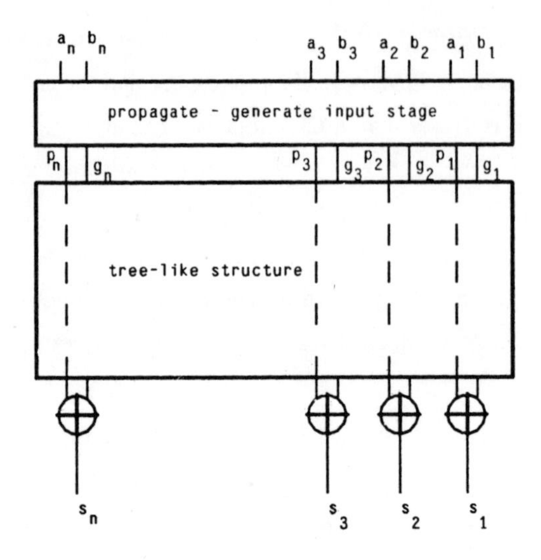

Figure 6-33: Block-scheme of the adder.

a_i and b_i are the i-th bits of the A operand and the B operand, respectively. The result s_i is:

$$s_i = p_i \text{ XOR } c_{i-1} .$$

The i-th carry output c_i can be written as:

$$c_i = G_i ,$$

with $c_0 = 0$. We also have:

$$G_i = g_1, \quad \text{for } i = 1 ;$$

$$G_i = g_i \vee (p_i \wedge G_{i-1}), \quad \text{for } 2 \le i \le n .$$

The internal cell of the tree-structure is a 4-input, 4-output cell and is shown in Fig. 6-34(a). Its boolean equations are:

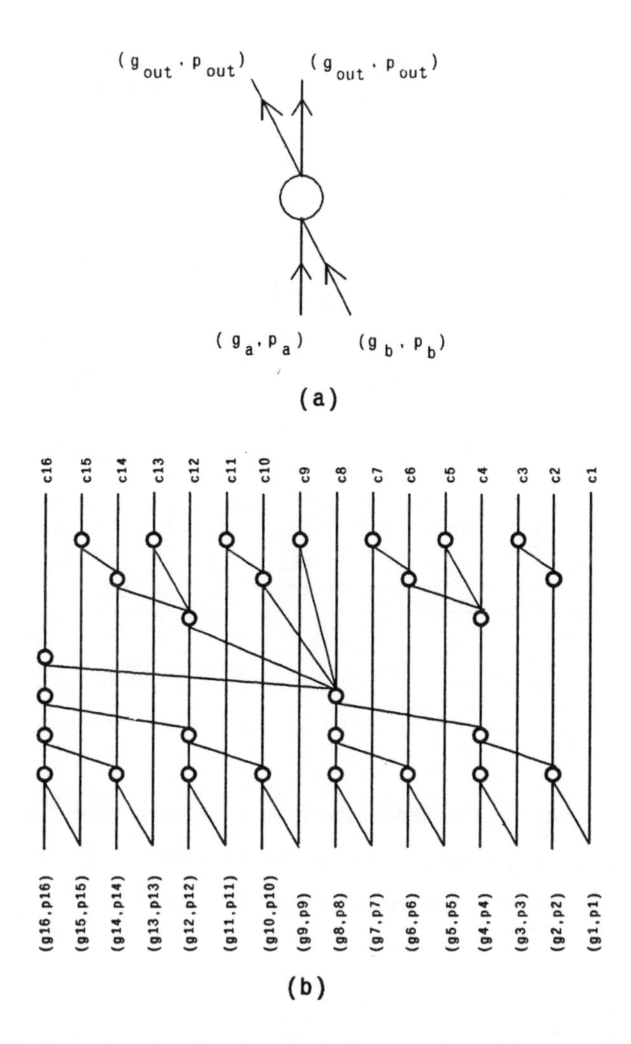

Figure 6-34: Internal cell and tree-structure for carry generation (n = 16).

$$g_{out} = g_a \vee (p_a \wedge g_b)$$

$$p_{out} = p_a \wedge p_b$$

Finally, Fig. 6-34 (b) shows the tree-structure for a 16-bit adder.

If we assume that each adder receives the input operands from a latch, it is easy to design a latch providing the output data in both polarities. In other words, we have both a and $\overline{a}$, b and $\overline{b}$. The scheme of the propagate/generate cell becomes straightforward.

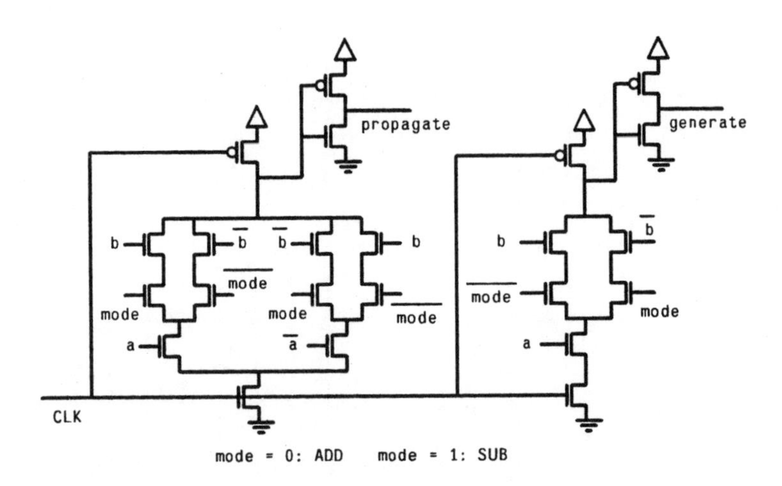

Figure 6-35: Scheme of the propagate/generate cell.

Fig. 6-35 shows the circuit diagram of the propagate/generate cell: a mode signal is provided to allow the execution of addition and subtraction. The area of such an adder can be expressed with the following *empirical* formula:

$$\text{Area} = L_w (42B\,\mu m \times 48\ln(B)\mu m) \text{ (one metal layer technology),}$$

where L_w is the minimum line width and B is the number of bits of each operand. The layout of a 32-bit adder will occupy an area of about 2700μm x 480μm (L_w = 2μm). Note that if more than one metal layer is available, the area can decrease considerably. Finally, the actual layout would fold the cells as shown in Fig. 6-36, resulting in more complex routing but significant saving

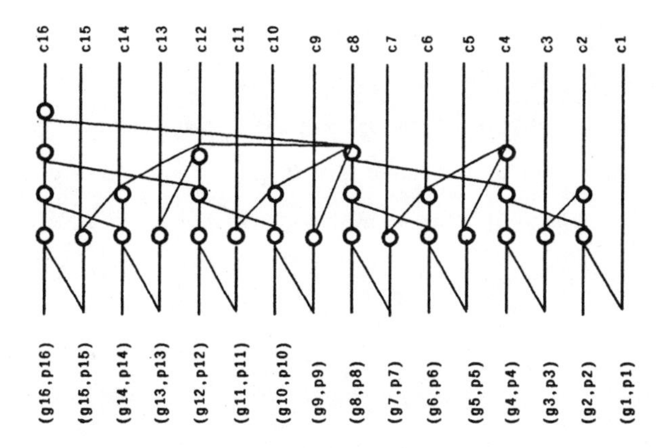

Figure 6-36: Folded tree for the actual layout.

6.6. Parallel Multiplier

The design of a hardware multiplier depends on the constraints that must be satisfied at the architectural level, i.e., whether high throughput rather than short latency is required[1] . This section deals with the design of a multiplier in which latency and circuit density have been considered the most important points of merit to be minimized. Therefore, serial-parallel and strictly serial multipliers cannot be used, because of their high latency. Among all-parallel schemes, the array multiplier is preferred for its extremely high regularity.

However, the basic array multiplier, like a Baugh and Wooley scheme for two's complement multiplication [2] takes too much area when it has to process operands with more than 16 bits. A 32-bit[2] array multiplier occupies about 5-6mm^2 (2μm line

---

[1]This section is derived from the paper: *The Design of a Booth Multiplier: nMOS vs. CMOS Technology*, by M. Annaratone and W.Z. Shen, Proc. of the 2nd International Symposium on VLSI Technology, Systems and Applications, Taiwan, 1985.

[2]Throughout the entire section, the expression "n-bit multiplier" will refer to a multiplier that processes two n-bit input operands.

width), depending on the logic design style (dynamic or static). However, we need a multiplier which is not only smaller but also faster. If we considered speed as the only parameter, there would be several schemes that could be used: Wallace trees and Dadda parallel counters are among the most well-known. On the other hand, it is also well-known that, due to their tree structure, they are not suitable for regular implementation, unless they can be heavily pipelined — with particular reference to Dadda multipliers [5].

As previously stated, we cannot afford massive pipelining because of its negative effects on latency. Moreover, the multiplier we are interested in is a multiplier that can efficiently process operands with up to 64 bits each, and, therefore, we cannot use more exotic approaches — e.g., those presented in [13] and [18] — that usually outperform more common schemes when operands with *more* than 64-128 bits are concerned.

Another important assumption is that the number system has been considered as given: a positional number system with radix r = 2 — that is, the usual binary system — is assumed. More precisely, two's complement has been adopted. Interesting results could be achieved by using different number systems, such as the residue number system [9], for instance. Nonetheless, a multiplier featuring regularity, short execution time, and small area, will be interesting also when non-positional number systems are concerned. The algorithm chosen is the "modified Booth algorithm." First, a review of the algorithm will be presented. This is the algorithm that has been used in other designs [25]. Then, we will discuss an improved version which handles the sign extension in a more efficient way and leads to a multiplier which is smaller than that implemented via the standard method.

6.6.1. The Design of a Multiplier Based on the Modified Booth Algorithm

The original Booth algorithm [3] did not deal with parallel multiplication, but was aimed at improving the speed of the add-and-shift algorithm. The Booth algorithm belongs to the class of "recoding" algorithms, i.e., algorithms that "recode" one of the two operands in such a way that the number of partial products to be added together decreases. For istance, a simple recoding scheme would consist of considering all the bits of one operand sequentially and skipping all its zeroes, because they do not contribute to the final result. However, this leads to a variable execution time of the multiplication. This can be interesting in self-timed systems, where a "done" signal can announce the completion of the operation, but it is useless in other timing strategies where the user has to trigger the system on the worst case, i.e., when the operand has all ones. Actually, the original Booth algorithm itself was not constant in time, although it was much more effective than the simple "skipping over 0's," because it dealt with *strings* of 0's and 1's properly recoded.

A slightly different algorithm, called "modified Booth algorithm," strictly considers groups of bits in one operand, rather than being able to skip over arbitrarily long strings. This leads to a longer, but constant, execution time. The original Booth algorithm is presented in [6] together with its modified version. A brief analysis of the algorithm will now be presented for the sake of completeness. As an example, an 8-bit multiplier that multiplies two operands, $A = a_7, ..., a_0$ and $B = b_7, ..., b_0$, will be considered.

The process of multiplying two n-bit binary numbers is equivalent, when using the "paper and pencil" method, to adding n-bit numbers, properly shifted, n times. This method is used in the array multipliers. The bit-pair recoding scheme of the Booth algorithm can be found, for instance, in [6], and is also reported in Table 6-1. The Booth algorithm decreases the number of rows that have to be added together,

Table 6-1: Booth algorithm: recoding scheme for bit-pair recoding.

Bit $i+1$	Bit i	Bit $i-1$	Recoded Digit
0	0	0	0
0	0	1	+1
0	1	0	+1
0	1	1	+2
1	0	0	−2
1	0	1	−1
1	1	0	−1
1	1	1	0

therefore speeding up the computation. The speed up depends on the number of bits that the algorithm considers in each step. If a bit-pair is considered in each step, the scheme is called "bit-pair recoding." The speed up, if compared to straightforward implementation via an array multiplier, also depends on how the two multipliers are implemented. If both of them use a carry look-ahead adder in the last row, the Booth multiplier will give a speed-up of 40%. Moreover, the area will decrease significantly (again, by about 40%).

Table 6-2: How each recoded digit influences the A operand.

Recoded Digit	Operation on A
0	Add 0 to the partial product
+1	Add A to the partial product
+2	Add 2 x A to the partial product
-1	Subtract A from the partial product
-2	Subtract 2 x A from the partial product

Let us now consider a bit-pair recoding scheme. Later, more complex schemes will be discussed. The algorithm operates on one of the two operands (B), analyzes pairs of bits, and converts them into a set of five signed digits, i.e., 0, +1, +2, −1, and −2. For instance, if the operand B is (least significant bit (LSB) on the right):

01101110,

the algorithm will generate the following recoded string:

+2 -1 0 -2.

Each recoded digit performs some processing on the other operand (A), according to Table 6-2. An example will clarify how the algorithm works. Let A = 10110101 and B = 01110010. By recoding the operand B, we have:

$$01110010 \rightarrow +2\ {-1}\ {+1}\ {-2}.$$

Complete multiplication is shown in Table 6-3. We now have to solve two basic problems:

1. Identify the functionalities we need *inside the array* in order to implement the algorithm.

2. Because of the sign extension, the shape of the multiplier is trapezoidal rather than rectangular (or romboidal). Therefore, we must reconduct the shape of the array to a rectangle by analyzing the problem of sign extension.

These two problems will now be addressed separately.

6.6.2. Basic Building-blocks Inside the Array

Let A be an n-bit number, $A = a_{n-1}\ a_{n-2} \ldots a_1\ a_0$. Its two's complement will be: $A_{tc} = \overline{A} + 1$. An n-bit, two's complement number can represent values that range from -2^{n-1} (1000...000) to $+2^{n-1} - 1$ (0111...111). During the multiplication process,

A = 1 0 1 1 0 1 0 1	
x B = 0 1 1 1 0 0 1 0	
0 0 0 0 0 0 0 0 1 0 0 1 0 1 1 0	(-2): take two's complement of A and shift left one position
1 1 1 1 1 1 1 0 1 1 0 1 0 1	(+1): add A shifted two positions (note sign extension)
0 0 0 0 0 1 0 0 1 0 1 1	(-1): take two's complement of A
1 1 0 1 1 0 1 0 1 0	(+2): shift left A one position
1 1 0 1 1 1 1 0 1 0 0 1 1 0 1 0	Final 16-bit result

Table 6-3: The complete multiplication process.

the partial products are computed by multiplying A by the proper recoded signed digit, which may be 0, $+1$, $+2$, -1, -2. In this case, the partial product may range from -2^n (i.e., $+2 \times -2^{n-1}$) to $+2^n$ (i.e., $-2 \times -2^{n-1}$). This means that at least $n + 2$ bits would be necessary to represent the partial product. However, when the two's complement of a number is taken, there is also an additional "add 1" operation to be performed. Therefore, the maximum number of bits that are actually needed to represent any partial product is $n + 1$.

According to the previous discussion, all the partial products can be generated by a structure which only requires multiplexing elements and an add operation at the least significant bit. Table 6-4 shows the relationship between the multiplicand $A = a_7 a_6 \ldots a_0$ and a partial product $PP = pp_8\ pp_7 \ldots pp_0$. Apparently, we should perform an extra addition to take into account the add operation on the LSB. However, it is possible to delay this operation by simply using a common carry-save technique and a carry look-ahead adder at the bottom of the array.

As appears evident from Table 6-4, any partial product can be produced by a multiplexer circuit and an adder. Therefore, we have a truth table of the kind shown in Table 6-5. The boolean equation that describes the multiplexer is:

Table 6-4: The relationship between partial product and recoded signed digits.

Multiplier Recoded Digit	Partial Products									add	Remarks
	pp_8	pp_7	pp_6	pp_5	pp_4	pp_3	pp_2	pp_1	pp_0		
0	0	0	0	0	0	0	0	0	0	0	
+1	a_7	a_7	a_6	a_5	a_4	a_3	a_2	a_1	a_0	0	
-1	$\overline{a}_7$	$\overline{a}_7$	$\overline{a}_6$	$\overline{a}_5$	$\overline{a}_4$	$\overline{a}_3$	$\overline{a}_2$	$\overline{a}_1$	$\overline{a}_0$	1	Invert A and add 1 to LSB
+2	a_7	a_6	a_5	a_4	a_3	a_2	a_1	a_0	a_{-1}	0	Shift A one position left
-2	$\overline{a}_7$	$\overline{a}_6$	$\overline{a}_5$	$\overline{a}_4$	$\overline{a}_3$	$\overline{a}_2$	$\overline{a}_1$	$\overline{a}_0$	$\overline{a}_{-1}$	1	Shift A one position left and add 1 to LSB

Table 6-5: Truth table for partial products.

Multiplier Recoded Digit	Output	Add Operation
0 (xO)	PP(i) = 0 for i=1,...,8	0
1 (xP1)	PP(i) = A(i) for i=1,...,8	0
2 (xP2)	PP(i) = A(i-1) for i=1,...,8	0
-1 (xM1)	PP(i) = $\overline{\text{A(i)}}$ for i=1,...,8	1
-2 (xM2)	PP(i) = $\overline{\text{A(i-1)}}$ for i=1,...,8	1

Note that A(8) = A(7) = sign bit of multiplicand.
PP(8) = sign bit of partial product.

$$PP(j) = xP1 \wedge A(i) \vee xP2 \wedge A(i\text{-}1) \vee xM1 \wedge \overline{A(i)} \vee xM2 \wedge \overline{A(i\text{-}1)},$$

while the logic for the "add 1" block is simply:

$$\text{add} = xM1 \vee xM2 .$$

The three basic cells we need inside the array to implement a Booth multiplier are shown in Fig. 6-37. A full-adder and two different combinational circuits — called Cl and C2 — are used. Cl is a multiplexer, and C2 is an "add 1" generator.

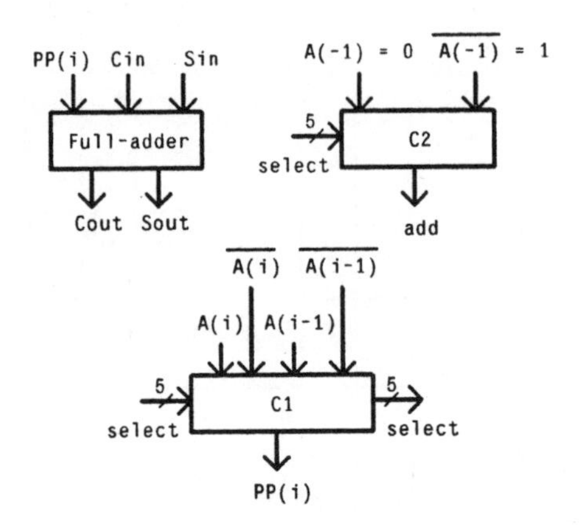

Figure 6-37: The three basic cells that are necessary inside the array of a Booth multiplier.

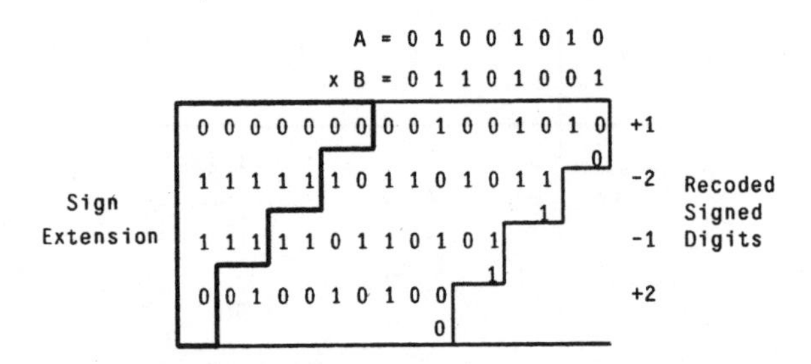

Figure 6-38: 8-bit multiplication with the necessary sign extension of the partial products to preserve the correctness of the final result.

6.6.3. The Problem of Sign Extension

From the previous discussion, we know that only nine bits and an add operation are necessary to represent any partial product, with the ninth bit being used simply to represent the sign. In order to reduce the array to a rectangle, the sign extension must now be considered. In this section we present two possible approaches to solve the problem of sign extension. For the sake of simplicity, the first approach will be referred to as the "sign propagate" method, while the second one will be referred to as the "sign generate" method. The "sign propagate" method has often been used: see, for instance, [14, 25].

6.6.3.1. The "Sign Propagate" Method

The partial products in an 8-bit multiplier can be generated by using a 9x4 adder array. In order to achieve a correct result, it is necessary to extend the sign of the partial products (see Fig. 6-38). This obviously leads to a multiplier that does not have a rectangular shape.

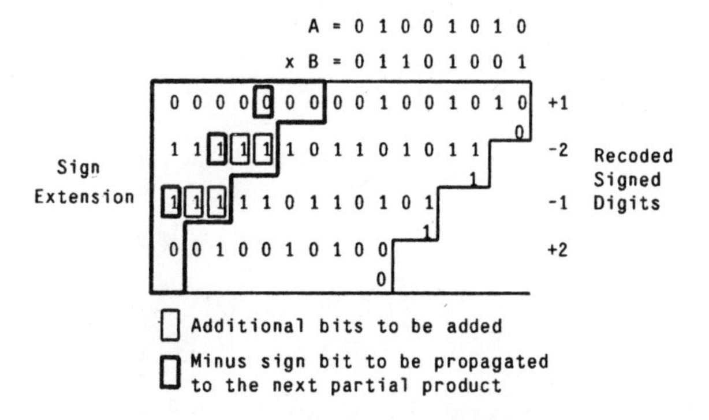

Figure 6-39: Three bits are necessary to handle the sign extension throughout the array.

The sign bits of the partial products are located two bits apart from each other. The second partial product in Fig. 6-38 must propagate to the third row, namely to the sign extension of the third partial product. The problem is shown graphically in Fig. 6-39, and it can be noted that a *third* bit is necessary, in order to propagate a "minus sign" to the next partial product.

The sign bit for the j-th partial product PP(j) is defined as:

$$S(j) = xM2(j) \wedge A(7) \vee xM1(j) \wedge A(7) \vee xP2(j) \wedge A(7) \vee xP1 \wedge A(7) .$$

We now define four terms:

Bit(2J):	sign extension to be added to the 8th bit of the next partial product
Bit(2J+1):	sign extension to be added to the 9th bit of the next partial product
M(J):	indicates whether there is a propagation of a "minus sign" from the previous partial product
M(J+1):	indicates whether there is a minus sign to be propagated to the next partial product

Therefore:

- If $M(J) = 0$ (i.e., no minus sign has occurred in the previous stage), then:

$$\text{if } S(j) = 0 \rightarrow \text{Bit}(2J) = \text{Bit}(2J+1) = M(J+1) = 0 ;$$

$$\text{if } S(j) = 1 \rightarrow \text{Bit}(2J) = \text{Bit}(2J+1) = M(J+1) = 1 .$$

- If $M(J) = 1$ (i.e., a minus sign has occurred in the previous stage), then:

$$\text{if } S(j) = 0 \rightarrow \text{Bit}(2J) = \text{Bit}(2J+1) = M(J+1) = 1 ;$$

$$\text{if } S(j) = 1 \rightarrow \text{Bit}(2J) = 0 \text{ and } \text{Bit}(2J+1) = M(J+1) = 1 .$$

We can derive the following boolean equations:

$$M(J+1) = M(J) \vee S(j) ;$$

$$\text{Bit}(2J) = M(J) \text{ XOR } S(j) ;$$

$$\text{Bit}(2J+1) = M(J) \vee S(j) = M(J+1) .$$

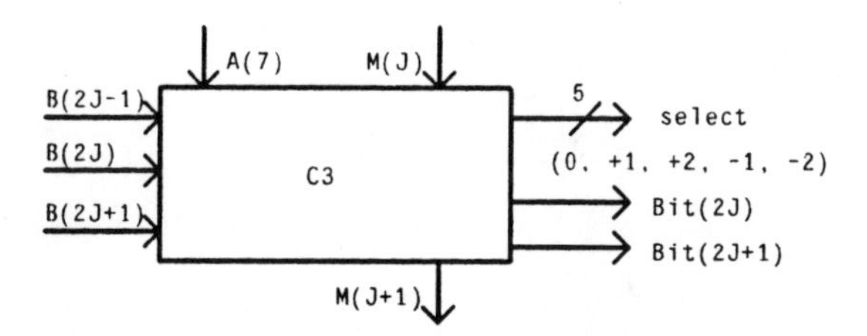

Figure 6-40: Block-scheme of the module implementing the recoding algorithm.

A module, called C3 (see Fig. 6-40), implements the above equations. In other words, C3 is the combinational logic that actually performs *both the recoding algorithm and the sign extension.* The sign "propagates" from one stage to the next one. This is the reason why this method has been called of the "sign propagate."

We finally have all the necessary submodules to build the Booth multiplier. Fig. 6-41 shows the straightforward implementation. The A operand, together with its one's complement — not shown in the figure — flow across the carry-save adder array. The B operand is simply processed by the C3 blocks that generate the five selection lines which correspond to the five signed digits. The Bit(J) terms properly handle the sign extension. A fast carry look-ahead adder is used to speed up the execution time further.

It is evident from Fig. 6-41 that the delay associated with this scheme is four full-adders plus the delay introduced by the carry look-ahead adder, *provided that the C3 logic is faster than a single full-adder.* It is also evident that the scheme shown in Fig. 6-41 has not been minimized, e.g., the first row has full-adders with one input only. A much more compact scheme is shown in Fig. 6-42, where only three rows are used. Moreover, the first one consists of half-adders. For an n-bit multiplier, this scheme has (n/2 - 1) rows of (n + 1) full-adders (or half-adders) each.

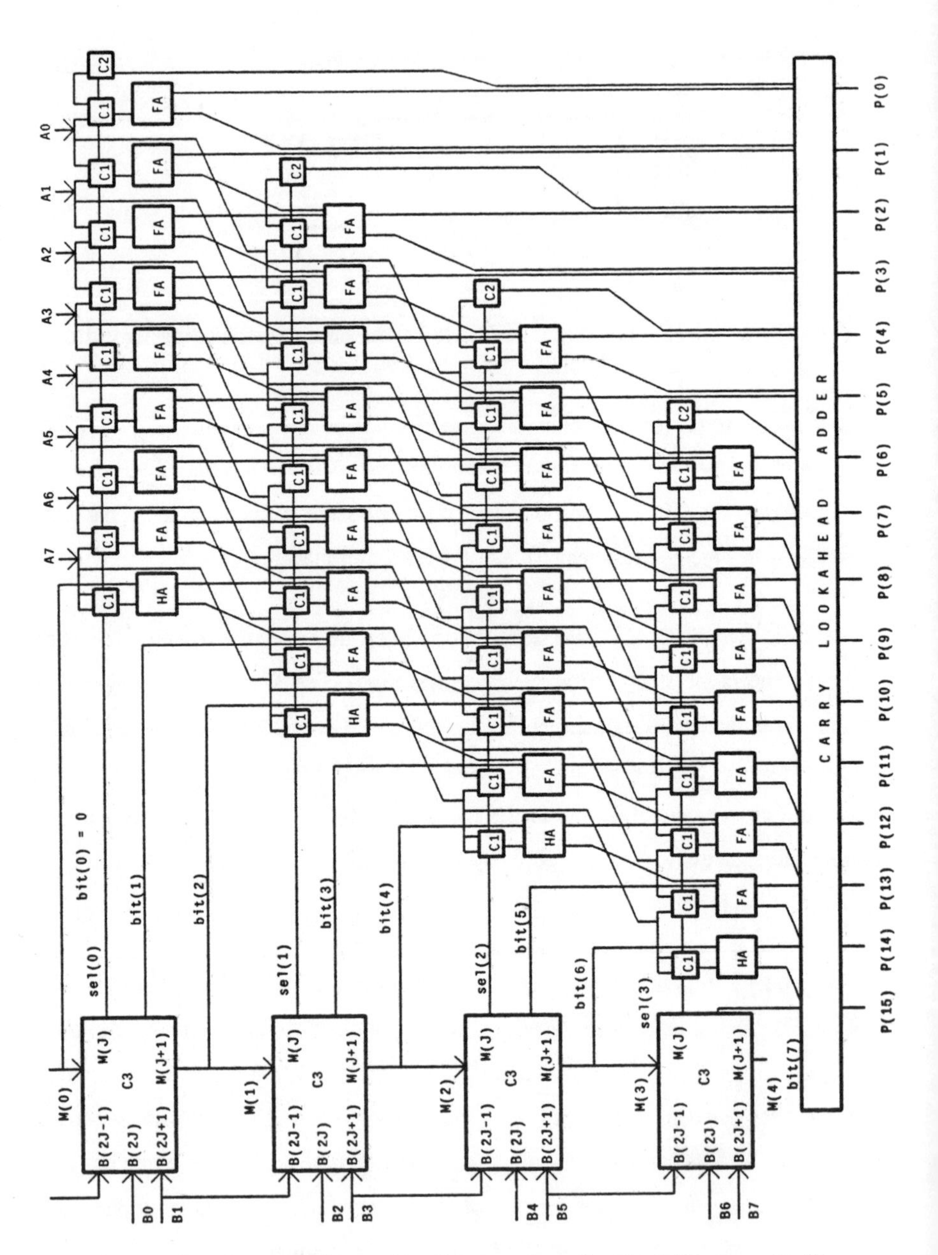

Figure 6-41: A straightforward implementation of an 8-bit Booth multiplier according to the "sign propagate" method.

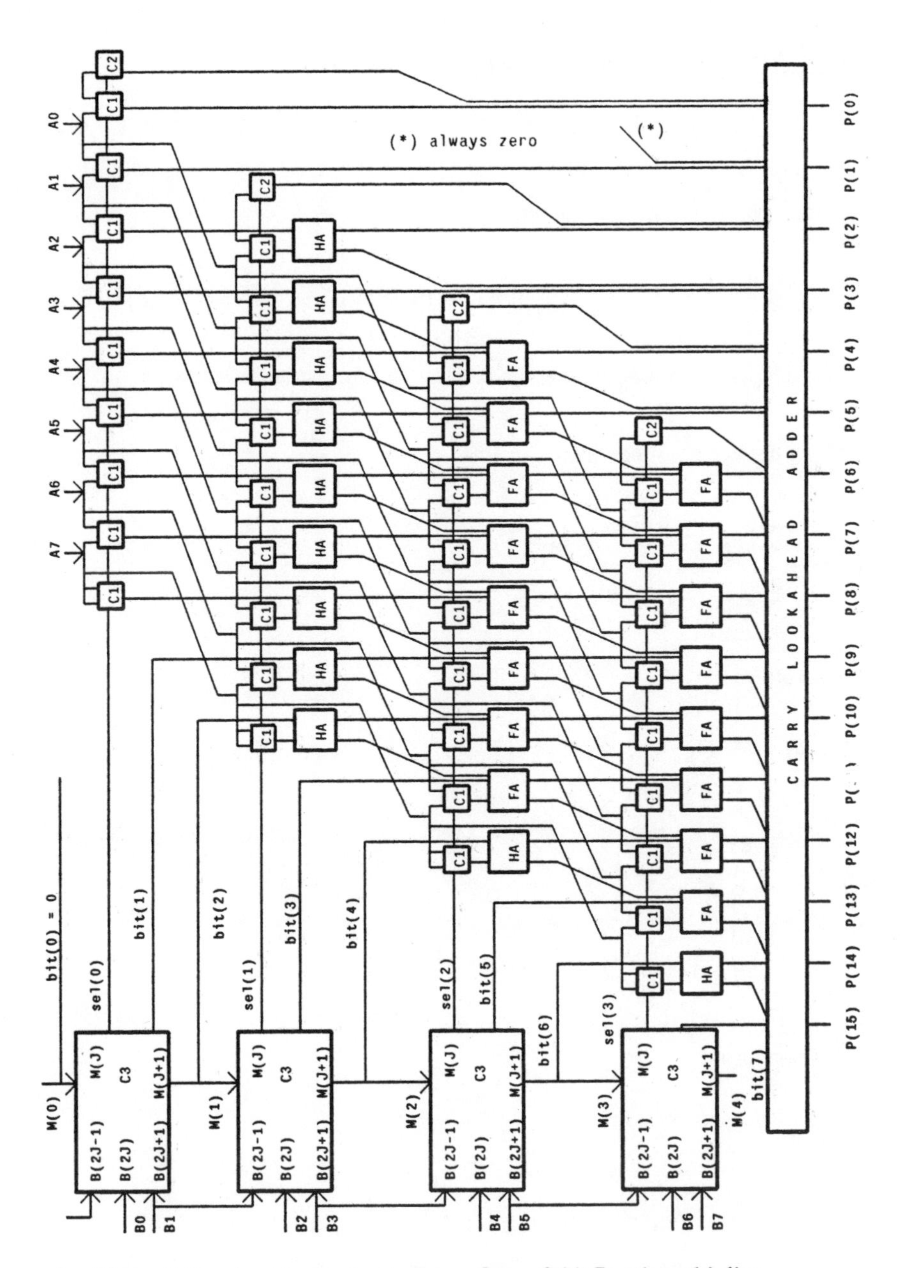

Figure 6-42: A more compact scheme for an 8-bit Booth multiplier according to the "sign propagate" method.

6.6.3.2. The "Sign Generate" Method

A different method for sign extension is now presented. The algorithm is discussed for an 8-bit multiplier, but it can easily be extended to operands with any word-length. We can write the sign bit of the *result* as:

$$S=S_0\sum_{i=8}^{15}2^i+(S_1\sum_{i=8}^{13}2^i)\times 2^2+(S_2\sum_{i=8}^{11}2^i)\times 2^4+(S_3\sum_{i=8}^{9}2^i)\times 2^6,$$

where S_0, S_1, S_2 and S_3 are the sign bits for the four partial products.

Using the two equivalences:

$$\sum_{i=j}^{k}2^i = 2^j(2^{k+1-j}-1) = 2^{k+1} - 2^j,$$

$$\overline{S_j} = 1 - S_j,$$

S becomes:

$$\begin{aligned} S &= (1-\overline{S_0})(2^{16}-2^8) + (1-\overline{S_1})(2^{16}-2^{10}) + (1-\overline{S_2})(2^{16}-2^{12}) + \\ &\quad + (1-\overline{S_3})(2^{16}-2^{14}) \\ &= [4-(\overline{S_0}+\overline{S_1}+\overline{S_2}+\overline{S_3})]2^{16} + \overline{S_0}2^8 + \overline{S_1}2^{10} + \overline{S_2}2^{12} + \\ &\quad + \overline{S_3}2^{14} - 2^8 - 2^{10} - 2^{10} - 2^{12} - 2^{16} \\ &= [3-(\overline{S_0}+\overline{S_1}+\overline{S_2}+\overline{S_3})]2^{16} + \overline{S_0}2^8 + \overline{S_1}2^{10} + \overline{S_2}2^{12} + \\ &\quad + \overline{S_3}2^{14} + 2^{16} - (2^8 + 2^{10} + 2^{12} + 2^{14}). \end{aligned}$$

As we have:

$$\begin{aligned} 2^{16} &= \sum_{i=0}^{15}2^i \\ &= \sum_{i=0}^{7}2^i + 1 + 2^8 + 2^9 + 2^{10} + 2^{11} + 2^{12} + 2^{13} + 2^{14} + 2^{15} \\ &= 2^8 + 2^8 + 2^9 + 2^{10} + 2^{11} + 2^{12} + 2^{13} + 2^{14} + 2^{15}, \end{aligned}$$

the sign bit of the result is:

$$S = [3 - (\overline{S_0} + \overline{S_1} + \overline{S_2} + \overline{S_3})]2^{16} + \overline{S_0}2^8 + \overline{S_1}2^{10} + \overline{S_2}2^{12} + \\ + \overline{S_3}2^{14} + 2^8 + 2^9 + 2^{11} + 2^{13} + 2^{15}.$$

The first term of S is the 17th bit and can be ignored; S can therefore be written as.

$$S = \overline{S_0}2^8 + \overline{S_1}2^{10} + \overline{S_2}2^{12} + \overline{S_3}2^{14} + 2^9 + 2^{11} + 2^{13} + 2^{15} + 2^8.$$

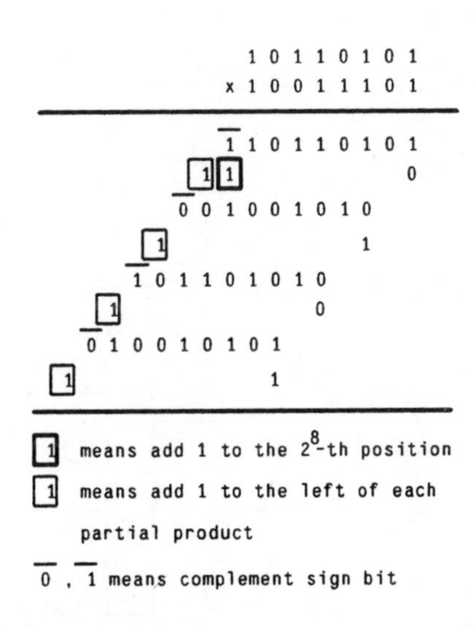

Figure 6-43: Example of the "sign generate" method.

The previous equation can be interpreted as follows:

1. Complement the sign bit of each partial product.

2. Add 1 to the left of the sign bit of each partial product.

3. Add 1 to the 9th bit of each partial product.

An example is shown in Fig. 6-43. In this approach, that we called of the "sign generate," the sign bit does not really propagate along the left edge of the array

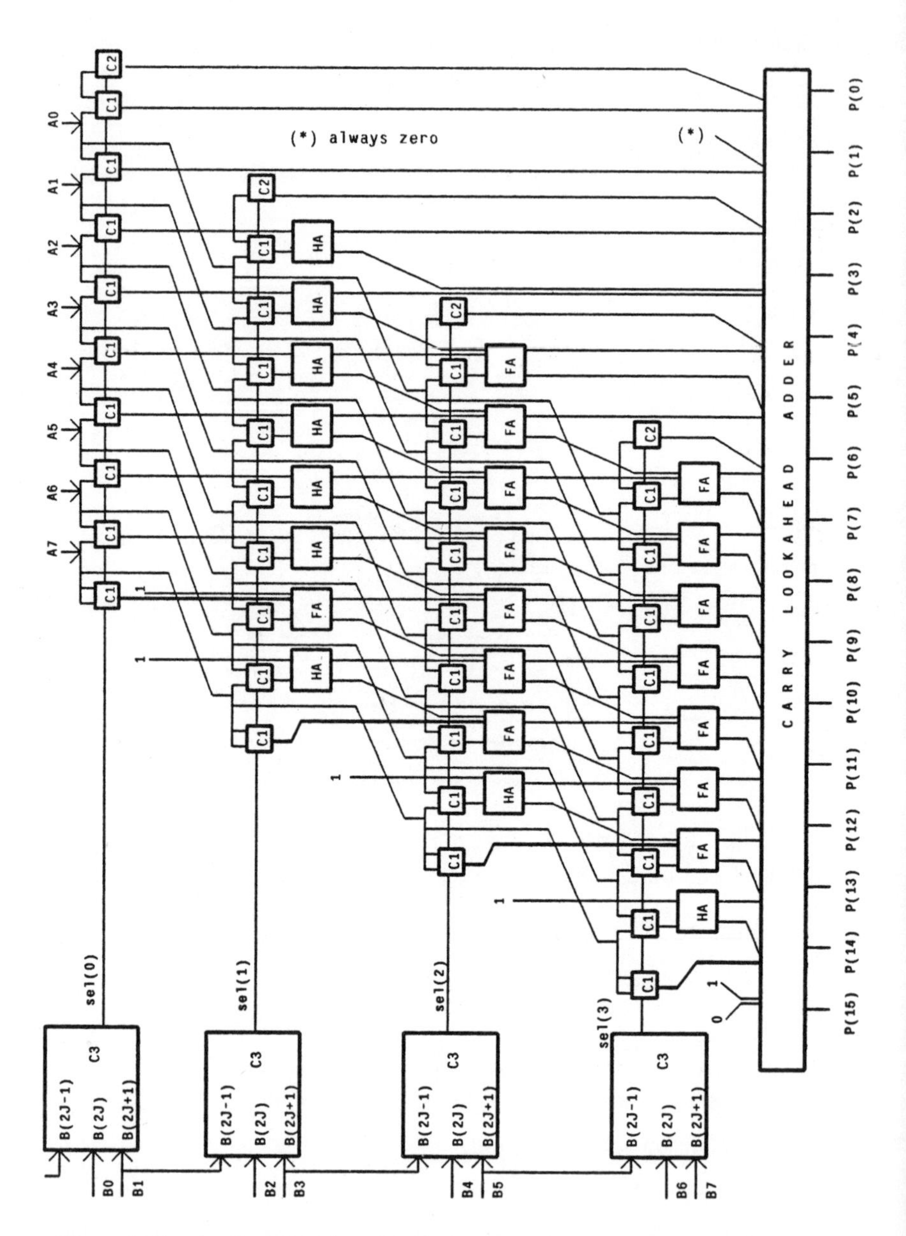

Figure 6-44: An 8-bit Booth multiplier using the "sign generate" method.

multiplier, but is "generated" statically. The hardware of a Booth multiplier that uses this method is different from the one shown in Fig. 6-42. The three operations that are necessary to generate the sign bits involve, as a first step, the negation of the sign bit. This can be accomplished easily by exchanging the A operand with its one's complement — and vice-versa — when they are input to the C1 logic. Fig. 6-44 shows the block-scheme of an 8-bit multiplier that uses the "sign generate" method. Note that only 8 full-adders are needed for each row. An n-bit multiplier will consist of (n/2 - 1) rows with n full-adders each. Moreover, the C3 logic becomes very simple, because the two terms Bit(2J) and Bit(2J + 1) are no longer required.

Finally, some considerations on the choice of a bit-pair recoding scheme are worth being made. Undoubtedly, a three-bit recoding scheme seems to be extremely promising, at least for medium-size and large multipliers — that is, 24-bit multipliers and larger. Further significant speed up could be achieved and area could be even smaller. There are, however, also some good reasons not to implement a three bit-recoding scheme. In a bit-pair recoding scheme, we need the A and 2 x A terms, together with their one's complement. The 2 x A is a simple shift, and the one's complement usually comes without additional circuitry, provided that both inverting and non-inverting latches are used at the input of the multiplier. If a three-bit recoding scheme were used, we would also need the 4 x A and the 3 x A term. The time it takes to compute the last term — especially if long operands are concerned — definitely discourages the use of more complex recoding schemes, at least in the VLSI field (arithmetic units inside mainframes have used up to four bit-recoding schemes). One solution, which is usually proposed to overcome these problems, is to compute the 3 x A term during *idle times*, e.g., during precharging. However, this solution does not seem to be easily applicable:

1. A three-bit recoding scheme does not make sense for small multipliers (up to 16-bit), because the increased complexity is not fairly balanced by significant gains both in area and in speed.

2. As far as larger multipliers are concerned, the time for a 32-bit addition — that is, what the 3 x A term consists of — cannot be compared to the precharging time, unless extremely fast adders — that occupy a large area and do not have regular layout — are used.

6.6.4. The Implementation of a 24-bit CMOS Booth Multiplier

Domino logic has been used to implement a 24-bit CMOS Booth multiplier. The only significant drawback of domino logic is that, being a non-inverting logic, an XOR gate is not provided. This does not affect the design of adders significantly, because carry look-ahead schemes can be almost completely implemented in domino logic — an XOR is necessary only in the last stage, as we saw in Section 6.5. However, the lack of an XOR gate makes the design of domino logic carry-save adders more problematic.

The scheme shown in Fig. 6-45 and Fig. 6-46 only features one static inverter inside the full-adder (besides the buffering inverters used in domino gates). The circuit is a straightforward implementation of the logic circuit in Fig. 6-45. However, the behavior of the circuit is critical, and a different implementation is necessary. In fact, the circuit can provide a wrong result when the inputs are ab=11 and cin=0. During precharging, node X is pulled down and node Y goes high. Therefore, transistor n1 conducts. This is not dangerous, because a, b, and cin are low. Let us now assume that during evaluation ab=11 and cin=0. We have cout=1 and sout=0. cout is computed correctly, and node X goes high. This might not be true for sout. In fact, before X goes high, Y goes low and n1 turns off, the two transistors n4 and n5 discharge node P through n1 to V_{ss}, and sout goes high. When Y finally goes low, it is too late, and the output is wrong.

Therefore, to insure the correct behavior of the adder, we must keep n1 off during

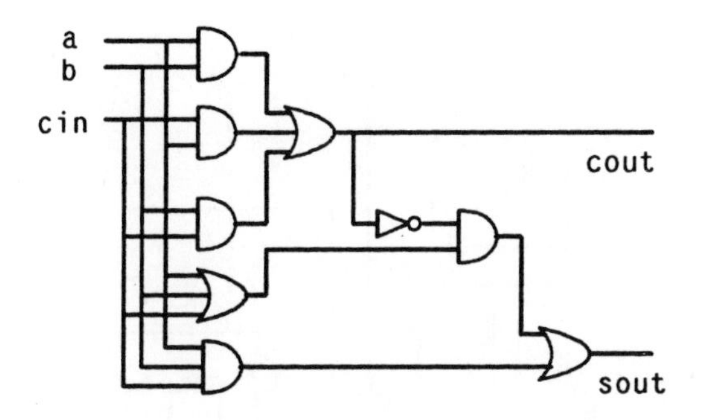

Figure 6-45: CMOS full-adder: logic scheme.

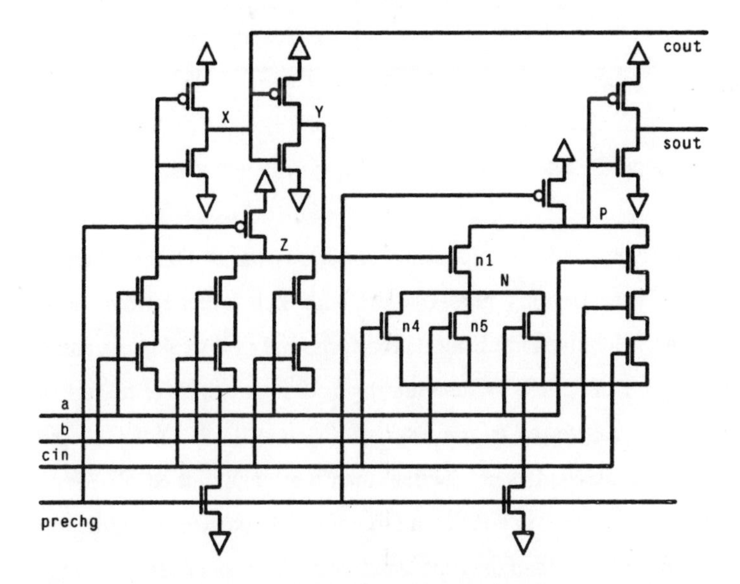

Figure 6-46: CMOS full-adder: a straightforward implementation that leads to unreliable behavior (see text).

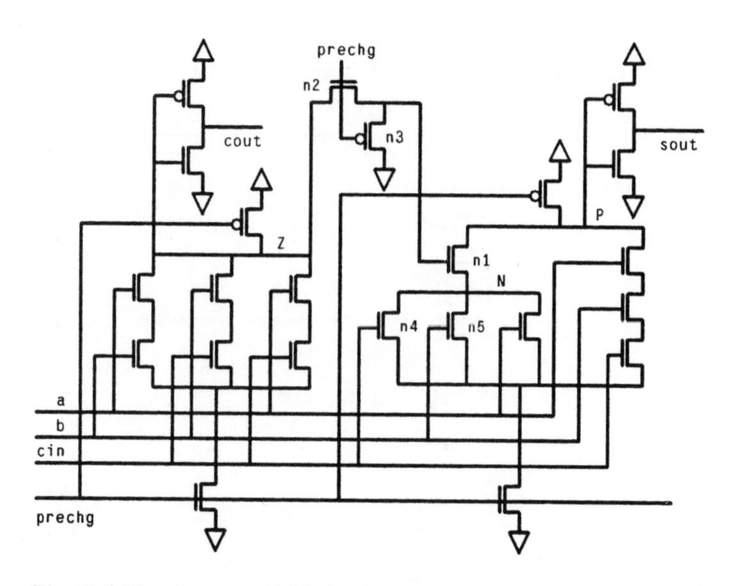

Figure 6-47: A more reliable implementation of the full-adder.

precharge. This can be accomplished in various ways. One of them is shown in Fig. 6-47. When prechg goes low, the node Z is isolated from transistor n1 by transistor n2. Moreover, transistor n3 conducts, and n1 is weakly off, enough to avoid the precharging of node N. When prechg goes high, transistor n2 conducts and the value on Z is transferred to transistor n1 (n3 is off). If node Z is pulled down, transistor n1 weakly enters conduction, then is strongly turned off. Some charge is transferred from node P to node N in this critical period. However, the capacitance at node P is much larger than the capacitance at node N, and charge sharing takes place but it is not destructive — moreover, n1 is not strongly conducting. If node Z remains high, transistor n1 enters conduction more strongly. (Note that it is not driven by a V_{dd} signal, because of the voltage drop across n2.) Finally, if the three inputs to the full-adder had been available in both polarities, different, smaller schemes could have been used. In the case of a multiplier of this kind — and, generally, with all the array multipliers — it does not seem to pay off to provide each full-adder with inputs in both polarities. As far as C3 logic is concerned, after a trivial rearrangement of scheme, a dynamic PLA-like circuit has been designed.

The technology used is bulk p-well CMOS, with a 3μm minimum feature size, one level of metal. The full-adder and selection logic take 131 x 272 μm^2, while the C3 logic takes 340 x 267 μm^2. The 24-bit carry look-ahead takes 579 x 3031 μm^2. The whole multiplier takes 4010 x 3520 μm^2. A similar 24-bit, nMOS multiplier implemented with same feature size and using static logic would have taken about 3000 x 2200 μm^2. Even dynamic logic cannot decrease the area significantly, if a comparison to nMOS is made. However, the decrease in area *is significant* if compared with a similar implementation in complementary logic: a 24-bit Booth CMOS multiplier would take in this case about 5500 x 3400 μm^2.

References

[1] Annaratone, M. *et al.*.
Extending the CMU Warp Machine with a Boundary Processor.
In William J. Miceli, Keith Bromley , Editors (editor), *Proc SPIE 564, Real Time Signal Processing VIII*, pages 56-65. August, 1985.

[2] Baugh, C.R. and B.A. Wooley.
A Two's Complement Parallel Array Multiplication Algorithm.
IEEE Trans. on Comput. C-22:1045-1047, 1973.

[3] Booth, A.D.
A Signed Binary Multiplication Technique.
Q.J. Mech. Appl. Math. 4:236-240, 1951.

[4] Brent, R.P. and H.T. Kung.
A Regular Layout for Parallel Adders.
IEEE Trans. on Computers :260-264, March, 1982.

[5] Cappello, P.R. and K. Steiglitz.
A VLSI Layout for a Pipelined Dadda Multiplier.
ACM Trans. on Computer Systems 1(2):157-174, May, 1983.

[6] Cavanagh, J.J.F.
Computer Science Series: Digital Computer Arithmetic.
McGraw-Hill Book Co., 1984.

[7] Childs, L.F. and R.T. Hirose.
An 18ns 4Kx4 CMOS SRAM.
IEEE Journal of Solid-State Circuits SC-19(5):545-551, October, 1984.

[8] Friedman, V. and S. Liu.
Dynamic Logic CMOS Circuits.
IEEE Journal of Solid-State Circuits SC-19(2):263-266, April, 1984.

[9] Garner, H.L.
The Residue Number System.
IEEE Trans. on Electronic Computers :140-147, June, 1959.

[10] Heller, L.G. and J.W. Davis.
Cascode Voltage Switch Logic.
In *ISSCC Dig. Tech. Pap.*, pages 16-17. 1984.

[11] Isobe, M. *et al.*.
An 18 ns CMOS/SOS 4K Static RAM.
IEEE Journal of Solid State Circuits SC-16(5):460-465, October, 1981.

[12] Linderman, R.W. *et al.*.
CUSP: A 2-μm CMOS Digital Signal Processor.
IEEE Journal of Solid-State Circuits SC-20(3):761-769, June, 1985.

[13] Luk, W.K.
A Regular Layout for Parallel Multiplier of O(log↑2 n) Time.
In Kung, H.T., R.F. Sproull and G.L. Steele (editor), *VLSI Systems and Computations*, pages 317-326. Computer-Science Department, Carnegie-Mellon Univeristy, Computer Science Press, Inc., October, 1981.

[14] Masumoto, R.T.
The Design of a 16x16 Multiplier.
VLSI Design (Lambda) First Quarter:15-21, 1980.

[15] Miyamoto, J.I. *et al.*.
A High-Speed 64K CMOS RAM with Bipolar Sense Amplifiers.
IEEE Journal of Solid-State Circuits SC-19(5):557-563, October, 1984.

[16] Mohsen, A. *et al.*.
The Design and Performance of CMOS 256K Bit DRAM Devices.
IEEE Journal of Solid-State Circuits SC-19(5):610-618, October, 1984.

[17] Ochii, K. *et al.*.
An Ultralow Power 8Kx8-bit Full CMOS Ram with a Six-Transistor Cell.
IEEE Journal of Solid-State Circuits SC-17(5):798-802, October, 1982.

[18] Preparata, F.P.
A Mesh-Connected Area-Time Optimal VLSI Integer Multiplier.
In Kung, H.T., R.F. Sproull and G.L. Steele (editors), *VLSI Systems and Computations*, pages 311-316. Computer Science Department, Carnegie-Mellon University, Computer Science Press, Inc., October, 1981.

[19] Shimohigashi, K. *et al.*.
An n-Well CMOS Dynamic RAM.
IEEE Journal of Solid-State Circuits SC-17(2):344-348, April, 1982.

[20] Sood, L.C. *et al.*.
A Fast 8Kx8 CMOS SRAM With Internal Power Down Design Techniques.
IEEE Journal of Solid-State Circuits SC-20(5):941-949, October, 1985.

[21] Taylor, R.T. and M.G. Johnson.
A 1-Mbit CMOS Dynamic RAM with a Divided Bitline Matrix Architecture.
IEEE Journal of Solid-State Circuits SC-20(5):894-902, October, 1985.

[22] Uya, M., K. Kaneko and J. Yasui.
A CMOS FLoating Point Multiplier.
IEEE Journal of Solid State Circuits SC-19(5):697-701, October, 1984.

[23] Walker, H.
A 4-Kbit Four-transistor Dynamic RAM.
Research Report CMU-CS-83-140, Computer-Science Department, Carnegie-Mellon University, June, 1983.

[24] Walker, H.
The Control Store and Register File Design of the Programmable Systolic Chip.
Research Report CMU-CS-83-133, Computer-Science Department, Carnegie-Mellon University, May, 1983.

[25] Ware, F.A. *et al.*.
64 Bit Monolithic Floating Point Processors.
IEEE Journal of Solid-State Circuits SC-17(5):898-907, October, 1982.

Chapter 7

Driver and I/O Buffer Design

CMOS I/O buffers and bus drivers — or any circuit which is to drive a significant load both on-chip and off-chip — have not received much attention in the literature, and a comprehensive treatment of them — delay minimization, power dissipation, second order effects, and layout techniques to minimize noise and maximize speed — is lacking. This chapter aims at filling this gap: both input and output buffers will be dealt with, from the stand-point of speed, power dissipation, noise robustness, degree of protection, etc. The design of on-chip drivers — such as bus drivers — can differ from the design of output buffers, because both transmitting and receiving stages are under the designer's control, and, therefore, a global optimization can be effectively carried out. Nonetheless, simple and reliable approaches are still implemented — for instance, scaled-up inverter chain. In this respect the design of on-chip drivers can be treated like the design of output buffers. Finally, on-chip driver design which optimize both transmitter and receiver is presented in Section 7.8.

The design of I-O pads[1] is a very important and critical task in CMOS technology, because protection against electrostatic discharge (ESD) is carried out in the I-O stage. If ESD goes to the internal circuitry unfiltered, it can fire latchup or even

---

[1] "I-O pad" means either an input or an output pad, while "I/O pad" means a bidirectional pad.

Plates referred to in this chapter are found in the PLATES section located between pages 172 and 173.

destroy the chip.

I-O pad design is a trade-off among many different parameters, such as speed, area, noise sensitivity, degree of protection, etc. Moreover, shrinking feature size not only provides the designer with higher circuit density and speed, but also with more delicate circuits (e.g., lower breakdown voltage). As a consequence, the protection circuitry associated with an I-O stage must be implemented carefully.

Finally, smaller feature sizes decrease the circuit delay, and, therefore, the influence of the I-O section on overall performance increases, because the off-chip load is much harder to be scaled down[1]. Optimizing the I-O section becomes crucial. In order to provide the designer with a more accurate — albeit still tractable — model for the dynamic behavior of the MOS transistor, it is necessary to depart from the simple optimization schemes presented in Chapter 5. Coming up with an accurate model for the fall-time and rise-time of a CMOS inverter is the first problem to be addressed. The same model can be applied to multiple input gates in a straightforward way.

7.1. CMOS Inverter Delay Estimation

This section aims to present an accurate model for the fall-time and rise-time of a CMOS inverter when driving a capacitive load C_L. Fall-time and rise-time are specified in different ways in the literature. For the time being, we shall consider the fall-time τ_n and rise-time τ_p for a certain "output reference voltage" V_o, as depicted in Fig. 7-1, where the bold curve is the output voltage, while the solid ramp is the input voltage. The subscript of τ indicates which device — that is, either the n-channel or the p-channel transistor — turns on during the transition.

Given a circuit as shown on the right-hand side of Fig. 7-2, the output voltage

---

[1]In fact, low-capacitance substrates, which replace common printed-circuit boards, are currently being actively investigated.

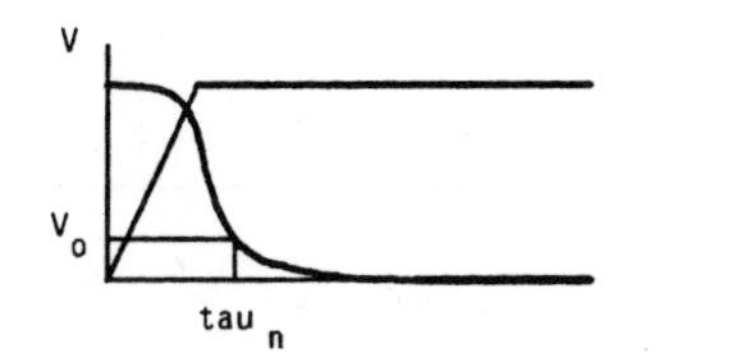

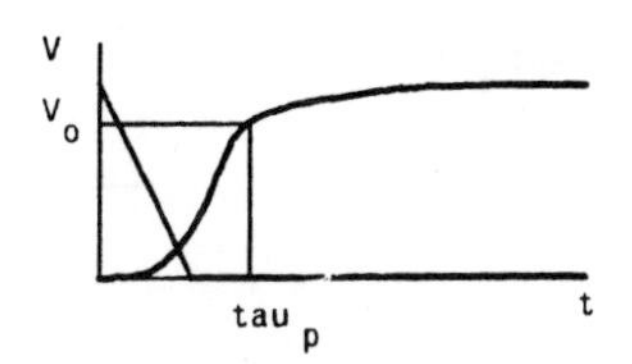

Figure 7-1: Fall-time (τ_n) and rise-time (τ_p).

$v_{out}(t)$ can be expressed as:

$$C_L \frac{dv_{out}(t)}{dt} = i_L(t) . \tag{7-1}$$

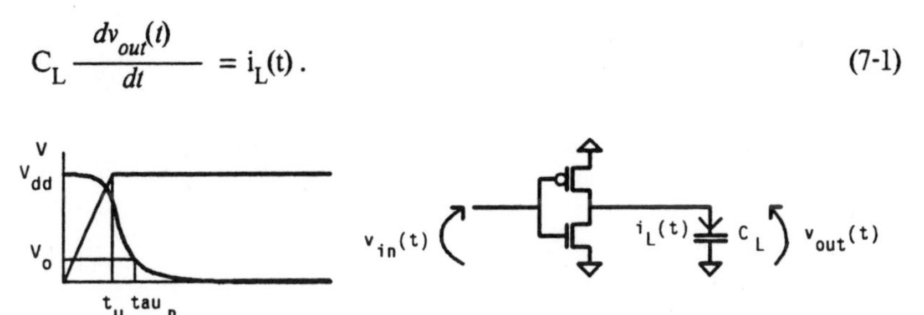

Figure 7-2: Parameters for delay estimation.

In order to compute the output voltage, and hence the delay, it is necessary to know the load current $i_L(t)$. From Kirchoff's current law we have the relationship:

$$i_L(t) = - i_{ds(n)}(t) - i_{ds(p)}(t) ,$$

where $i_{ds(n)}$ and $i_{ds(p)}$ are the drain-to-source current of the n-channel and of the p-channel device, respectively. This equation would lead to an exact[1] estimate of the load current. Unfortunately, such an approach rapidly becomes intractable, given the complexity of the equations that model the two drain-to-source currents, and even simplified equations lead to integrals that are hard to solve analytically.

---

[1]That is, with an accuracy which only depends on the accuracy of the equations describing the drain-to-source current of both devices.

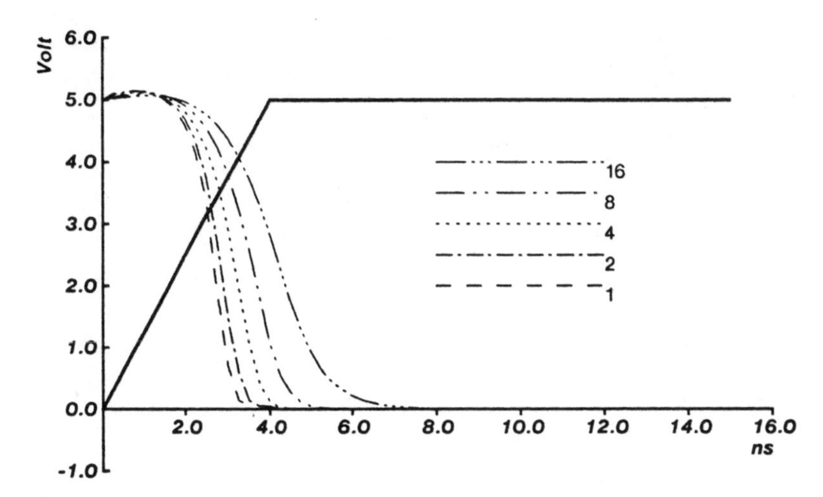

Figure 7-3: Input (bold) and outputs for different values of load capacitance: fall-time.

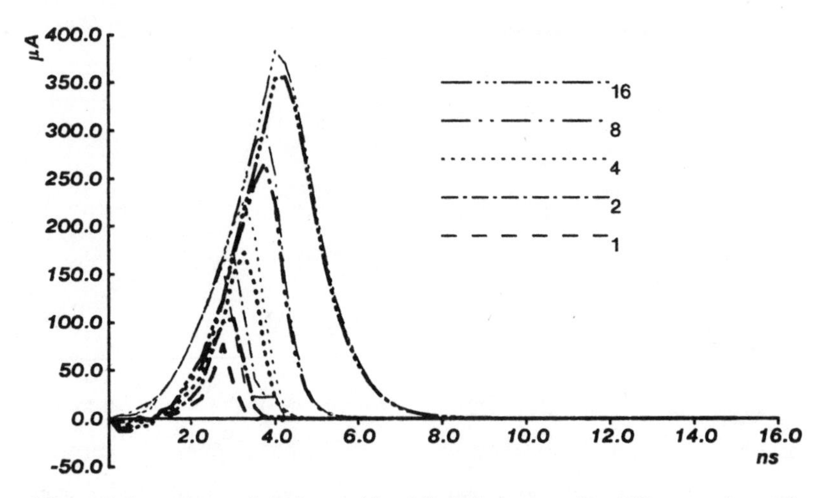

Figure 7-4: n-channel device and load (bold) currents for different values of load capacitance.

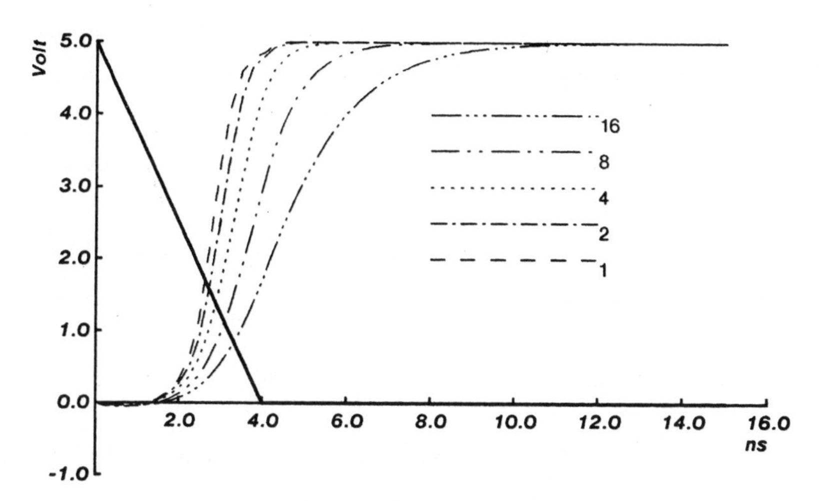

Figure 7-5: Input (bold) and outputs for different values of load capacitance: rise-time.

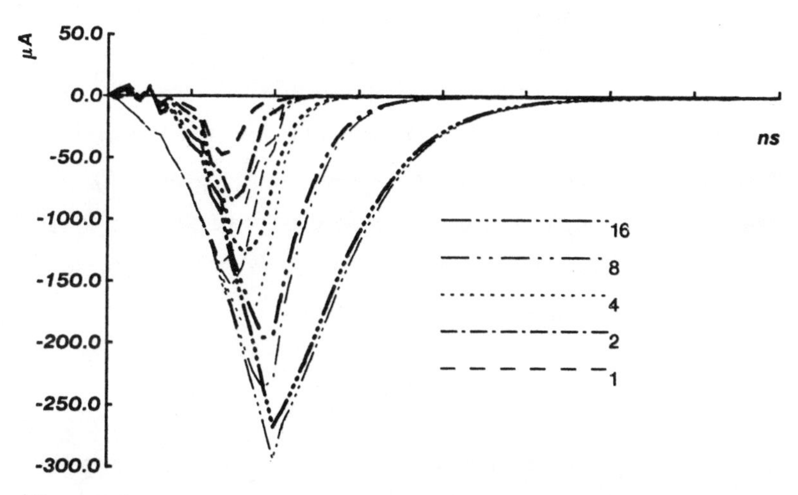

Figure 7-6: p-channel device and load (bold) currents for different values of load capacitance.

However, an important simplification is possible if we look at the dynamic behavior of a CMOS inverter as shown in Figs. 7-3 through 7-6. Fig. 7-3 shows the input signal to a CMOS inverter and five output voltages. The five output curves refer to five different loads; the number 8 in Fig. 7-3 means that the output capacitance is 8 times the driving capacitance. Throughout the entire chapter, "driving capacitance" indicates the capacitive value of the driving transistor, i.e., the one which turns on. This value is expressed by the usual formula:

$$\text{driving capacitance} = \frac{\varepsilon_{ox}\varepsilon_{o}}{t_{ox}} W_{n/p} L_{n/p} = C_{o}\,.$$

Fig. 7-4 corresponds to Fig. 7-3, but shows the *current* flowing through the n-channel (i.e., $i_{ds(n)}(t)$, in bold face in Fig. 7-4) and through the load (i.e., $i_L(t)$). Notice that when the load is moderately large (e.g., load capacitance eight times the driving capacitance), the two currents are almost identical in absolute value. Significant differences arise when load capacitance is small.

Figs. 7-5 and 7-6 correspond to Figs. 7-3 and 7-4 but refer to a high-going output transition, that is, they consider the relationship between $i_{ds(p)}(t)$ and $i_L(t)$. Again, if we equate the load current to the drain-to-source current of the p-channel, the error is small for a load capacitance / driving capacitance ratio (R_c) larger than, say, eight. We shall therefore assume that:

$$i_L(t) \approx i_{ds(n/p)}\,, \tag{7-2}$$

and choose $i_{ds(n)}(t)$ when computing the fall-time and $i_{ds(p)}(t)$ when computing the rise-time. This assumption is commonly adopted in the literature [7, 9].

The output response of a CMOS inverter depends on many parameters. Most of them are fabrication process parameters — such as thin oxide thickness, substrate doping concentration, etc. — or physical constants — such as SiO_2 dielectric constant — while others *can* be considered as constants, at least in a first-order approximation (e.g., majority carrier mobility, thin-oxide capacitance per unit area, etc.). Note that the output response of a CMOS inverter depends on the *dynamic characteristics* of the input waveform. This dependence has been neglected in the literature because it makes the model more complex to deal with. The typical assumption [7] considers the input signal as an ideal pulse of zero rise- or fall-time. Since the input signal is an ideal step-wise function with infinite slew-rate, the signal $v_{in}(t)$ — which is also the gate voltage of the n-channel transistor — is substituted by V_{dd} in the two (simplified) classical equations of the MOS device, that is:

$$i_{ds(n)} = \frac{\beta_n}{2} (v_{in} - V_{Tn})^2 \text{ (saturation region) ;}$$

$$i_{ds(n)} = \beta_n \left[(v_{in} - V_{Tn})\, v_{out} - \frac{v_{out}^2}{2} \right] \text{ (linear region) .}$$

With such an assumption, fall-time (or rise-time) delay τ_n (τ_p) can be computed by solving the following system:

$$C_L \int_{V_{dd}}^{V_{dd} - V_{Tn}} dv_{out} = -\beta_n (V_{dd} - V_{Tn})^2 \int_0^{t_s} dt \, ;$$

$$C_L \int_{V_{dd} - V_{Tn}}^{xV_{dd}} \frac{dv_{out}}{2(V_{dd} - V_{Tn})v_{out} - v_{out}^2} = -\beta_n \int_{t_s}^{\tau_n} dt \, ;$$

and, therefore:

$$\tau_n = \frac{2L_n t_{ox} C_L}{\mu_n W_n \varepsilon_{ox} \varepsilon_o} \left[\frac{V_{Tn}}{(V_{dd} - V_{Tn})^2} + \right.$$

$$+ \frac{1}{2(V_{dd} - V_{Tn})} \ln \frac{(2 - \chi)V_{dd} - 2V_{Tn}}{\chi V_{dd}} \Bigg] . \qquad (7\text{-}3)$$

t_s is the time in which the n-channel transistor leaves the saturation region and enters the linear region. χV_{dd} $(0.0 < \chi < 1.0)$ is the output level at which we want to compute the delay τ_n. The application of this model to the optimization of an inverter chain — in the sense of delay minimization — is presented as a reference in Appendix D.

Although this assumption makes the delay estimation straightforward and simple, it also produces serious inaccuracies. Not only do these inaccuracies lead to an output voltage curve quite different from the actual one, but the estimated delay at critical voltage levels — such as at the inverter threshold voltage, which is $V_{dd}/2$ for a balanced inverter — is also different from the correct one for many R_c values. When the model is used in optimization programs which determine the optimum size of devices, the results are — at best — of dubious reliability. Fig. 7-7 shows the fall-time of a 3μm p-well inverter as a function of R_c and the input signal slew-rate; both axes are on logarithmic (base 10) scale. The inverter has channel widths of 5μm (n) and 10μm (p), channel length of 3μm for both devices, threshold voltages of 0.844V (n) and -0.826V, mobilities of 220 cm^2/Vs (n) and 100 cm^2/Vs (p), and substrate doping concentrations of 1.0×10^{16} cm^{-3} (p-type) and 2.6×10^{15} cm^{-3} (n-type). The input signal is a ramp function starting at time zero from zero volt and reaching V_{dd} (5V) at 2ns or 4ns. Two voltage levels have been considered to compute the delay: 2.5V and 0.4V. The former is the threshold of a balanced inverter, while the second is the low reference voltage of most TTL families.

The two bold lines show the delay as predicted by the model ([7] and Eq. (7-3)) for 2.5V and 0.4V. The other four curves refer to SPICE [11] simulations of the same inverter. Fig. 7-8 shows the relative error of the model compared to SPICE simulations: when the relative error is negative, the model predicts a shorter delay than SPICE's. Accordingly, it is evident that input voltage characteristics have to be included in the model. This does not actually increase the complexity of the computation as much as one might think and will provide us with a more accurate model.

An interesting issue concerns the *range of accuracy* of the model. Should the model be accurate even for "unreasonably" large values of R_c, such as 1000 or more? The answer is yes. The rationale for having such a wide range of accuracy is two-fold:

- The optimization method used to determine the "best" sizes of the transistors in a chain is based on the delay model we are going to present. All these methods have problems of convergence — or unbearably slow convergence — for ill-behaved functions. Serious inaccuracies even for very large values of R_c might drive the algorithm into minima that do not have any physical correspondence or are not a physical minimum all together, thus leading to inaccurate results.

- If we want to use an optimization criterion based on the delay model to design output buffers, we have to realize that an "optimization" in the mathematical sense can bear no relationship with reality, because it might lead to a driver with channel widths as wide as 10,000μm, a size no one can afford. Optimization, as applied to output buffers, therefore means something different from optimization of gates driving on-chip loads. We will return to this issue later in this chapter; right now, it suffices to say that even for very large values of R_c our model should give us results which are close to reality, otherwise any optimization method might proceed right away toward values that are, from a physical/engineering point of view, unacceptable — for instance, the model might heavily underestimate the delay for very large R_c values.

We now present a model describing the low-going and high-going output transitions

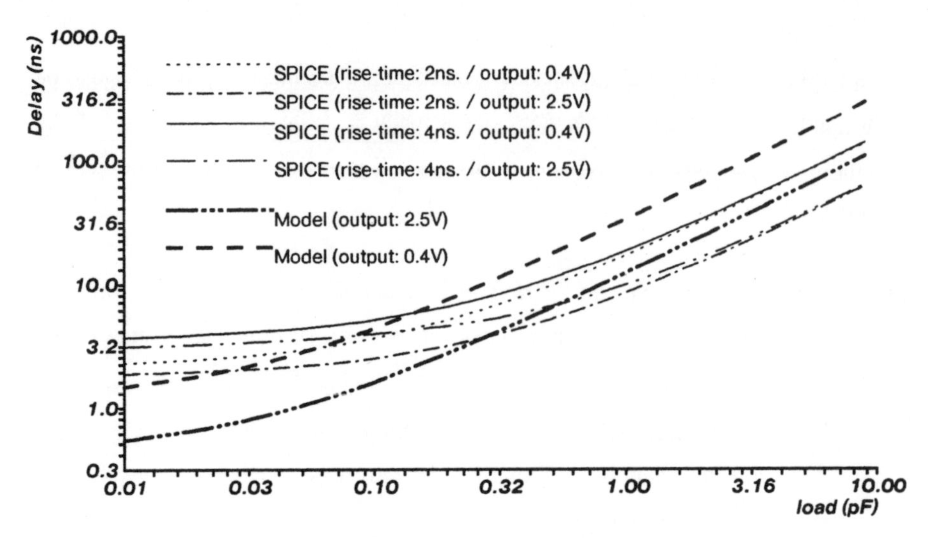

Figure 7-7: Fall-time vs. output load and input signal slew-rate for two output voltages of 2.5V and 0.4V: SPICE results and (bold) results obtained from the model presented in [7]. $R_c \approx$ load x 100.

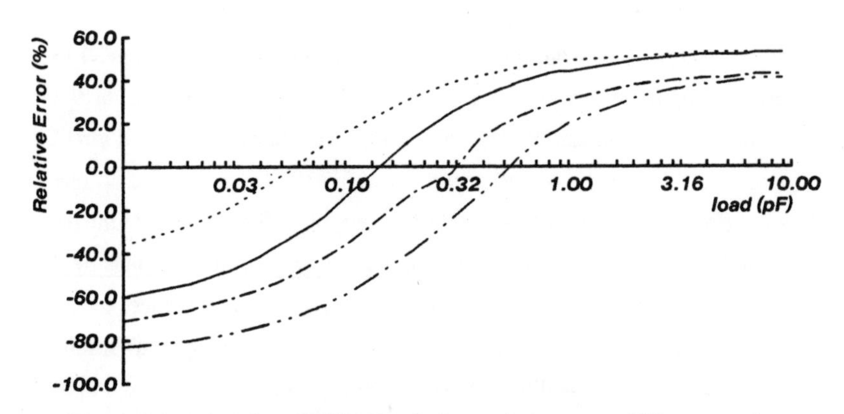

Figure 7-8: Model vs. SPICE simulation: relative error. When negative, the model underestimates the actual delay (i.e., model delay < SPICE delay). $R_c \approx$ load x 100.

of a CMOS inverter; in Section 7.5 we shall discuss the application of the model to the optimization of an inverter chain which drives a known capacitive load. This application is useful for the design of output pads and bus drivers.

7.1.1. Fall-time Delay Estimation

To compute the fall-time delay τ_n, the following assumptions have been made:

$$i_L(t) = -\, i_{ds(n)}(t)\,;$$

$$v_{in}(t) = \frac{V_{dd}}{t_u}\, t, \qquad 0 \le t \le t_u\,; \tag{7-4}$$

$$v_{in}(t) = V_{dd}, \qquad t > t_u\,.$$

As input signal, we consider a ramp with slew-rate V_{dd}/t_u (see Fig. 7-2). The output voltage can be expressed from Eq. (7-1) as:

$$C_L \frac{dv_{out}(t)}{dt} = -\, i_{ds(n)}\,.$$

V_o is the output reference voltage which corresponds to a delay of τ_n. The transistors of the inverter will go through two different operating regions, namely:

- *Region 1*: the n-channel device is operating in saturation and the p-channel device is operating in the linear region. Region 1 is characterized by the relationship:

$$V_{ds(n)} \ge V_{gs(n)} - V^e_{Tn}\,,$$

that is,

$$v_{out}(t) \ge v_{in}(t) - V^e_{Tn}\,.$$

- *Region 2*: the n-channel device operates in its linear region, while the p-channel device is in saturation. Region 2 is characterized by the relationship:

$$V_{ds(n)} < V_{gs(n)} - V^{e}_{Tn} ,$$

that is,

$$v_{out}(t) < v_{in}(t) - V^{e}_{Tn} .$$

V^{e}_{Tn} is the *effective threshold voltage* of the n-channel device and will be discussed below. We shall now discuss both regions separately.

7.1.1.1. Region 1: n-channel Device in Saturation

From the simplified theory of the MOS transistor, we have the following equation for the drain-to-source current in a saturated device:

$$i_{ds(n)} = \frac{\beta_n}{2} (v_{in} - V^{e}_{Tn})^2 , \tag{7-5}$$

and we can express the delay in this region with the following equation:

$$C_L \int_{V_{dd}}^{V_S} dv_{out} = -\frac{\beta_n}{2} \int_{0}^{t_s} \left[\frac{V_{dd}}{t_u} t - V^{e}_{Tn} \right]^2 dt ; \tag{7-6}$$

$$V_S = \frac{V_{dd}}{t_u} t_s - V^{e}_{Tn} .$$

As we already said, V^{e}_{Tn} is the effective threshold voltage. We have:

$$V^{e}_{Tn} = V_{Tn} - K\sqrt{2\phi_F} ;$$

$$K = \sqrt{\frac{2\varepsilon_{Si} q}{\varepsilon_o}} \frac{t_{ox}}{\varepsilon_{ox}} \sqrt{N_B} ;$$

$$\phi_F \approx \frac{kT}{q} \ln \frac{N_B}{n_i} .$$

β_n includes the *effective* width and length (i.e., not drawn) of the device. We have:

$$C_L \int_{V_{dd}}^{V_S} dv_{out} = -\frac{\beta_n}{2} \int_0^{t_S} \left[\frac{V_{dd}^2}{t_u^2} t^2 + V_{Tn}^{e\;2} - 2 \frac{V_{dd}}{t_u} V_{Tn}^e \right] dt ;$$

$$C_L \int_{V_{dd}}^{V_S} dv_{out} = -\frac{\beta_n}{2} \left[\frac{V_{dd}^2}{3t_u^2} t^3 + V_{Tn}^{e\;2} t - \frac{V_{dd}}{t_u} V_{Tn}^e t^2 \right]_0^{t_s} ;$$

$$C_L \left[\frac{V_{dd}}{t_u} t_s - V_{Tn}^e - V_{dd} \right] =$$

$$= -\frac{\beta_n}{2} \left[\frac{V_{dd}^2}{3t_u^2} t_s^3 + V_{Tn}^{e\;2} t_s - \frac{V_{dd} V_{Tn}^e}{t_u} t_s^2 \right] ;$$

$$t_s^3 - \frac{3t_u V_{Tn}^e}{V_{dd}} t_s^2 + \left[\frac{3t_u^2 V_{Tn}^2}{V_{dd}^2} + C_L \frac{6t_u}{\beta_n V_{dd}} \right] t_s +$$

$$- \frac{6t_u^2 C_L}{\beta_n V_{dd}} + \frac{6t_u^2 V_{Tn}^e C_L}{\beta_n V_{dd}^2} = 0 .$$

Given:

$$a_2 = -\frac{3t_u}{V_{dd}} V_{Tn}^e ,$$

$$a_1 = \frac{3t_u^2}{V_{dd}^2} V_{Tn}^{e\;2} + C_L \frac{6t_u}{\beta_n V_{dd}} ,$$

$$a_0 = -C_L \left[\frac{6t_u^2}{\beta_n V_{dd}} + \frac{6t_u^2}{\beta_n V_{dd}^2} V_{Tn}^e \right] ,$$

we have a cubic equation in t_s:

$$t_s^3 + a_2 t_s^2 + a_1 t_s + a_0 = 0\,.$$

Let q and r be:

$$q = \frac{a_1}{3} - \frac{a_2^2}{9}\,; \qquad r = \frac{1}{6}\,(a_1 a_2 - 3a_0) - \frac{a_2^3}{27}\,;$$

$$q = C_L \frac{2t_u}{\beta_n V_{dd}} > 0\,; \qquad r = 3C_L \frac{t_u^2}{\beta_n V_{dd}} - \frac{t_u^3}{2V_{dd}^3} V_{Tn}^{e\;3}\,.$$

Because $q > 0$, the cubic equation has one real root and a pair of complex conjugate roots. The real root is:

$$t_s = s_1 + s_2 - \frac{a_2}{3}\,;$$

$$s_1 = \left[r + \sqrt{q^3 + r^2}\, \right]^{1/3};$$

$$s_2 = \left[r - \sqrt{q^3 + r^2}\, \right]^{1/3}$$

Eq. (7-6) represents a large class of situations, but does not cover all the possibilities. There are some exceptions, as when R_c is very large. In this case, some adjustment is necessary. In fact, for small values of R_c, the output voltage behaves as shown in Fig. 7-9(a); the n-channel transistor stays in the saturation region (the thickest bold part corresponds to the n-channel device in saturation), while the input voltage is rising. On the other hand, when R_c is very large, the output voltage behaves as shown in Fig. 7-9(b). Again, the thickest bold part of the curve corresponds to the n-channel device in saturation. The input voltage reaches V_{dd} *before the n-channel transistor leaves the saturation region.* Eq. (7-6) assumes, however, that the input voltage keeps increasing at a V_{dd}/t_u rate. In this case, Eq. (7-6) is substituted by the following two equations:

$$C_L \int_{V_{dd}}^{V_X} dv_{out} = -\frac{\beta_n}{2} \int_0^{t_u} \left[\frac{V_{dd}}{t_u} t - V^e_{Tn} \right]^2 dt \, ; \tag{7-7}$$

$$C_L \int_{V_X}^{V_{dd} - V^e_{Tn}} dv_{out} = -\frac{\beta_n}{2} \int_{t_u}^{t_s} (V_{dd} - V^e_{Tn})^2 \, dt \, . \tag{7-8}$$

We shall compute V_X from Eq. (7-7) and plug it into Eq. (7-8) to obtain t_s.

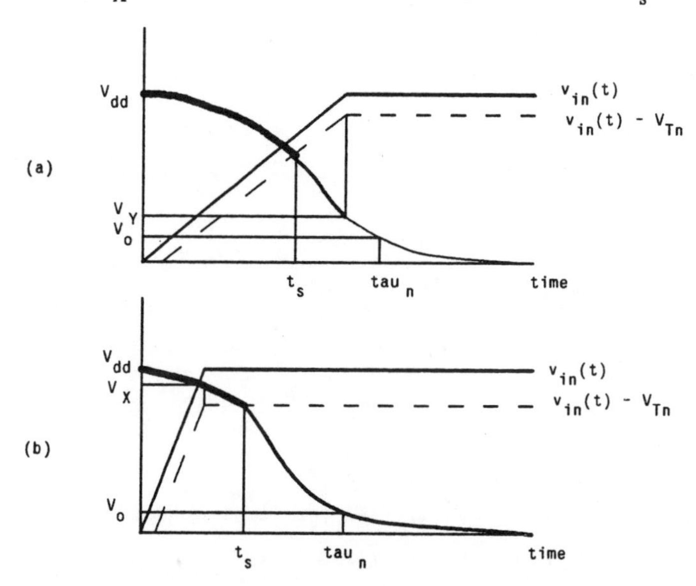

Figure 7-9: Bold curve: n-channel device in saturation. Solid curve: n-channel device in linear region. Two input signals with different slew-rate.

7.1.1.2. Region 2: n-channel Device in Linear Region

From the simplified theory of the MOS transistor we have the following relationship for the device's drain-to-source current:

$$i_{ds(n)} = \beta_n \left[(v_{in} - V^e_{Tn})\, v_{out} - \frac{v^2_{out}}{2} \right].$$

When computing the output voltage in this region, we shall often divide the computation in two separate contributions; the first considers the input signal still rising, while the second considers the input signal as settled at V_{dd}. We have, therefore:

$$C_L \int_{V_L}^{V_Y} dv_{out} = -\beta_n \int_{t_s}^{t_u} \left[\left(\frac{V_{dd}}{t_u} t - V^e_{Tn}\right) v_{out} - \frac{v^2_{out}}{2} \right] dt\,; \qquad (7\text{-}9)$$

$$C_L \int_{V_Y}^{V_o} dv_{out} = -\beta_n \int_{t_u}^{\tau_n} \left[(V_{dd} - V^e_{Tn}) v_{out} - \frac{v^2_{out}}{2} \right] dt\,; \qquad (7\text{-}10)$$

$$V_L = \frac{V_{dd}}{t_u} t_s - V^e_{Tn}.$$

Eq. (7-9) can be ignored if $t_s \geq t_u$. In this case, Eq. (7-10) becomes:

$$C_L \int_{V_{dd} - V^e_{Tn}}^{V_o} dv_{out} = -\beta_n \int_{t_s}^{\tau_n} \left[(V_{dd} - V^e_{Tn}) v_{out} - \frac{v^2_{out}}{2} \right] dt. \qquad (7\text{-}11)$$

From the system of Eqs. (7-9) and (7-10), we compute V_Y from Eq. (7-9) and, plugged into Eq. (7-10), we determine τ_n. V_Y is shown in Fig. 7-9(a). We now solve Eqs. (7-9), (7-10) and (7-11). Eq. (7-9) becomes:

$$C_L \int_{V_L}^{V_Y} dv_{out} = -\beta_n (t_u - t_s) \left[\left(\frac{V_{dd}}{2t_u} (t_u + t_s) - V^e_{Tn} \right) v_{out} - \frac{v^2_{out}}{2} \right];$$

$$C_L \int_{V_L}^{V_Y} \frac{dv_{out}}{\left[\frac{V_{dd}}{2t_u} (t_u + t_s) - V^e_{Tn} \right] v_{out} - \frac{v^2_{out}}{2}} =$$

$$= -\beta_n (t_u - t_s).$$

The above integral belongs to the class of integrals:

$$\int \frac{dx}{ax^2 + bx + c},$$

which has three different solutions depending on whether $b^2 - 4ac$ is greater than zero, less than zero, or equal to zero. In our case, because $c = 0$, $b^2 - 4ac$ can only be zero or greater than zero. Let us assume $b = 0$; we have:

$$\frac{V_{dd}}{2t_u} (t_u + t_s) - V^e_{Tn} = 0;$$

$$\frac{V_{dd}}{2} + \frac{V_{dd}}{2} \frac{t_s}{t_u} = V^e_{Tn}.$$

This requires V^e_{Tn} to be greater than $V_{dd}/2$, which is clearly impossible; b^2 is, therefore, strictly greater than zero. In this case the general solution of the integral is:

$$\int \frac{dx}{ax^2 + bx + c} = (b^2 - 4ac)^{-1/2} \ln \left| \frac{2ax + b - \sqrt{b^2 - 4ac}}{2ax + b + \sqrt{b^2 - 4ac}} \right|,$$

which leads to:

$$f(V_Y) = \frac{1}{H} \left[\ln \frac{-v_{out}}{K} \right]_{V_L}^{V_Y} = -\beta_n \frac{t_u - t_s}{C_L}, \qquad (7\text{-}12)$$

where:

$$H = \frac{V_{dd}}{2t_u}(t_u + t_s) - V^e_{Tn} ;$$

$$K = -v_{out} + \frac{V_{dd}}{t_u}(t_u + t_s) - 2V^e_{Tn} .$$

Eq. (7-12) is a transcendental equation in V_Y. We solve it with a Newton-Raphson iterative procedure:

$$V_{Y(q+1)} = V_{Y(q)} - \frac{f(V_{Y(q)})}{f'(V_{Y(q)})} .$$

Eq. (7-10) becomes:

$$\int_{V_Y}^{V_o} \frac{dv_{out}}{\left[(V_{dd} - V^e_{Tn})v_{out} - \frac{v^2_{out}}{2}\right]} = -\beta_n \frac{\tau_n - t_u}{C_L} .$$

Because $V_{dd} - V^e_{Tn} > 0$, we have:

$$\frac{1}{V_{dd} - V^e_{Tn}} \left[\ln \left| \frac{-v_{out}}{-v_{out} + 2(V_{dd} - V^e_{Tn})} \right| \right]_{V_Y}^{V_o} = -\beta_n \frac{\tau_n - t_u}{C_L} ,$$

and the equation can be solved directly for τ_n. As far as Eq. (7-11) is concerned, it is similar to Eq. (7-10), where $V_{dd} - V^e_{Tn}$ substitutes V_Y in the left hand-side integral. We have:

$$\tau_n = t_u + \frac{C_L}{\beta_n(V_{dd} - V^e_{Tn})} \ln \left| \frac{-V_o + 2(V_{dd} - V^e_{Tn})}{-V_o} \right| .$$

Fig. 7-10 shows the sequence of steps necessary to compute the fall-time delay for any value of V_o, including values very close to V_{dd}. For instance, after computing t_s, it is necessary to check whether t_s is less than or equal to the rise-time of the input signal. All the equations can be found in Appendix E and are modifications of the basic Eqs. (7-6) through (7-11).

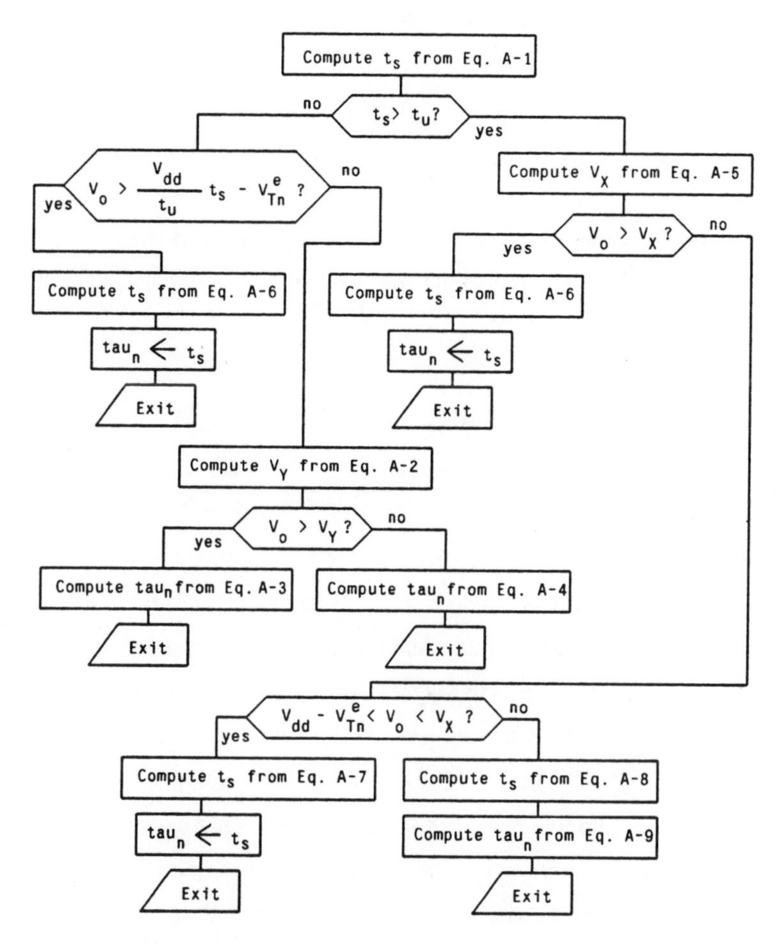

Figure 7-10: Flow-chart for computation of the fall-time (rise-time). The equations are shown in Appendix E.

7.1.2. Rise-time Delay Estimation

The same equations presented in the previous section can be used to compute the rise-time of a CMOS inverter, substituting all the "n" subscripts with "p." When using the previous equations to compute the rise-time, it is important to introduce the following two changes:

1. The threshold voltage of the p-channel, which has a negative value, must be included in the equation with opposite sign (i.e., always positive).

2. V_o is the complement to V_{dd} of the actual V_o we need to compute. If, for instance, we want to compute the rise-time delay to 4.0V, V_o will be actually $V_{dd} - V_o = 1.0V$ (for $V_{dd} = 5V$).

7.1.3. Refining the Model

The model presented in the previous section provides the designer with higher accuracy than the simplified model of Eq. (7-3). Nonetheless, some inaccuracy is still present. More precisely, the inaccuracy is directly proportional to the output load and inversely proportional to the input signal rise-time. Medium values of R_c — such as 80, or more — combined with short input rise-times or large values of R_c, combined with longer input rise-times, will trigger this behavior in the model. In this case, the model overestimates the delay and produces results that are comparable to the ones given in Eq. (7-3). The reason for this behavior is quite simple to understand if the equations of the model presented in the previous section are compared to Eq. (7-3). When input voltage reaches V_{dd}, there is no difference between Eq. (7-3) and the model presented; both assume that the input has a constant value of V_{dd}. In this case, the output follows a logarithmic decay in a sort of "free-fall," at a rate which *linearly* depends on C_L, the output load.

The fact that the model is inaccurate means that equating the load current to the n-channel transistor current — that is, for an output low-going transition — is no

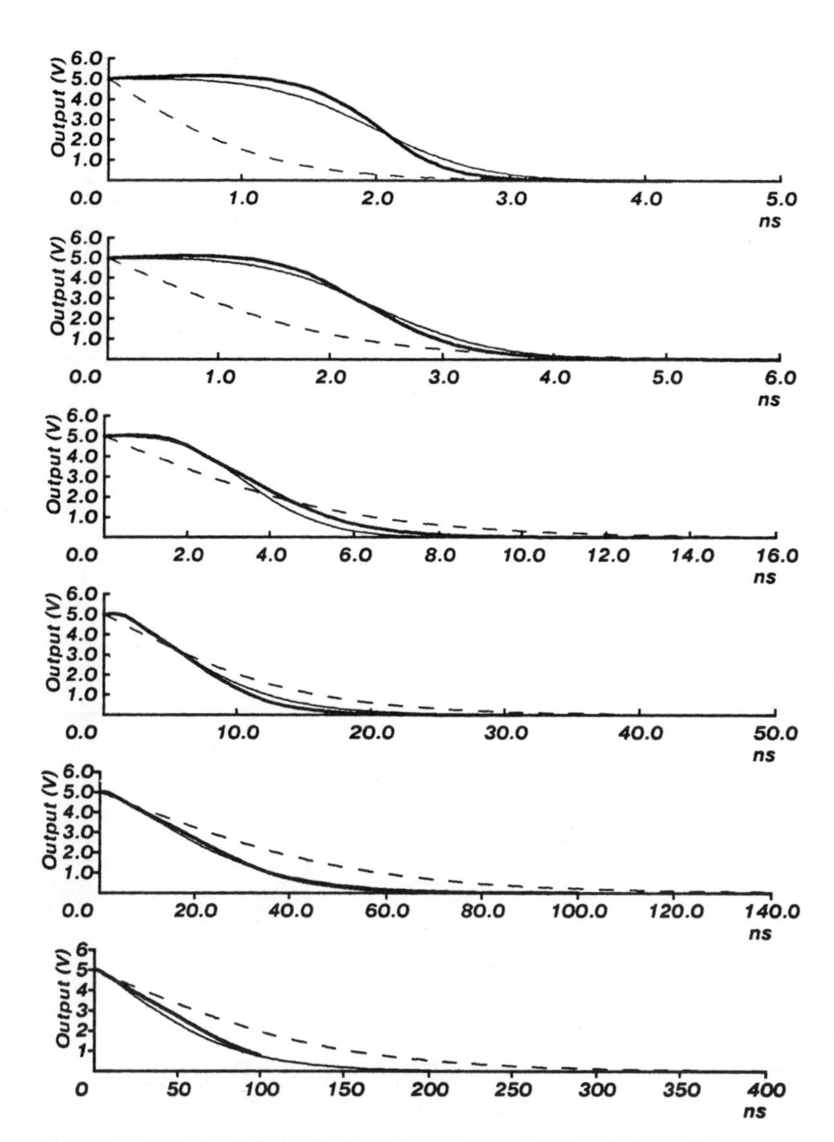

Figure 7-11: SPICE output (bold), model output (solid) and Eq. (7-3) (dashed) for 2ns rise-time input signal. From top to bottom: R_c = 3, 8, 30, 80, 300, 800.

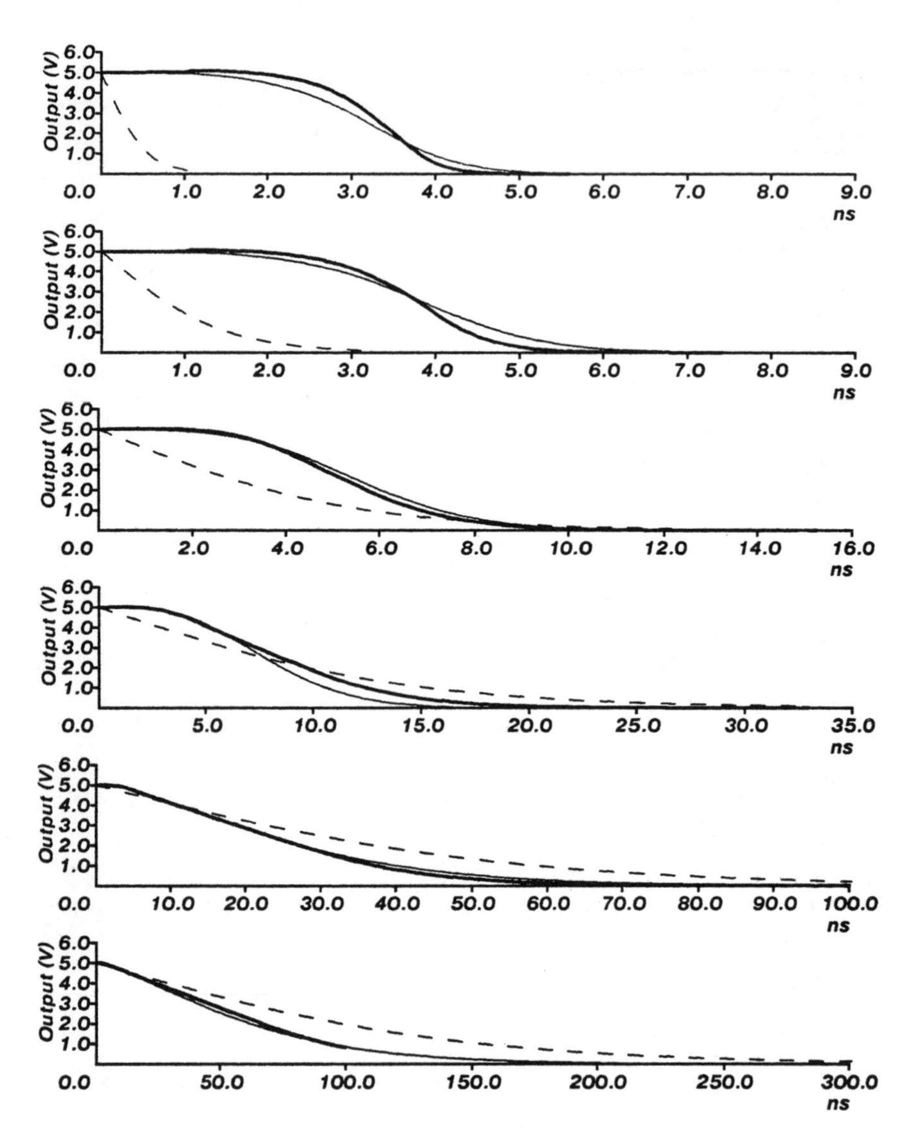

Figure 7-12: SPICE output (bold), model output (solid) and Eq. (7-3) (dashed) for 4ns rise-time input signal. From top to bottom: R_c = 3, 8, 30, 80, 300, 800.

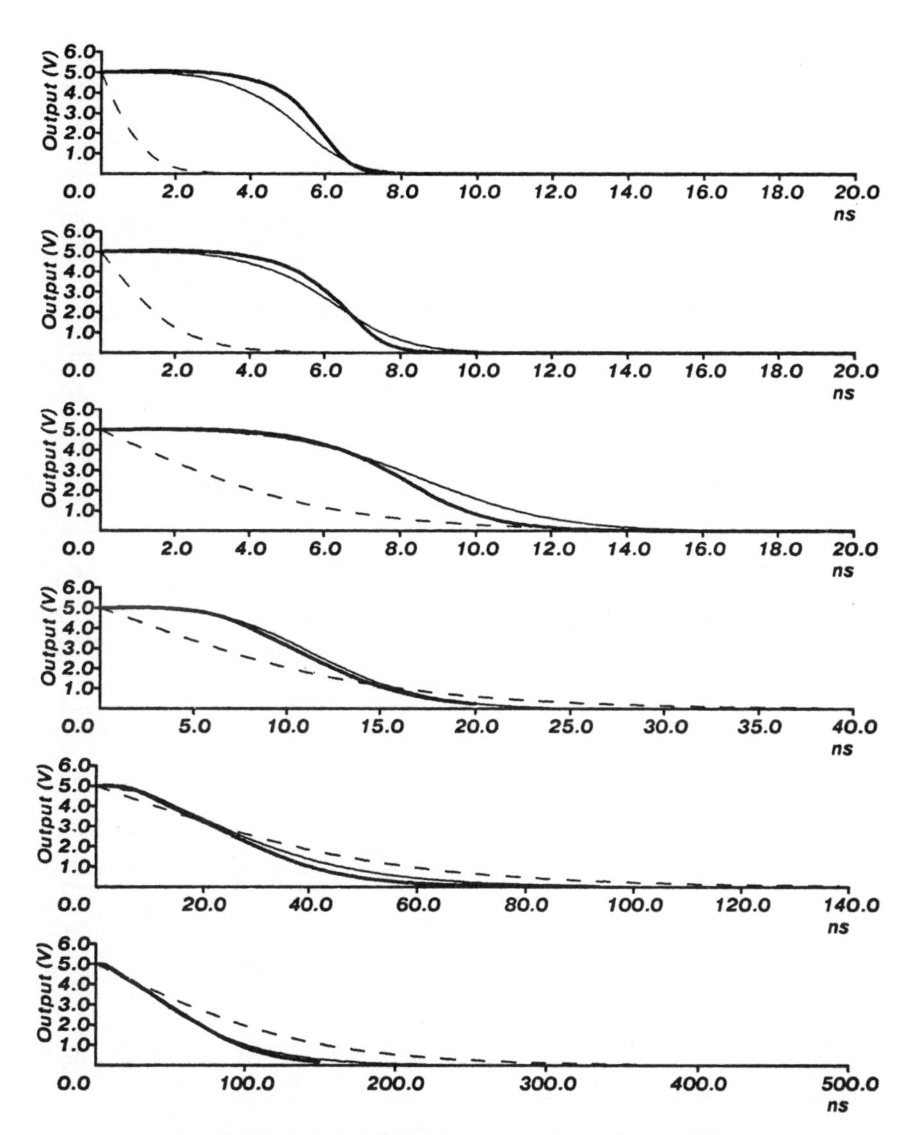

Figure 7-13: SPICE output (bold), model output (solid) and Eq. (7-3) (dashed) for 8ns rise-time input signal. From top to bottom: R_c = 3, 8, 30, 80, 300, 800.

longer valid. However, the following observation has been made, which helped to solve this problem of inaccuracy: *the decrease in accuracy of the model starts when the time t_s in which the n-channel transistor stays in the saturation region becomes larger than the rise-time of the input signal.* In other words, so long as the "saturation period" is shorter than the rise-time of the signal, the model accurately tracks the output waveform computed by SPICE. The larger the saturation time becomes with respect to input rise-time, the less accurate the model is.

The solution to the problem consists of providing the inverter with a drastic input overdrive, a sort of very large bootstrapping at the gate. From a mathematical point of view, this consists of letting the input signal rise indefinitely without setting it at V_{dd} after t_u seconds. This is done only when t_s is larger than t_u. When t_s is smaller, the equations presented in the previous sections and summarized in Appendix E are used, and the computation flow shown in Fig. 7-10 can be used.

Some results will now be presented. The bold curves show the SPICE output, the solid curves show the output provided by the model, and the dashed curves refer to the model represented by Eq. (7-3). Three different sets of curves are shown; they are characterized by three different input signal rise-times. Each set presents six output waveforms, each one referring to a different value of R_c. Values of R_c from 1 to 1000 have been studied and compared to SPICE results. More precisely the following R_c values have been considered: 1, 2, 3, 4, 5, 6, 7, 8, 9, 10, 20, 30, 40, 50, 60, 70, 80, 90, 100, 200, 300, 400, 500, 600, 700, 800, 900, 1000. Only the results for six of these values are presented in the next figures: the values are 3, 8, 30, 80, 300, 800. Note that these values refer to the load/driving ratio as defined in this document. An R_c value of 3 roughly corresponds to an inverter driving an identical inverter. Fig. 7-11 shows the results for an input signal with rise-time of 2ns; Fig. 7-12 shows the same curves for an input signal with 4ns rise-time, and, finally, Fig. 7-13 shows the results for an input signal with 8ns rise-time. The accuracy of the model is well within 10% of SPICE's, for rise-times from 1ns to 10ns and R_c values from 1 to 1000. Note that while 3, 8, and 30 result from the model described in the previous sections (for all three rise-times considered), curves 80, 300, and 800 in Fig. 7-11, curves 300 and 800 in Fig. 7-12, and curve 800 in Fig. 7-13 are obtained through the gate "overdrive" solution explained above.

7.2. Input Buffer

An input buffer can be omitted when the input signal is already CMOS compatible and the current provided is sufficient (this is not true for protection, which must be present anyway, as we shall see later). However, when a "TTL compatible input buffer" is required, some logic circuitry has to be implemented. A simple approach is to use an inverter, as Fig. 7-14 shows. Note that a direct connection between the input of the first logic gate — say, an inverter — and the output of an off-the-chip circuit which delivers TTL logic levels would be very dangerous and unreliable. While a logic TTL "0" can be considered low enough to switch the CMOS gate, a TTL logic "1" (say, 2.7V) is too close to the inverter threshold voltage — which is 2.5V for a balanced inverter when V_{dd} = 5V — to guarantee reliable operation. Moreover, even though correct operation could be achieved, a TTL logic "1" is so close to the inverter threshold voltage that both p- and n-channel devices would operate in the saturation region, with very high static power dissipation. It is clear, therefore, that an input buffer has to be interposed between the TTL input and the internal circuitry. This section will focus only on the TTL compatible input buffer, because a "CMOS compatible input buffer" can be a simple wire — if the current provided by the off-the-chip circuitry is sufficient and we ignore the protection section — or a series of properly scaled-up inverters to increase current capability. In this case, the design of such a chain follows the methodology for output buffer or bus driver design which will be presented in Section 7.5.

An inverter, acting as a level shifter, basically requires an "unbalanced" design, that is, it consists of an inverter with a threshold voltage different from $V_{dd}/2$. In

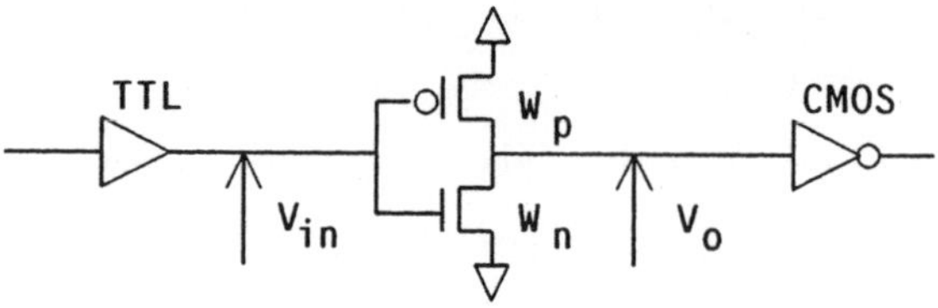

Figure 7-14: A simple level shifter for a TTL compatible input buffer.

particular, an inverter which provides 0V output for, say, 2.4V input and 5V output for 0.4V input can be considered adequate for our purposes. We already know from basic CMOS theory that this would require W_n to be much larger than W_p — with both having the same length (see Fig. 7-14). Any accurate model of the MOS device, such as that of Section 2.6, can be used to determine the W_n/W_p ratio required to achieve a *statically* balanced inverter. If precision is not needed, conservative figures, such as an n-channel 5 to 7 times wider than the p-channel (both having the same channel length), will usually work.

Fig. 7-15 and 7-16 show a comparison of three different inverters, which have the following characteristics (the curve numbers refer to the enumeration below):

1. p-channel width: 15μm, n-channel width: 100μm;

2. p-channel width: 15μm, n-channel width: 15μm;

3. p-channel width: 100μm, n-channel width: 100μm.

As has already been pointed out, a large "balanced inverter" would work with both devices close to saturation and, as curve 3 in Fig. 7-16 shows, power dissipation would be dangerously high (note the high "plateau" in the center of the figure). Curve 1 in both figures shows the "best" inverter, that is, a level shifter with the n-channel much wider than the p-channel. Note that neither solution 2 nor solution 3 can deliver a

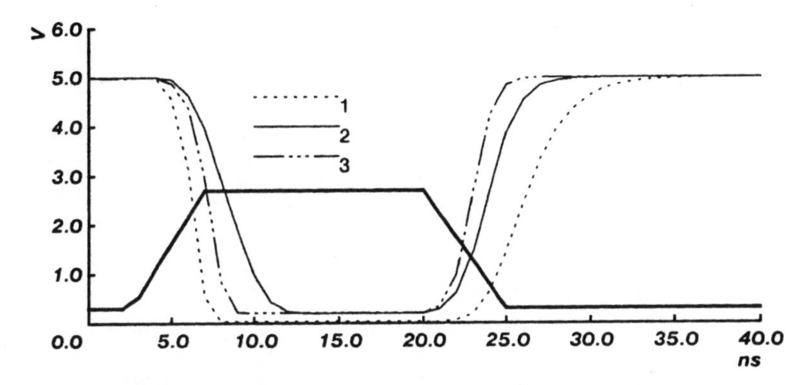

Figure 7-15: V_{in} (bold line) and three output voltages (see text).

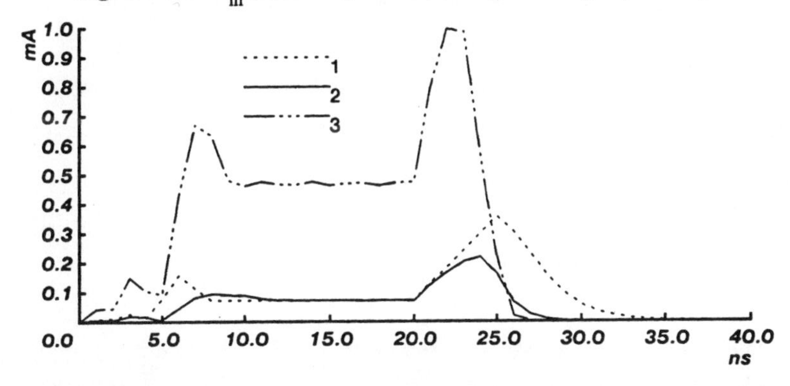

Figure 7-16: Dynamic currents flowing through the level shifter for the same cases shown in Fig. 7-15.

low logic level of zero volts. The need for having a zero volt logic level does not stem from being able to correctly switch the next gate — either zero volts or, say, 1V would be just fine, for that matter. Zero volts are necessary because of noise margin requirements, which are even more crucial here because the gate receives a signal from the environment and is more subject to noise disturbances. If we compare the different solutions, we can conclude that:

- Gate 2 does not deliver a low logic level of zero volts and is sensitive to variations in the process parameters.

- Gate 3 does not deliver a low logic level of zero volts, it is sensitive to variations of the process parameters and features very high power dissipation.

- Gate 1 is the right compromise: power dissipation in steady state is comparable to that of gate 2. Dynamic power dissipation (switching), which becomes important if the number of input pads is high, is comparable to that of gate 2. Finally, speed is comparable to that of gate 2 — faster fall-time, slower rise-time — and a low logic level equal to zero volts is delivered.

If the level shifter is to drive a significant load inside the circuit, two different approaches can be used:

1. Implementation of one very wide level shifter: the level shifter would be driven in turn by outside circuitry, capable of delivering high current levels (TTL logic, for instance).

2. Implementation of a small level shifter, followed by a scaled-up chain of inverters.

The two approaches differ with respect to speed and power dissipation. The former maximizes speed, because it exploits the higher current capability of bipolar, off-the-chip circuitry. However, power dissipation is high as well, because a very wide n-channel has to be used, and a large current spike will take place during a low-going transition at the output of the level shifter. The second approach does not exploit the fact that the level shifter is driven by circuitry able to provide high current at high speed. However, power dissipation is lower than that of the first solution. Which methodology is to be adopted depends on parameters such as total number of input buffers (i.e., power dissipation constraints), speed requirements, noise considerations, etc.

7.3. Output Buffer

The output pad has to deliver sufficient current to charge or discharge the load capacitance, which consists of bonding wire, pin, conductor(s) on the printed-circuit board, and input capacitance of all the gates — and pins — of other chips that are connected to the output pad. This capacitance is a specification and usually ranges from 10pF to 200pF. Full specifications call for a minimum rise- and fall-time at the output of the chip for a certain load over a specified temperature range. For example, a specification can be:

$$\frac{\tau_f + \tau_r}{2} \leq 8\text{ns} \;@\; 15\text{pF [commercial range]}$$

The rise- and fall-time can be defined in various ways; Fig. 7-17 shows one definition of τ_f and τ_r. In this example, rise- and fall-times are computed between $0.1V_{dd}$ and $0.9V_{dd}$.

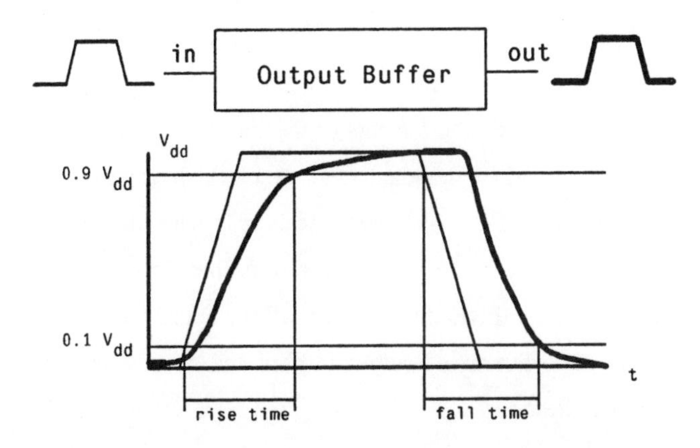

Figure 7-17: A definition of rise-time (τ_r) and fall-time (τ_f).

The nMOS output buffer is usually designed with a "quasi-complementary" configuration in the last stage, with two n-channel devices driven by a signal and its complement, in a typical push-pull configuration, see Fig. 7-18(a). This approach pays in terms of performance — two n-channel devices, no depletion transistor — but especially in terms of power dissipation, because no static path between V_{dd} and V_{ss} is present. This approach is not used in CMOS because of the availability of the p-channel device. The final stage in the static output buffer is a simple CMOS inverter (see Fig. 7-18 (b)).

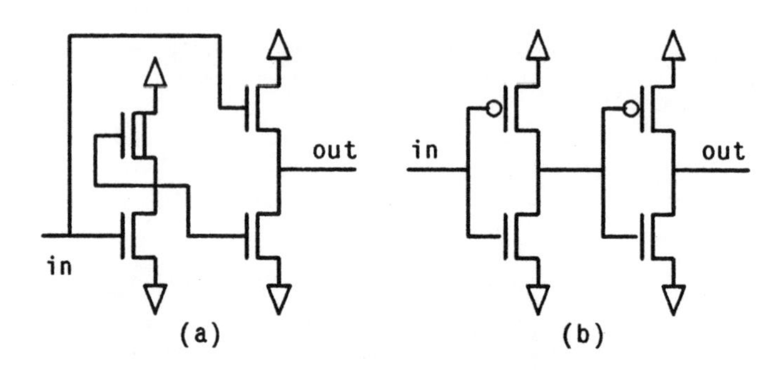

Figure 7-18: nMOS (a) and CMOS (b) pre-driver and driver.

A static CMOS output buffer consists of a chain of inverters that are scaled *down* until they meet the loading requirements of the internal circuitry, that is to say, the last stage (driver)[1] should be designed first. A block diagram of a CMOS output buffer is shown in Fig. 7-19. This section deals only with the design of the driver, because the preceding stages are normal inverters that do not usually require the special treatment that the driver does, as we shall see below. Section 7.5 will present an approach for the synthesis of output buffers and/or bus drivers based on the model developed in Section 7.1.

The driver of a CMOS output buffer features, like all the other CMOS gates,

the capability of a full swing between V_{dd} and V_{ss}. If the output specifications call

---

[1]The driver is the last stage which is connected directly to the load, while the pre-driver, as the name implies, is the stage preceding the driver.

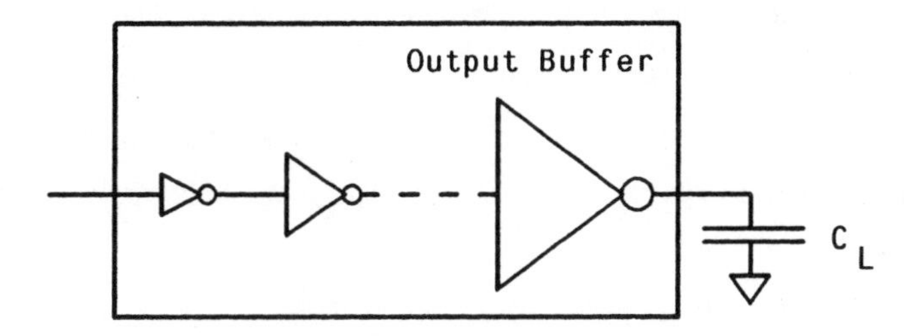

Figure 7-19: A simple output buffer is a chain of properly scaled inverters.

for a "CMOS compatible" output buffer, no precautions have to be taken, as far as logic levels. This is also true when the specification calls for a "TTL compatible" output buffer. Because the output swings from V_{dd} to V_{ss}, there is no problem in reaching the TTL "high" and "low" logic levels. The only constraint calls for the driver to sink and source the necessary amount of current in both logic states. The TTL output specifications are different depending on the logic family and are shown in Table 7-1.

Table 7-1: TTL output specifications for various families.

	Standard	54LS/74LS	54S/74S	25LS
High	40μA @2.4V	20μA @2.7V	50μA @2.7V	20μA @2.7V
Low	-1.6mA @0.4V	-0.36mA @0.4V	-2.0mA @0.5V	-0.36mA @0.4V

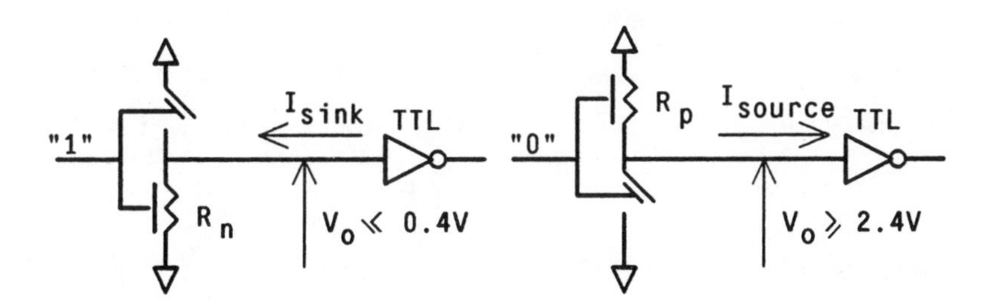

Figure 7-20: Requirements for standard TTL compatible output driver.

Let us assume that the output buffer has to interface with the TTL Standard logic family: Fig. 7-20 shows the problem. To correctly interface with a TTL gate, when the driver generates a TTL low logic level ($\leq$ 0.4), the n-channel device must sink 1.6mA. We can express the channel resistance of the MOS transistor in the linear region as:

$$R_{channel} = \frac{L}{W\mu_{n/p}C_{ox}(V_{gs} - V_{Tn/p})} . \tag{7-13}$$

By combining Eq. (7-13) with the above constraints we have:

$$\frac{L}{W\mu_n C_{ox}(V_{gs} - V_{Tn})} = \frac{V_o}{I_{sink}} = \frac{0.4}{1.6x10^{-3}} .$$

Assuming t_{ox} = 400Å and μ_n = 400cm^2/Vs we have ($V_{gs} - V_{Tn}$ = 4.2V):

$$\frac{L}{W} = 36.2x10^{-3} \approx \frac{1}{27}$$

The n-channel device width must be at least 27 times larger than the n-channel device length; this is not a difficult requirement to meet, because the n-channel transistor of the output driver will always be much wider than 27 times its channel length in order to provide enough current to the load capacitance. However, the above constraint assumes a fan-out of one. When the fan-out is greater than one, the constraint between the n-channel device width and its length should be more carefully considered. For instance, a fan-out of fifteen would require a channel width about 420 times greater than the channel length and this constraint would become more difficult to satisfy.

When the output buffer performs a high-going transition, the p-channel device must source 40μA. We can write:

$$R_p = \frac{V_{dd} - V_o}{I_{source}} = \frac{2.6}{40\mu A} = 65K\Omega \,.$$

Assuming $\mu_p = 160cm^2/Vs$, we have:

$$\frac{L}{W} \approx 3 \,.$$

Hence, any ratio will be acceptable, as long as the width of the p-channel transistor is *wider* than one-third the channel length.

I-O buffers are among the first blocks included in libraries. Often, the output buffer starts with a minimum feature size inverter — to offer the lowest input load to the internal circuitry — followed by a chain of scaled up inverters. This approach is useful to cut down design costs. However, it should also be noted that this "library output buffer" has a serious drawback, that is, it has been designed for any possible circuit, and, therefore, it will not be able to offer *optimized* performance for every circuit. This is not to say that "library I-O buffers" are a wrong approach, but simply to point out that, when special situations arise — for instance, severe input or output speed constraints — it is mandatory to design buffers tailored to the specific implementation. On the other hand, a full characterization of them — in terms of noise robustness, driving capability, latchup insensitivity — takes significant design effort, which is an argument in favor of standardization of I-O buffers.

Sizing an output pad must start from an assumption that may sound somewhat unusual, that is: *we already know that, as far as speed is concerned, the output driver might not be optimized.* Delay minimization for very large load capacitance values can be physically impossible to achieve. Let us assume a specification calling for an output load of 100pF. Whatever the approach may be, the output driver cannot be sized to achieve minimum delay. If a 4:1 ratio were used (driving capacitance 4 times smaller than load capacitance), we should have an n-channel device ($L = 2\mu m$, $t_{ox} = 400Å$) roughly 1.5cm (centimeter) wide. In reality, the device will have a much narrower channel width and, therefore, a much smaller driving capacitance. Given the enormous disparity between load capacitance and driving capacitance, even a significant increase in the size of the transistors does not pay off as much as one might think. Fig. 7-21 and 7-22 show the output fall-time and rise-time of the output buffer driver for different sizes; the load is 100pF. Note how little the curves differ. Even doubling the width of the transistors does not drastically speed up the charging/discharging of the load capacitor.

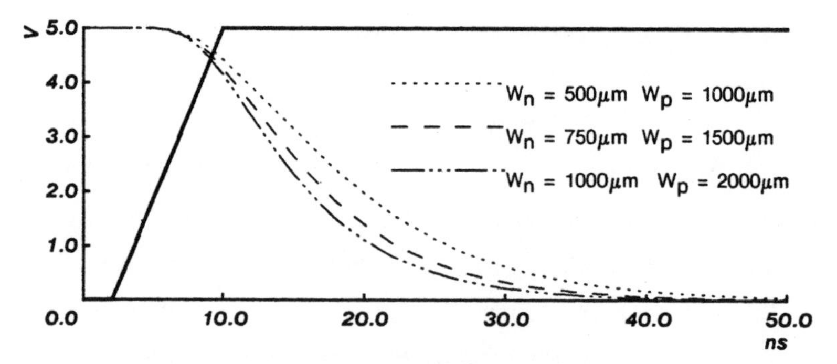

Figure 7-21: Low-going transition of the output driver for different transistor sizes.

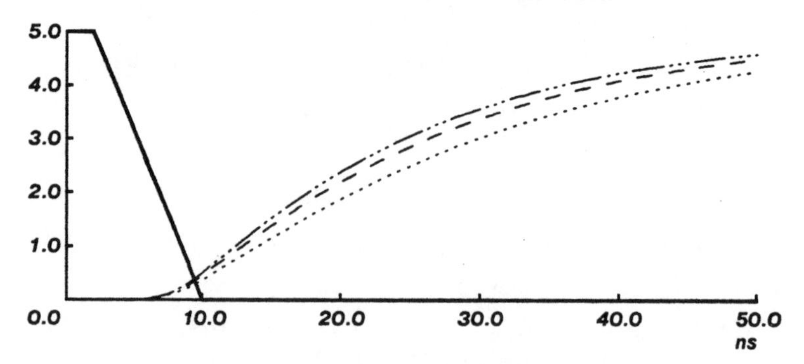

Figure 7-22: High-going transition of the output driver for different transistor sizes.

It is worth remembering that the output signal is also influenced by the *shape of the input signal*, as we already pointed out in Section 7.1. This characteristic is usually neglected during the design of gates in the internal circuitry, because this feature, given the comparatively small dimensions of the devices, is not as notable as it is in large buffers or bus drivers. Fig. 7-23 shows one example. The same driver is driven by two inputs that cross the inverter threshold at the same time but have different rise-times; the output corresponding to the steepest input has a fall-time shorter than the output corresponding to the slower input. It is therefore important, when designing output buffers, to provide the driver with a fast, steep input waveform. This can usually be done, because the pre-driver is much smaller than the driver, and, therefore, can be truly optimized. This also implies that the pre-driver ought to be slightly oversized to guarantee a fast, "clean" output signal.

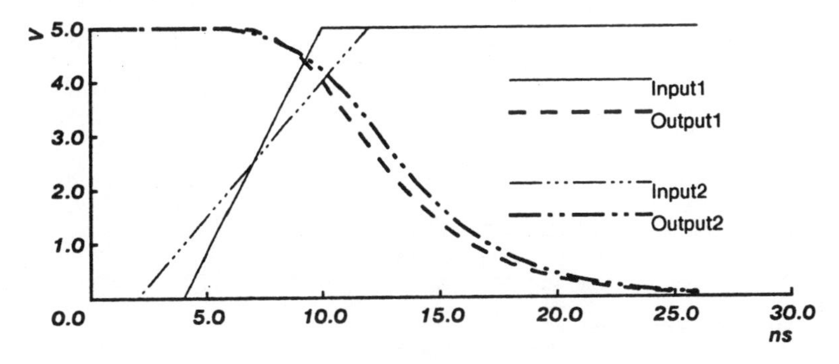

Figure 7-23: The shape of the input signal to a driver does influence the performance of the buffer.

The size of the output driver is determined by four fundamental constraints: *speed*, *area*, *power dissipation*, and *noise*. These parameters are not independent of one another. The area of the output driver depends on the size of the transistors and the layout style, which is crucial to achieve high performance. As far as noise is concerned, it is important to realize how difficult it is to deal with it in an accurate way, given the fact that noise is also influenced by layout geometry of V_{dd} and V_{ss}

power lines.

Let us assume we have 32 output buffers working synchronously, that is, a change in output occurs simultaneously in all of them. If we assume (worst-case) that all these buffers keep switching from V_{dd} to V_{ss}, and back, with a cycle time of 100ns, the current (switching) flowing in the V_{dd} and V_{ss} lines can reach peaks of up to 2A. Assuming that the metal line is capable of delivering 0.7mA/μm, a 1400μm-wide metal power supply line should be laid out. Moreover, this switching current would operate at 10MHz. This high current, coupled with the inductance of the power lines, creates noise that, being generated on-chip, cannot be easily filtered. This noise can be coupled to adjacent logic lines and can create glitches, decrease the circuit speed, and even cause latchup. Therefore, it is absolutely necessary to apply both layout and design techniques to minimize this noise. What has been presented seems to be an ideal case, because the 64 devices will never be in saturation at the same time, even though they all switch "at the same time." There will always be some skew in the switching, and this peak current ought to be lower than the worst case. Nonetheless, the problem remains and has to be considered seriously when output drivers are designed. In order to minimize the noise, we can:

- Decrease the current requirements for the output pads by making the output drivers smaller.

- Decrease the overload on the power lines by making them very wide.

- Filter possible spikes with capacitors between V_{dd} and V_{ss}. This can be accomplished either by external capacitors and/or by proper layout techniques, when more than one metal layer is available. This is shown in Plate X, where the second metal is used for V_{dd}, while the first metal is used for V_{ss}. The two metal layers run overlapped, with the insulator between them acting as the capacitor dielectric. This creates a capacitance between second and first metal and, therefore, between V_{dd} and V_{ss}.

- Decrease the inductive coupling factor by proper layout techniques, such as avoiding closed rings in the pattern of the power lines, although this often requires multiple V_{dd} and V_{ss} pads. Note that using multiple V_{dd} and V_{ss} pads has become an acceptable procedure in advanced commercial chips.

The first point interests us immediately. It calls for a minimization of the sizes of the devices in the driver. Therefore, if no output specifications are given, the designer should compromise between speed and noise induction, when the size of the driver has to be decided. It is difficult to give "typical" figures for this size; we can say that most of the p-channel devices in drivers have widths ranging from 600μm to 2000μm with few exceptions below and above these values.

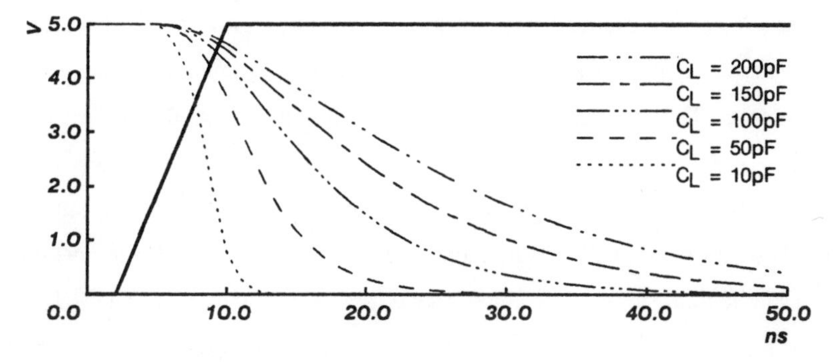

Figure 7-24: Low-going transition of the same driver for different output loads.

The load capacitance also plays a crucial role, as Fig. 7-24 shows. Achieving high performance for large loads requires careful optimization of all parameters, including optimization of the input signal to the driver, minimization of parasitic output capacitances, and consideration of other effects such as RC gate delays. Decreasing the output capacitance by shortening interconnections on the board and using sockets with low values of pin/contact capacitance does help in decreasing the load capacitance, without calling for a logic redesign — for instance, decreasing the fan-out

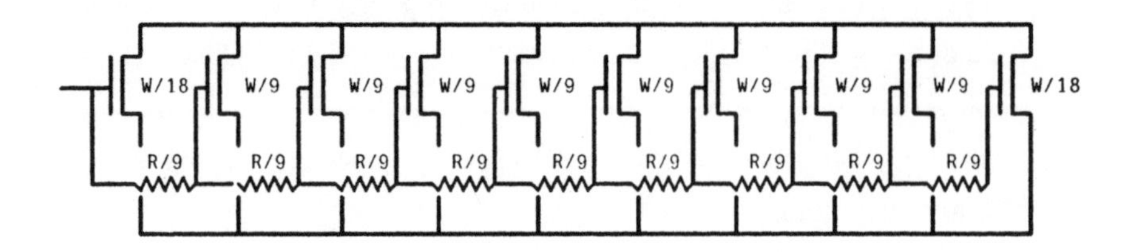

Figure 7-25: Model that takes into account the RC gate delay in very wide devices.

— of the system. Up to now, we have dealt with "ideal" drivers, that is, transistors that have the same characteristics of their much narrower counterparts, but linearly scaled up. However, for very wide channel devices, second order effects such as RC gate delay [13] start playing an important role. Polysilicon features a delay which increases quadratically with the length of the interconnect, because both resistance and capacitance contribute to the overall delay, unlike low resistivity metal-like interconnect that features a delay which is linearly proportional to the length of the interconnect. This quadratic build-up of delay heavily affects the performance of drivers — and pre-drivers, to a lesser extent — because the input signal to a very wide device is not "instantly" propagated to every point on the gate electrode, but, rather, flows through the gate with considerable delay. Modeling very wide channel devices with a simple transistor can introduce large errors, even of the order of 45% [13]. The influence of the wide polysilicon gate can be taken into account by simply carrying out the same procedure adopted for polysilicon interconnects, that is, using a T-ladder or π-ladder model, with the interconnect capacitance being the gate capacitance. It has been shown [13] that a 9-step ladder, like the one shown in Fig. 7-25, leads to an error of less than 1%. In Fig. 7-25, W is the total gate width, and R is the sheet resistance of the polysilicon gate. A typical R-value is about 30Ω/□. The RC gate delay build-up becomes significant for channel widths as short as 100μm. The use of gate material with low resistance, such as polycide, significantly decreases this phenomenon.

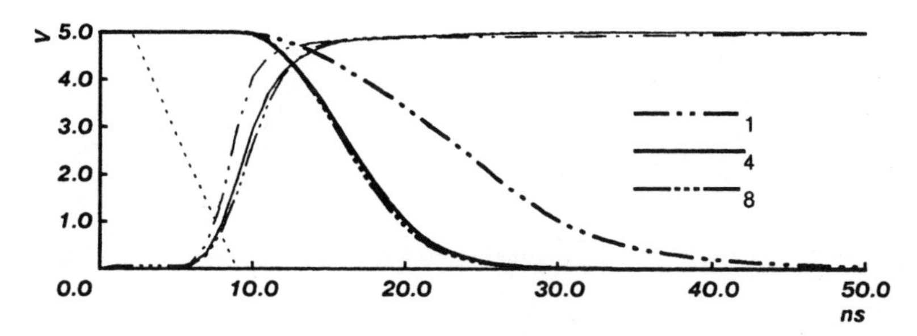

Figure 7-26: Different implementations of the same driver: low-going transition.

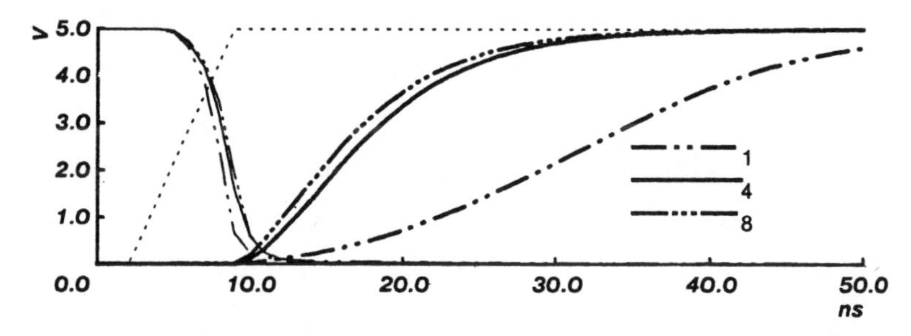

Figure 7-27: Different implementations of the same driver: high-going transition.

The presence of the RC gate delay suggests that better performance could be achieved if many small devices were put in parallel, instead of using one very wide device. The speed up that can be obtained by this approach is shown in Figs. 7-26 and 7-27: the same pre-driver precedes a driver, implemented by one very large device (curve 1 in Figs. 7-26 and 7-27), four narrow devices in parallel (curve 4), and eight narrow devices in parallel (curve 8), respectively. In all three cases the model shown in Fig. 7-25 has been used. The dotted line represents the input to the pre-driver, the same lines show the output of the pre-driver and the output of the driver (bold). Note that the output of the pre-driver is almost identical for all the implementations. The difference between the single device implementation and the "four-in-parallel" implementation is outstanding, while going from four devices in parallel to eight does not significantly speed up the driver.

7.4. Tri-state Output Buffer and I/O Buffer

The most common tri-state output pad is shown in Fig. 7-28. The last gate is the typical C^2MOS gate driven by an "enable" ($\overline{EN}$) signal which can put the output in tri-state. This gate suffers from charge sharing problems, as we saw in Section 4.3. Therefore, the interconnections must be laid out correctly. Moreover, the performance of the driver is degraded compared to that of an output pad, because the total channel length of the pull-up and pull-down stage increases due to the two devices in series. It helps to make the devices much wider than the corresponding devices in an output pad driving the same load, but the price in terms of power dissipation and area may be so high that this approach is seldom carried out. Therefore, tri-state pads with such topology feature larger delays than similar output pads driving identical loads.

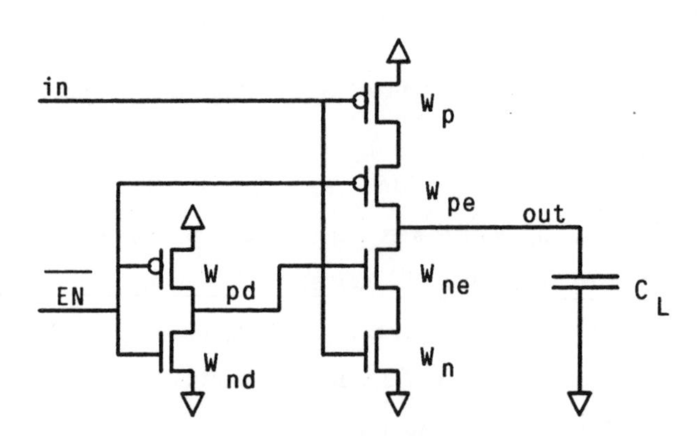

Figure 7-28: Tri-state output driver.

The enable signal needs current amplification to be able to control the driver. Moreover, both polarities shall be generated. This requires an extra stage, as shown in Fig. 7-28. If the enable signal comes from outside and has TTL logic levels, proper level shifting techniques — similar to what is done in TTL compatible input pads — are necessary. In fact, a TTL signal cannot directly drive the driver in Fig. 7-28, even though both polarities are available. This significantly affects the area of the pad.

The topology shown in Fig. 7-29 helps to solve the problem of long channel length. When $\overline{EN}$ is high, the output of the NOR gate is low, and the n-channel transistor is off. At the same time, the output of the NAND gate is high, and the p-channel device is off as well. The driver is in tri-state. When $\overline{EN}$ goes low, both NOR and NAND act as an inverter, and the input signal is sent to the gates of both transistors. This allows one to have a tri-state driver which does not suffer from having two devices in series in both the pull-up and pull-down stage.

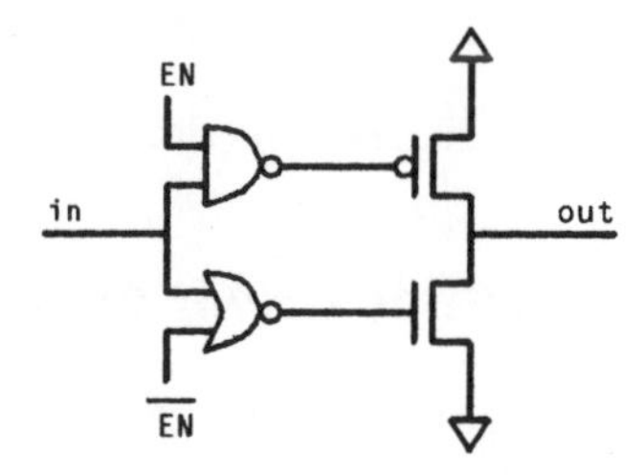

Figure 7-29: Tri-state pad featuring a driver with shorter total channel length.

The I/O pad circuit shown in Fig. 7-30 shows an output section conceptually similar to that of Fig. 7-29. When CNTRL is high, the pad behaves like an output pad; when CNTRL is low, the pad becomes an input pad. The input pad has some degree of protection given by the two drain junctions of the p-channel and n-channel devices in the driver. A series resistance is added to protect against overcurrents. This resistance — diffused or via interconnect, depending on speed vs. degree of protection requirements — is shared with the output section, and some performance degradation in the output swing has to be expected.

It is recommended that the two driver devices be kept as far apart as possible, both for latch-up avoidance purposes and because the two drain junctions are used as protection diodes when the structure works as an input pad. This can be done by laying out the p-channel and the n-channel transistors on opposite sides of the pad.

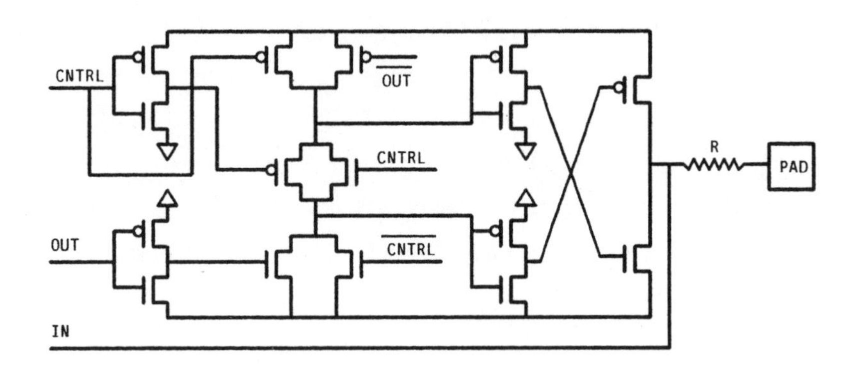

Figure 7-30: I/O pad circuit diagram.

7.5. Output Buffer and Bus Driver Design Optimization

Output pads can be designed according to two different strategies: delay minimization and delay*power minimization. Peak power minimization with constrained delay is an alternative to delay*power minimization. The rationale for having two different objective functions lies in the fact that delay minimization can be impossible, in the physical sense, when large loads are driven. In this case it is more appropriate — and useful, from the designer's point of view — to include peak power dissipation in the objective function; when the number of output pads is very high, delay*power minimization is commonly used. The next two sections will present two approaches for unconstrained delay minimization and peak power dissipation minimization in presence of delay constraints, respectively. Delay*power minimization is an unconstrained version of the latter and is useful when a specific requirement for the delay is lacking.

7.5.1. Unconstrained Delay Minimization

We want to use the model presented in Section 7.1 to derive an optimization algorithm which solves the following problem:

1. Given a known capacitive load;

2. given the size of the first inverter;

3. given all the fabrication process parameters;

4. *determine*:

   - The number of stages in the inverter chain, and

   - the size of each device in the chain,

   which minimize the delay.

This is a typical problem of constrained non-linear optimization (minimization) in which an analytical representation of the first and second partial derivatives of the objective function is not available. The objective function we are to minimize is the inverter chain delay, which is defined in this chapter as follows:

Definition: Inverter chain delay. The delay of an inverter chain is the average of both transitions, one corresponding to a rising input voltage and one corresponding to a falling input voltage. The delay of each transition is computed as the difference between the time in which the input signal starts to change and the time in which the output of the inverter chain reaches a voltage value chosen by the designer. This output voltage can be $V_{dd}/2$, when the design of a bus driver is carried out (for both transitions), or 2.7V and 0.4V, when a TTL compatible output pad driver is being designed.

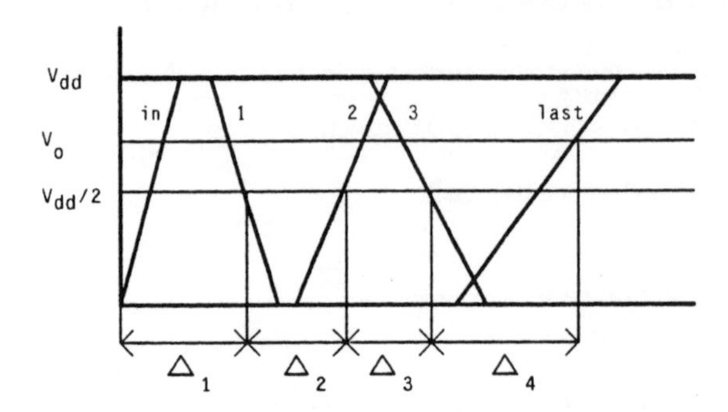

Figure 7-31: Different contributions in the inverter chain delay.

An example will show how the inverter chain delay is actually computed. Let us assume that the inverter chain has four stages. The delay $\Delta\uparrow$ for a rising input voltage is given by (see Fig. 7-31):

$$\Delta\uparrow = \Delta_1 + \Delta_2 + \Delta_3 + \Delta_4,$$

where:

- Δ_1 is the interval between the time in which the input starts to rise and the output voltage of the first stage reaches $V_{dd}/2$.

- Δ_2 is the interval between the time in which the output voltage of the first stage reaches $V_{dd}/2$ and the output voltage of the second stage reaches $V_{dd}/2$.

- Δ_3 is the interval between the time in which the output voltage of the second stage reaches $V_{dd}/2$ and the output voltage of the third stage reaches $V_{dd}/2$.

- Δ_4 is the interval between the time in which the output voltage of the third stage reaches $V_{dd}/2$ and the output voltage of the last stage reaches the output reference voltage chosen by the designer (for instance, 2.7V).

$\Delta\downarrow$ is expressed in the same way. Finally, we have:

$$\Delta = \frac{\Delta\uparrow + \Delta\downarrow}{2} \, .$$

In Section 7.1 we have presented a model which incorporates the information on the input voltage signal; in particular, it assumes that the input voltage is a ramp with constant slew-rate. We shall apply the same model to all the stages of the inverter chain. The output of each stage is therefore linearized, as shown in Fig. 7-32, its slew-rate is computed and the signal is fed to the next stage.

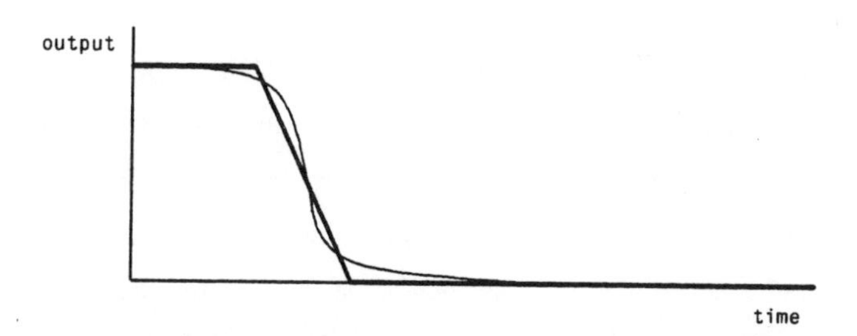

Figure 7-32: Inverter output and its linearized replica.

Finally the load that each stage sees is the sum of two contributions:

1. The input gate capacitance of the next stage (if the stage is not the driver) or the output load (if the stage is the driver).

2. The load of the two drain junction capacitances of the gate itself. This value is voltage dependent, as we saw in Section 2.2; however, for practical purposes, it is possible to consider this value as constant by computing it for an "average voltage value" (e.g., $V_{dd}/2$). We can define two values, C_{jp} and C_{jn}, which are the p-channel and n-channel device

drain-substrate junction capacitance *per unit area*, respectively.

Therefore, the i-th stage in the inverter chain will see a load C_{Li} equal to:

$$C_{Li} = (L_n W_{n(i+1)} + L_p W_{p(i+1)})C_{ox} + D(C_{jn} W_{n(i)} + C_{jp} W_{p(i)}),$$

where W is the channel width of the device, L is the channel length, D is the diffusion length — a parameter which depends on the layout and the design rules — and C_{ox} is the gate capacitance *per unit area.* Note that

$$(L_n W_{n(i+1)} + L_p W_{p(i+1)})C_{ox}$$

is substituted by C_L in the driver.

Among the various methods for non-linear optimization in the presence of non-differentiable objective functions, the variable metric algorithm [5] has been chosen. This algorithm is suitable for small and medium sized problems such as the one we are dealing with. The objective function we have to minimize is $\Delta(\mathbf{W})$, where **W** is the vector of channel widths of all the devices in the inverter chain. A four-stage inverter chain will have eight variables. Because it is unlikely that a chain will be longer than six or seven stages, the number of variables will never be larger than, say, fourteen, which qualifies our problem as a small to medium size optimization.

The algorithm consists of the following steps:

1. Choose an initial $\mathbf{W}_0$ and a positive definite symmetric matrix (for example, the identity matrix); set $\mathbf{S}_0 = -\mathbf{I}\mathbf{G}_0$, where $\mathbf{G}_j$ is the gradient of $\Delta(\mathbf{W})$ computed for $\mathbf{W} = \mathbf{W}_j$.

2. Compute $\mathbf{W}_{q+1} = \mathbf{W}_q + \alpha_q \mathbf{S}_q$, where α_q minimizes $\Delta(\mathbf{W}_q + \alpha_q \mathbf{S}_q)$.

3. Compute $\mathbf{H}_{q+1} = \mathbf{H}_q + \mathbf{M}_q + \mathbf{N}_q$, where:

$$Y_q = G_{q+1} - G_q ;$$

$$M_q = \alpha_q \frac{S_q S_q^T}{S_q^T Y_q} ;$$

$$N_q = -\frac{(H_q Y_q)(H_q Y_q)^T}{Y_q^T H_q Y_q} .$$

4. Compute $S_{q+1} = - H_{q+1} G_{q+1}$ and return to step 2.

The method has proved to converge reliably toward the minimum even when it starts quite far from it. The algorithm has also proved to be very sensitive to the α_q chosen to update the Hessian-type matrix H_q. Values of α_q that do not make $S^T_{i+1} G_{i+1}$ *very* close to zero will slow down the convergence. In this case, that is when α_q does not make $S^T_{i+1} G_{i+1}$ very close to zero, it is recommended not to update H_q and to re-use the old one. As long as the objective function decreases — and H_q is a positive definite symmetric matrix — the process will not diverge. Finding the value of α_q which minimizes $\Delta(X + S_q)$ is a problem of one-dimensional optimization. However, lacking a first derivative on α_q — the function does belong to C^1 but the derivative is very cumbersome to compute — methods such as quadratic interpolation have to be used, and these methods do not always find stationary points. Sometimes they get trapped into local minima; sometimes they provide values quite far from zeroing $S^T_{i+1} G_{i+1}$. Finally, since the objective function is non-separable (see, for instance, [12]), a slow rate of convergence is to be expected.

A result of this algorithm is presented in Fig. 7-33. The inverter chain is four stages long; the load is 50pF, and the first inverter has fixed size: $W_{n(1)}$ is 5μm and $W_{p(1)}$ is 10μm. μ_n and μ_p are 220cm^2/Vs and 100cm^2/Vs, respectively. The inverter chain output reference voltages have been fixed at 4.0V and 0.4V. The "best" inverter chain features a total delay of 28.08ns, while the SPICE simulation of the same chain — for the same output reference voltage — has produced a delay of 28.35ns. How well optimized is this result, in practical terms? The answer to this question is two-fold:

- From a mathematical point of view, a test such as

$$\frac{\mathbf{G}_q^T \mathbf{H}_q \mathbf{G}_q}{|\Delta(\mathbf{W}_q)|} < \varepsilon$$

  gives us the difference between the current iteration and the actual minimum and represents a measure of confidence for the optimality of the solution. ε is a small value chosen by the designer.

- Even though the above inequality tells us that the solution is very close to a global minimum, we have to remember that the entire algorithm simply optimizes *a model*, and, therefore, the result can only be as reliable as the model on which the algorithm is based. According to Section 7.1, we can consider the model to be reliable for a wide range of situations. Note that the model has been compared to SPICE simulations and not to physical circuits, which means that any inaccuracy in the SPICE model also affects the model presented.

We can conclude that the result is close to optimal, especially if we take into account the large variations of fabrication process parameters, which make a very precise optimization useless. This method can be used iteratively with different inverter chain lengths to find the "best" number of stages. Note that the usual specification calls for an inverting (or non-inverting) output buffer, and, therefore, only odd (even) lengths, such as 3, 5, 7 (2, 4, 6), are considered.

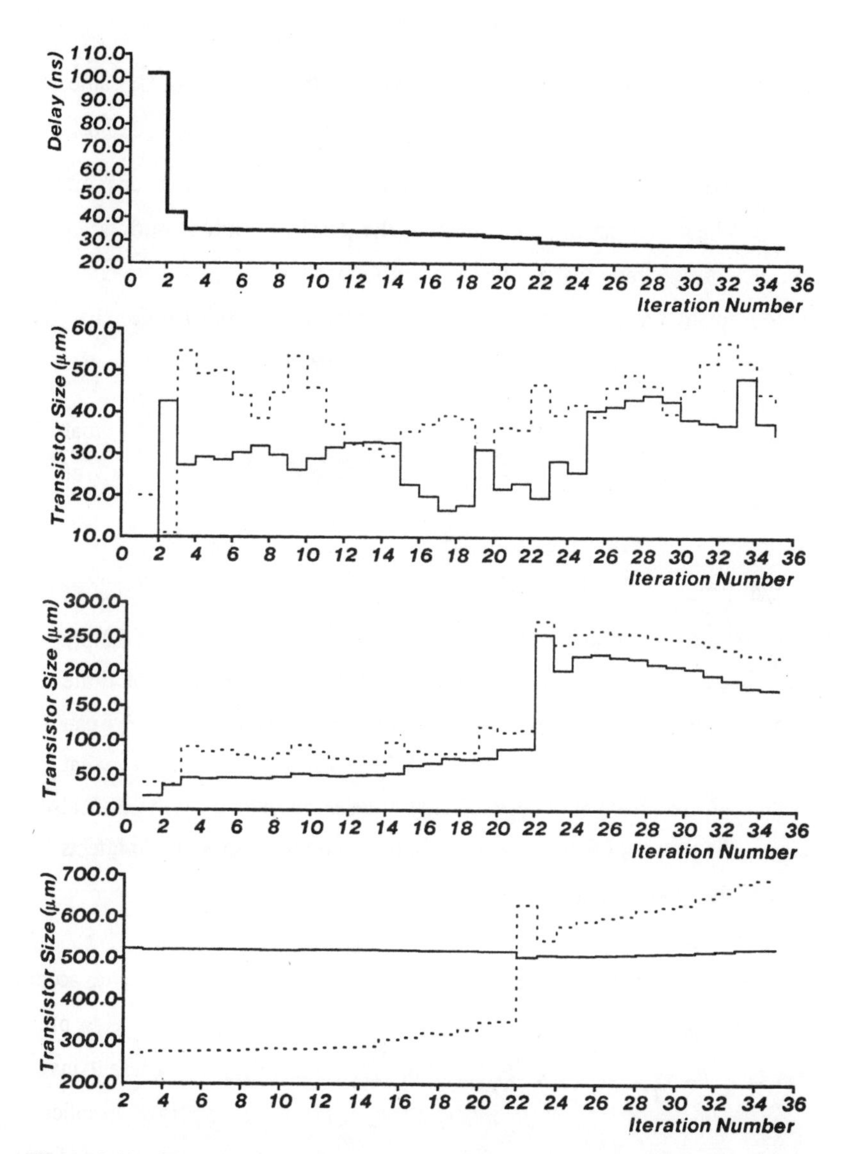

Figure 7-33: From top to bottom: inverter chain delay vs. number of iterations; second stage n-channel (solid) and p-channel (dots) device widths vs. number of iterations; third stage n-channel and p-channel device widths vs. number of iterations; last stage n-channel and p-channel device widths vs. number of iterations.

7.5.2. Constrained Delay Minimization

The optimization we dealt with in the previous section put much emphasis on speed. As we saw at the beginning of the chapter, speed is only one of many parameters which play an important role in the design of output buffers and bus drivers. Power dissipation is another very important parameter which actually can become dominant when a large number of drivers and/or output pads are necessary inside the chip. A typical problem that the designer has to face is the following:

Design an output buffer which provides a delay of less than Xns and minimizes power dissipation.

While the problem treated before was only theoretically one of constrained minimization — e.g., all the widths of the devices had to be non-negative — but it could be dealt with as an unconstrained problem, the constraints of the design problem presented above must be explicitly taken into account.

This problem has been studied in [18], by using a simplified model of the MOS transistor. The results of that study show that "minimum dissipation design" is fully specified by the following four equations:

$$\beta_N = \frac{C_N}{\tau_N} A\,; \tag{7-14}$$

$$A = \frac{1}{V_{dd} - V_{Tn}} \left[\frac{2V_{Tn}}{V_{dd} - V_{Tn}} + \ln \frac{2(V_{dd} - V_{Tn}) - V}{V} \right]; \tag{7-15}$$

$$\frac{\beta_{N-1}}{\beta_N} = \frac{A}{\tau\beta_\square}(1+b)C_{ox}L_{nN}[L_{nN}(1+\mu_r a^2) - \Delta L_{nN} - \mu_r a \Delta L_{nN}]\,; \tag{7-16}$$

$$\left[\frac{\beta_N}{\beta_{N-1}}\right]^N = \frac{C_N}{C_{ox}}\,. \tag{7-17}$$

N is the number of stages, β_N is the beta of the last stage (the beta of the n-channel and p-channel are identical and equal to β), threshold voltages are identical (in absolute value) and equal to V_{Tn}, and the rise-time τ_r and fall-time τ_f are identical and equal to τ. L_n and L_p are the channel lengths of the two devices, while ΔL_n and ΔL_p are the drawn channel lengths minus the effective channel lengths. μ_r is the mobility ratio, C_N is the load capacitance of the last stage, and V is the reference voltage at which the delay is computed. "a" and "b" are constants:

$$a = \frac{L_p}{L_n} ;$$

$$b = \frac{i-th\ stage\ parasitic\ (output)\ capacitance}{load\ capacitance\ of\ i-th\ stage} .$$

Finally, $\beta_\square$ is the beta of a device with equal length and width. Eqs. (7-14), (7-16), and (7-17) allow us to determine N, the number of stages, and the width of the transistors, by using the following formula:

$$W_n = (L_n - \Delta L_n)\frac{\beta_N}{\beta_{n\square}} , \tag{7-18}$$

and

$$W_p = \frac{\mu_n}{\mu_p} W_n . \tag{7-19}$$

The above formulae are used as follows. From Eq. (7-15) we compute $A \approx 0.8$. Then, we compute $\beta_{n\square}$ and $\beta_{p\square}$. C_N is the sum of the load capacitance and the parasitic output capacitance of the last stage. At this point, we determine β_N from Eq. (7-14). Given β_N, $\beta_{n\square}$, and $\beta_{p\square}$, we can determine W_{nN} and W_{pN} for the last stage from Eq. (7-18) and (7-19). The scaling factor is determined from Eq. (7-16). Finally, from Eq. (7-17) we derive the number of stages. The scaling factor allows us to determine the widths of all devices in the inverter chain, knowing the width of the p-channel (W_{pN}) and n-channel (W_{nN}) devices in the last stage.

The approach presented above suffers from a limited range of accuracy; for large loads, or slow devices — or both cases — the results can be incorrect. If we use the same parameters used in [18], but we increase the load to 50pF, and we allow a rise-time/fall-time of 10ns — two constraints that are quite usual — we have:

$$V_{Tn} = -V_{Tp} = 1V; \qquad V_{dd} = 5V; \qquad V = 0.05V_{dd};$$

$$\beta_{p\square} = 14\mu A/V^2; \qquad \beta_{n\square} = 42\mu A/V^2;$$

$$C_{ox} = 70nF/cm^2; \qquad C_o = 100fF; \qquad A \approx 1;$$

$$L_n = 2.5\mu m; \qquad \Delta L_n = 0.5\mu m; \qquad \Delta L_p = 1.0\mu m;$$

$$a = 3.0/2.5; \qquad b \approx 0.1;$$

$$C_N = C_L + C_{parasitic} \approx (1 + b)C_L = 55pF;$$

$$\beta_N = 5.5mA/V^2;$$

$$W_{nN} = 262\mu m; \; W_{pN} = 786\mu m;$$

$$\frac{\beta_N}{\beta_{N-1}} = 23.7 \rightarrow N \approx 2.$$

A two-stage inverter with the above parameters cannot discharge 50pF in 10ns.

When power dissipation has to be factored in the design, another approach is to minimize the delay*power product objective function. The two approaches, that is, minimization of the power dissipation objective function constrained by an assigned minimum acceptable delay, and unconstrained minimization of the delay*power product function, can coexist, because the first one requires an explicit specification of the delay, while the second does not.

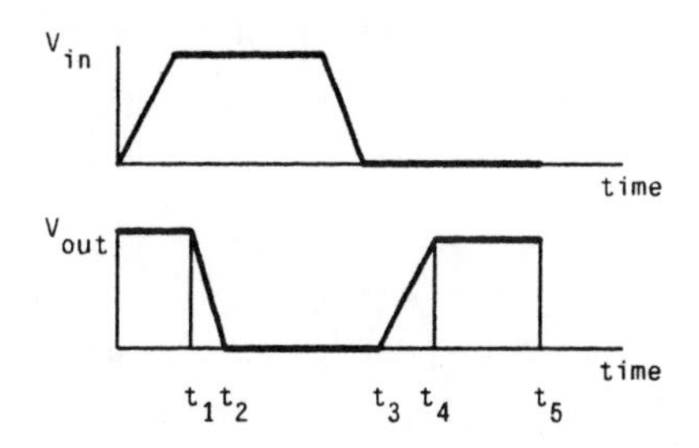

Figure 7-34: Input and output of CMOS inverter with parameters for power dissipation estimation.

Both problems can be solved in a similar fashion to the one in the previous section, that is, by using an accurate model for both delay and power dissipation and then solving the minimization problem by numerical methods. Moreover, this frees some unnecessary constraints, such as $V_{Tn} = |V_{Tp}|$, $\tau_r = \tau_f$, and so on, drastically improves accuracy, and widens the range of validity of the results.

Once an objective function for the power dissipation has been designed, minimizing power dissipation constrained by a pre-assigned maximum delay can be solved using a Lagrange technique. On the other hand, minimizing the power*delay product can be obtained with any technique for unconstrained minimization. The first problem can be restated in a formal way as follows:

$$\text{minimize } P_c = f(W_{n1}, \ldots, W_{nN}, W_{p1}, \ldots, W_{pN}, V_{dd}, \text{fab. proc. params.})$$

with the constraint:

$$\Delta = g(W_{n1}, \ldots, W_{nN}, W_{p1}, \ldots, W_{pN}, V_{dd}, \text{fab. proc. params.}) \leq X$$

We have already defined g(.) in the last section; let us derive f(.). To do that, we need to compute a relationship among power dissipation, geometrical parameters, and process parameters. Fig. 7-34 shows the input and output of an inverter with the parameters necessary in this computation. We can express the *short-circuit component of the power dissipation of the inverter* as follows:

$$P_{sc} = \int_{0}^{t_1} V_{dd} i(t)^* dt + \int_{t_1}^{t_2} \frac{V_{dd} t_2 - V_{dd} t}{t_2 - t_1} i(t) dt +$$

$$\int_{t_2}^{t_3} V_{ss} i(t) dt + \int_{t_3}^{t_4} \frac{V_{dd} t - V_{dd} t_3}{t_4 - t_3} + \int_{t_4}^{t_5} V_{dd}\, i(t)^*\, dt\ , \qquad (7\text{-}20)$$

where i(t) is the current flowing through the inverter. Assuming $V_{ss} = 0$ and $i(t)^* \approx 0$, only the two intervals t_1-t_2 and t_3-t_4 contribute to the power dissipation.

The other power dissipation component is the charge/discharge of energy accumulated in parasitic capacitors. In the case of an inverter chain, these capacitors are the input gate capacitance of the next inverter stage and the drain and source junction capacitances of the current inverter. The former depends on the size of the transistors in the next stage, as we already saw, while the latter depends on the junction area, that is, the transistor width and junction length. This contribution is:

$$P_{ds} = C_L V_{dd}^2 F\,. \qquad (7\text{-}21)$$

C_L is the total output capacitance, i.e., input gate capacitance of the next stage and junction capacitances, and F is the charging/discharging frequency. Total dynamic power dissipation depends on the frequency of the charge and discharge, in exactly the same way short-circuit dissipation depends on the switching frequency of the gate. A worst-case analysis could be carried out by setting:

$t_1 = 0;$

$$t_3 - t_2 = 0;$$

$$t_5 - t_4 = 0.$$

This means the output signal has a triangular shape. This analysis is very conservative, but can be useful to establish an upper bound for the power dissipation. Once Eq. (7-20) and Eq. (7-21) are expressed in terms of geometrical dimensions of all the devices in the inverter chain, either a delay*power minimization or a constrained minimization problem can be carried out. We will obtain widths and lengths of all the devices in the chain for which the objective function is optimized. We can conclude that:

$$f(.) = P_{ds} + P_{sc}.$$

7.6. Input Protection

Electrostatic discharge (ESD) is one of the principal causes of both soft and hard failures in electronic devices, and MOS technology is extremely vulnerable to ESD. The likelihood of ESD-related failures has been studied extensively, and it is acknowledged that device failures due to ESD happen often enough to recommend the adoption of proper precautions, both at the design level and when packages are handled [4].

ESD can take place through transfer of charges from the human body to the device, through transfer of charges from a charged device to a ground potential, or by static induction [3], when uncharged conductors are charged while grounded in the proximity of an electrostatic field. Masks, wafers, and IC's can suffer from this phenomenon.

The human body is a typical carrier of very high charges, accumulated during normal working activities; walking on a carpet can create voltages of up to 15,000V in the human body [2], and such charges can be transferred to electronic equipment via ESD. Moreover, ESD consists of very short pulses — in the hundred nanosecond range — and, therefore, power spikes of the order of kilowatts can be delivered to electronic circuits. Protection against ESD is very important in factories, too. Circuits must be protected against voltage pulses caused by personnel performing necessary operations — such as handling unpackaged dies, bonding, and so on. Note that while ESD of 400-500V is likely to induce failures in MOS circuits, the human body detects ESD only above 3000V; most of the time, therefore, ESD will damage a device without being detected by the laboratory personnel that are the source of the discharge.

Protection against ESD must be carried out as soon as the signal enters the device and is an essential part of the input pad circuitry. It is useful to divide ESD effects into *overvoltages* and *overcurrents*. In fact, both voltage-induced failures and current-induced failures are possible from ESD. There are protection schemes which better protect from one source of damage and are less effective for the other source, and combinations of both are used.

Current density, which is generated by the very high voltages and short duration time of ESD's, is more likely to affect the chip than the actual current value. This high current density can cause metallization or polysilicon vaporization of the wire which carries the input signal to the internal circuitry. We will analyze some protection schemes later; it suffices to say that all of them include an input resistor, which can be implemented in different ways. The connection between the pad and the input resistor can be a short metal line — likely to vaporize during ESD. If the resistor is a polysilicon line and is attached to the pad directly (that is, via a metal/polysilicon contact), either vaporization of the polysilicon line or metal/contact failure can take place. If the resistor is diffused and connected to the pad via a metal-diffusion contact, the contact can easily heat up and make the aluminum to alloy through the junction [17] (junction short). Electromigration, which results in open metal wires, also depends on current density. If this is too high, electrons push forward aluminum atoms and make them accumulate, thus devoiding a section of wire of aluminum atoms. This section eventually opens. Among other parameters, electromigration depends on interconnect temperature. Electromigration can also cause the opposite effect, that is wire shorts.

Run vaporization and junction short require different solutions. Run vaporization requires line width to be increased in order to sustain higher current density, while junction shorts require the use of many *evenly distributed* contacts. Fig. 7-35 shows an example. In both cases the same number of contacts have been used, but in Fig. 7-35(a) large current stress is put on contact A ("current crowding," [17]), and such a layout does not mitigate the problem of junction short. Fig. 7-35(b) shows the correct layout which evenly distributes the current stress to all the metal/diffusion contacts.

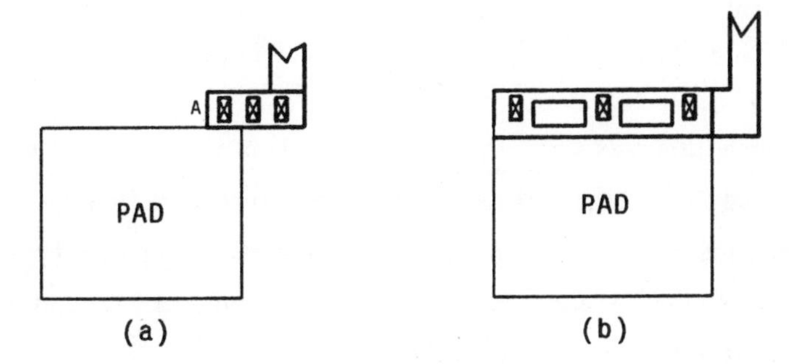

Figure 7-35: Incorrect layout of metal/diffusion contacts (a) and correct layout (b).

Voltage ESD overstresses, exceeding the breakdown voltage of the thin gate oxide, cause gate oxide pin-holes; another effect is metal/diffusion shorts. When diffused resistors underpass power lines, arcs may occur between diffusion and metal [17]. Although the breakdown voltage of glass separating the two conductors is roughly two orders of magnitude higher than the thin gate oxide, ESD can exceed such levels, and the zap will punch through the glass. It is generally a good design practice to avoid routing power lines above a diffused resistor. As far as the thin gate oxide is concerned, if we assume that the breakdown voltage of SiO_2 is roughly $80x10^6$V/cm, a 400Å gate oxide can sustain less than 32V. A voltage higher than 32V applied to the gate will likely cause gate pin-holes. Gate oxide used in current state-of-the-art technologies is about 200Å thick and, therefore, even more likely to break because of ESD.

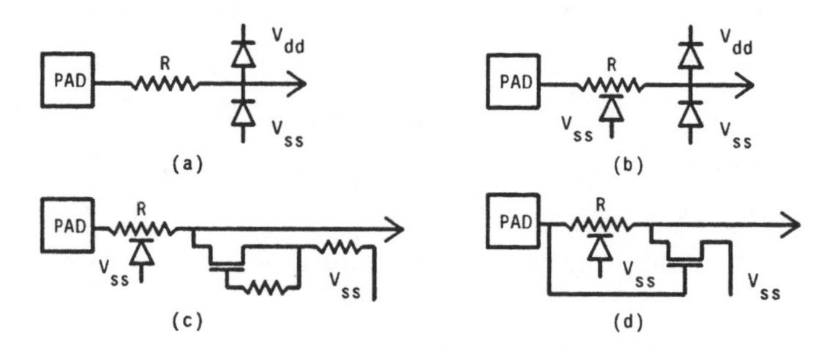

Figure 7-36: Different types of input protection circuits.

The input protection circuitry is commonly built around two concepts: spike filtering and clamp diodes. Some of these types of protection are shown in Fig. 7-36 [8]. The trade-off in designing protection circuits is between degree of protection and speed degradation, because most protection circuits introduce RC input structures in the signal path.

The most common protection circuitry is shown in Fig. 7-36 (a): a resistance R, associated with the capacitance of the reverse-biased clamp diodes, results in a low-pass filter which prevents sudden spikes in the input signal from reaching the internal logic. The two diodes, on the other hand, protect against overvoltages. If the input signal goes above (V_{dd} + 0.7) or below (V_{ss} − 0.7), they become forward-biased and route the signal either to V_{dd} or V_{ss}. Both diodes must be fabricated so that their reverse breakdown voltage is lower than the breakdown voltage of the thin gate oxide,

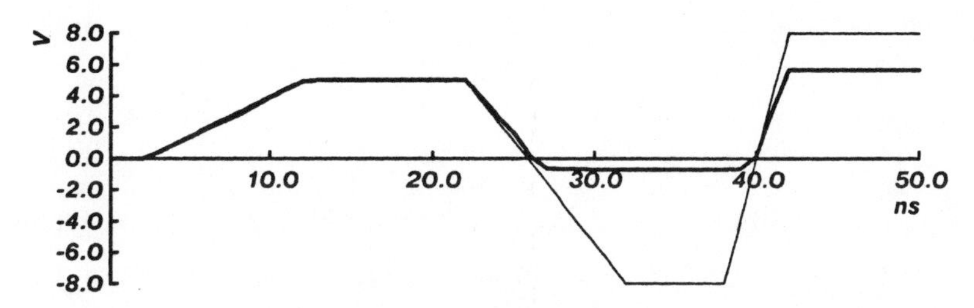

Figure 7-37: The input signal is "clipped" by the protection circuitry shown in Fig. 7-36 (a) when it goes below V_{ss} - 0.7 or above V_{dd} + 0.7.

to assure reliable protection. The resistor limits the current that reaches the internal circuitry and the protection diodes. Attention must be paid to assure that current density due to ESD does not cause vaporization, if the resistor is implemented with polysilicon. The resistor's value is between 1KΩ and 3KΩ. By using formulae available in the literature, it is possible to determine the size of the junction that, together with the series resistance, produces the required -3dB point in the frequency response of the low-pass filter. Typical -3db points range from 30MHz to 200MHz. Fig. 7-37 shows the effect of the protection on an input signal going outside the V_{dd}-V_{ss} range. The protection allows a voltage drop equal to the threshold voltage of the junctions (about 0.7V) either below V_{ss} or above V_{dd}.

Another approach consists of using diffused resistors which also include a clamp diode in the same structure. This technique is shown in Fig. 7-36(b). This circuit features higher current handling because of the two diodes in parallel. Fig. 7-36(c) shows a form of protection called "thin oxide punch-through transistor" [2]. The diffused region lies adjacent to another diffused area held at the substrate potential. When the diode junction is reverse-biased, the depletion region increases and for sufficient high bias voltage it extends to the nearby diffusion region. The approximate voltage which causes this effect according to [2, 16] is:

$$V_{pt} = \frac{L^2 q N_B}{\varepsilon_{Si}} ,$$

where L is the channel length.

Effective protection employs a *thick oxide transistor (metal gate)*, as shown in Fig. 7-36(d). The structure resembles the one discussed before, in that a series diffused resistor is placed near another diffusion which is electrically grounded. This area is then covered with thick oxide, and metal is deposited — forming a metal-gate thick-oxide transistor. The gate of the transistor(s) is then connected directly to the pad. For positive voltages higher than the threshold voltage of the thick oxide (e.g., 30V) the transistor will turn on and route to ground the overvoltage. Very high overvoltages can cause punch-through. Negative overvoltages are handled by the diode-resistor structure in parallel with the clamp diode.

Almost all protection circuits presented in the literature use a combination of the basic protection schemes described above [15, 6]. Finally, it is important to point out that layout is very critical in such circuits; thick-oxide based protection schemes use the parasitic bipolar transistors intrinsic in the structure to provide extra discharge paths, for instance. The characteristics of such bipolar transistors are *also* influenced by geometrical dimensions, and, consequently, the layout influences protection characteristics.

The realization of all the necessary components for protection circuits is now presented; both the layout and a cross-section are shown. The most important features of a protection resistor are current density handling and heat sinking capability. As far as the implementation of the resistor is concerned, two approaches are currently used, if the resistance does *not* have to be merged with a clamp diode:

1. Well-resistor (see Plate XII): the input signal is sent through the (p-type) well, via two metal-p+ diffusion contacts. Well resistance is usually about 1KΩ/□, which allows fabrication of very compact resistors. The

dashed area in the cross-section of Plate XII is the field oxide.

2. Poly-resistor (see Plate XIII): this approach is also used. Polysilicon has a resistance much lower than that of the well, and the resistor occupies more area. Usually, the polysilicon is covered with metal — preferably, second metal — for heat dissipation purposes.

The resistor-diode structure is implemented through a diffusion area, as shown (for a p-well technology) in Plate XIV. An n+ diffusion is formed inside the well. The signal flows through the metal interconnect on the left, goes through the diffusion (resistance), and continues through the metal on the right. The V_{ss} contact of the well is also shown in the plate. The well and the n+ diffusion form the diode.

Non-resistive diodes are implemented as shown in Plates XV and XVI (again, for a p-well technology). They are usually surrounded by guard rings for extra protection. The V_{ss} diode is shown in Plate XIV, and is formed on the secondary substrate (p-well). The junction consists of the p-well and the n+ diffusion, with a p+ guard ring tied to V_{ss}. The p-well is also connected to V_{ss}. The figure shows both the cross-section and top view of the diode implementation. The V_{dd} diode has a complementary structure, as shown in Plate XVI, and is implemented through an n-substrate-p+ junction, where the n-substrate is connected to V_{dd}. The guard ring is an n+ diffusion connected to V_{dd}.

The two diodes should be kept as far from each other as possible, either laying them out on opposite sides of the pad or, even better, on opposite corners of the pad. Note that the diode pair forms an SCR structure, and the use of guard rings is recommended. This also explains the need to put the two diodes as far apart as possible. The diodes should be sized so that they can sustain the maximum admissible current — which is a specification — due to ESD.

While the thin-oxide transistor protection technique employs a normal transistor, the thick-oxide transistor deserves some further comments. The structure of the transistor is shown in Plate XVII: a metal plate overlaps two regions which, in the case shown in Plate XVII, are p+ regions. This structure is *not self-aligned* like a polysilicon gate transistor, and it is necessary to draw a metal plate largely overlapping the two regions. Second, the source and drain should be separated significantly, to avoid p+ region to p+ region punch-through. There is another reason to separate them: in normal fabrication processes, the active areas are patterned and then doped accordingly. *The area below the polysilicon gate is not doped because the (thick) polysilicon protects the substrate*; this is no longer true in this case, because the transistor has a metal gate, rather than a polysilicon gate. If we patterned one active area, the net effect would be to have the entire area doped, resulting in one large junction. To physically separate the source from the drain, it is necessary to *explicitly* separate them in the layout.

7.7. Output Protection

A simple scheme of output protection is shown in Fig. 7-38 (a), where two diodes are connected in parallel with the driver's two devices. If the output signal goes above (V_{dd} + 0.7V), diode d1 becomes forward-biased. The output voltage is then "clipped" and cannot go above this value. Diode d2 operates in the same manner, but does so for signals going below (V_{ss} − 0.7V). Note that both devices are usually reverse-biased, and extra capacitance loads the driver. As usual, the trade-off is between current handling of the diode junction — which would require a large diode to be implemented to avoid junction punch-through — and output capacitance — which would require the diode to be as small as possible. The layout of diodes includes the usual protections, such as guard rings and separation between the two junctions for latchup avoidance. Fig. 7-38 (b) shows a slightly different approach, used in the 810 circuit — a CMOS chip which includes 1K RAM, programmable I/O ports and timers [19]. This is a standard output protection in CMOS circuits. The resistance acts as a current limiter and increases the RC constant at the output. This filters dangerous spikes, but, on the other hand, increases the output load of the driver.

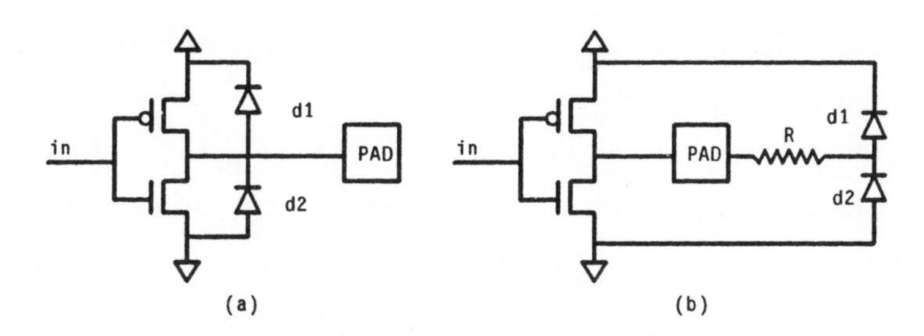

Figure 7-38: Two simple output protections.

7.8. Driving Large On-chip Loads

When both the transmitter and the receiver ends are on-chip, new opportunities become available to reduce the delay on long signal lines. Actually, different chips or sections of chips feature different optimizations schemes: for instance, output pads are a typical example of circuit with transmitter-only optimization. There is not much which the chip designer can do to optimize the receiver, because it is off-chip. On-chip clock drivers can be another example of transmitter-only optimization [10, 14].

Sense amplifiers in memory chips are a typical example of receiver-only optimization. In this case, the transmitter is the memory cell, which must be as small as possible. Buses are a typical example of transmitter *and* receiver optimization: both ends do not suffer from very hard constraints of area, and both ends are under the designer's control. In the previous sections dealing with output pads we have considered a transmitter-only optimization; in this section we present transmitter and receiver optimization. Receiver-only optimization, which is peculiar to memory design, has been dealt with in Section 6.4. Note that the design of an input pad could be considered as a case of receiver-only optimization, but other factors — for instance, insensitivity to noise and on-chip driving capability — play a more fundamental role.

One way of reducing the bus delay is to reduce the overall line capacitance by laying out shorter lines and/or using materials with low capacitance and/or resistance. These approaches concern layout and fabrication process engineers, respectively. As far as the logic designer is concerned, increasing the driving capability of the transmitter end while keeping the input capacitance of the receiver end at the smallest possible value is of little help, if real improvement is seeked. One interesting opportunity in CMOS is to use signals with low voltage swing.

To simplify matters, let us assume that the receiving end is a balanced inverter. The inverter threshold is roughly $V_{dd}/2$. Let us also represent the line capacitance with a single capacitor. It is evident that there is no need to charge this capacitor (that is, the line) to V_{dd} and discharge it to V_{ss}. Theoretically, one could charge the capacitor to $V_{dd}/2 + \kappa$ and discharge it to $V_{dd}/2 - \kappa$, where κ is a small positive value. This would be sufficient to switch the output of the receiving inverter and would drastically decrease the delay. Dynamic power dissipation would decrease as well, but at the expense of an increase in static power dissipation. While the capacitor is no longer charged to V_{dd} or discharged to V_{ss}, the receiving gate always has the devices operate in saturation, or close to it. But we can safely assume that the gate is minimum size.

This extreme approach would work for the ideal case, because noise, temperature, and fabrication process parameter variations, etc., create either glitches or a shift in the inverter threshold, or both. The receiving inverter would switch erroneously as a consequence of even the smallest noise voltage induced on the line by adjacent circuitry. In other words, the receiver behaves almost randomly.

In what follows, we present some circuits that are based, in different ways, on the concept of reducing voltage swing on long lines. We already saw one example in Section 4.1.4, where CVS logic was presented. A significant speed-up in this logic is obtained by reducing output voltage swing of the gate (transmitter), while keeping some devices close to their threshold voltage (receiver).

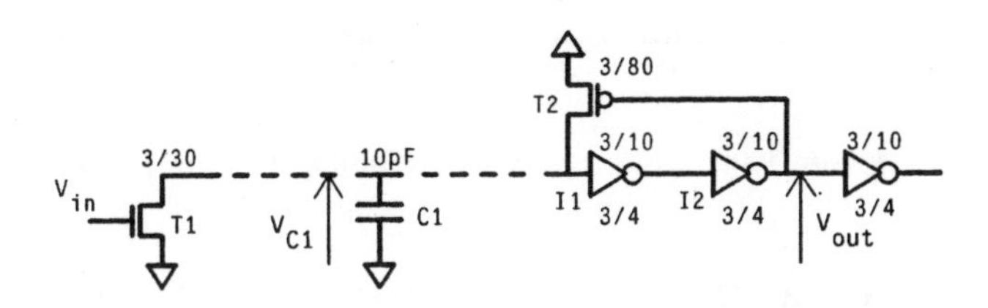

Figure 7-39: Circuit scheme which provides short delay when driving long, highly capacitive lines.

Fig. 7-39 [1] shows a static circuit which can drive large on-chip loads with very short delay. When V_{in} is low, the transistor T1 does not conduct, while the transistor T2 precharges the line. When the voltage across C1 goes above $V_{dd}/2 + \kappa$ (κ depends on fabrication process parameters, temperature, pull-up|pull-down ratio of the inverter I1, etc.), the input voltage to the gate of T2 goes high and turns T2 off. C1 remains charged to $V_{dd}/2 + \kappa$, rather than being pulled-up to V_{dd}. When V_{in} goes high, T1 turns on, and the capacitor is discharged. At the same time, T2 turns on. T1 and T2 represent a voltage divider, and, therefore, a proper relationship between the size of T1 and T2 must be used to limit the discharge of C1. With proper ratio between T1 and T2 it is possible to reduce the voltage swing across C1 drastically, as Fig. 7-40 shows. This circuit allows us to drive highly capacitive lines with very short delay. The bold curve in Fig. 7-40 shows that the delay is about 8ns (C1 = 10pF). The same capacitance driven by a single inverter with an n-channel device with the same size of T1 and with a p-channel device with the same size of T2 would drive the same capacitance with a delay of about 16ns, that is twice as long.

Compared to an inverter chain, the circuit of Fig. 7-39 decreases both the delay *and*

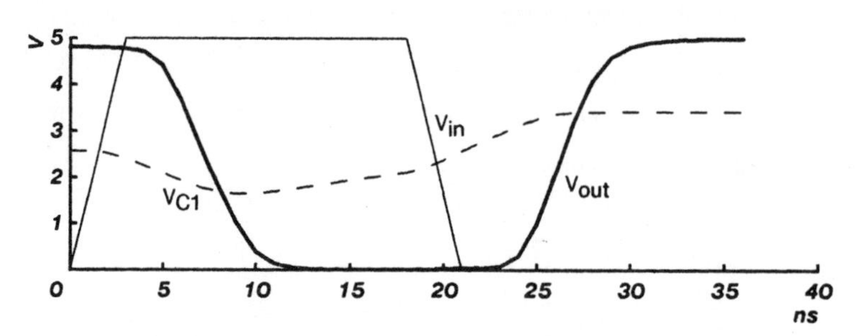

Figure 7-40: Voltage waveforms for the circuit in Fig. 7-39.

the area significantly. The only drawback is its static power dissipation, because both T1 and T2 are conducting at the same time. However, both T1 and T2 need not be very wide, and the amount of static power dissipation is not a major concern. Moreover, clocking schemes can be used to eliminate the static component of the power dissipation and make the T1-T2 pair ratioless [1]. Finally, the circuit in Fig. 7-39 would seem to have a poor noise margin. In practice, this is not true, if some precautions are taken. Noise induction from capacitive or inductive coupling between adjacent lines is proportional to the voltage swing on these lines, and it is not a major problem in this case, because we assume that all the lines are driven by the same circuit topology. However, all these lines should be laid out far apart from *other* sources of noise, typically power lines, output drivers, clock drivers, and so on. If these precautions are taken, the reduced voltage swing on the lines does not affect overall noise margin.

One way of reducing the influence of adjacent circuitry or power lines on reduced-swing interconnections is to use a differential configuration, as is done in CVSL. In this case, the symmetry of the circuit guarantees that both lines carrying opposite logic values are affected by noise in the same way. A differential receiver, with its high common mode signal rejection ratio, can assure reliable and fast operation even in the presence of noise. The drawback of this approach is an increase in area, because two lines are required to carry each signal.

References

[1] Bakoglu, H.B. and J.D. Meindl.
CMOS Driver and Receiver Circuits for Reduced Interconnection Delays.
In *Proc. of the 1985 International Symposium on VLSI Technology, Systems and Applications - Taiwan*, pages 171-175. May, 1985.

[2] Bhar, T.N. and E.J. McMahon.
Electrostatic Discharge and Control.
Hayden Book Co., 1983.

[3] Chemelli, R.G., B.A. Unger and P.R. Bossard.
ESD By Static Induction.
In *Proc. EOS/ESD Symposium*, pages 29-36. Reliability Analysis Center, September, 1983.

[4] Dangelmayer, G.T.
ESD - How Often Does It Happen?
In *Proc. EOS/ESD Symposium*, pages 1-5. Reliability Analysis Center, September, 1983.

[5] Richard L. Fox.
Optimization Methods for Engineering Design.
Addison-Wesley Publishing Co., 1971.

[6] Fujishin, E. *et al.*.
Optimized ESD Protection Circuits for High-Speed MOS/VLSI.
IEEE Journal of Solid-State Circuits SC-20(2):594-596, April, 1985.

[7] Kang, S.M.
A Design of CMOS Polycells for LSI Circuits.
IEEE Trans. on Circuits and Systems CAS-28(8):838-843, August, 1981.

[8] Keller, J.K.
Protection of MOS Integrated Circuits from Destruction by Electrostatic Discharge.
In *Proc. EOS/ESD Symposium*, pages 73-80. September, 1980.

[9] C.M. Lee and H. Soukup.
An Algorithm for CMOS Timing and Area Optimization.
IEEE Journal of Solid-State Circuits SC-19(5):781-787, October, 1984.

[10] Mohsen, A. *et al.*.
The Design and Performance of CMOS 256K Bit DRAM Devices.
IEEE Journal of Solid-State Circuits SC-19(5):610-618, October, 1984.

[11] Nagel, L.W.
SPICE 2: A computer program to simulate semiconductor circuits.
Technical Report ERL-M510, Electronics Research Lab., University of California, Berkeley, May, 1975.

[12] Donald A. Pierre.
Optimization Theory with Applications.
John Wiley & Sons, Inc., 1969.

[13] Sakurai, T. and T. Iizuka.
Gate Electrode RC Delay Effects in VLSI's.
IEEE Journal of Solid-State Circuits SC-20(1):290-294, February, 1985.

[14] Shimohigashi, K. *et al.*.
An n-Well CMOS Dynamic RAM.
IEEE Journal of Solid-State Circuits SC-17(2):344-348, April, 1982.

[15] Soden, J.M., H.D. Stewart and R.A. Pastorek.
ESD Evaluation of Radiation-hardened, High Reliability CMOS and MNOS ICs.
In *Proc. EOS/ESD Symposium*, pages 134-146. Reliability Analysis Center, September, 1983.

[16] Sze, S.M.
Physics of Semiconductor Devices.
John Wiley & Sons, New York, 1969.

[17] Turner, T.E. and S. Morris.
Electrostatic Sensitivity of Various Input Protection Networks.
In *Proc. EOS/ESD Symposium*, pages 95-103. September, 1980.

[18] Veendrick, H.J.M.
Short-Circuit Dissipation of Static CMOS Circuitry and Its Impact on the Design of Buffer Circuits.
IEEE Journal of Solid State Circuits SC-19(4):468-473, August, 1984.

[19] Wilson, D.
ESD Sensitivity of Complex ICs.
In *Proc. EOS/ESD Symposium*, pages 128-133. Reliability Analysis Center, September, 1983.

Appendix A

Layout

Poor layout affects the area of the chip and its performance (speed, power dissipation, etc.), and — ultimately — it can affect the correct behavior of the chip. Although layout is mainly driven by two-dimensional, geometrical constraints — such as aspect ratio of the circuitry and physical location of the terminals — performance issues and considerations on design reliability influence layout, as well. Poor layout can lead to very long interconnections which slow down the chip. Laying out power lines on diffused resistors may generate arcs which permanently short the two layers.

Given the large number of parameters involved, it is impossible to present layout techniques in a well-structured, "scientific" way. Usually, layout techniques are presented as a list of tips that, in *most* situations — but not in *all* situations — are indeed effective. This chapter makes no exception to this: some general techniques of layout are presented, and emphasis is placed on their range of applicability. Layout techniques which increase the chip's robustness against latchup are discussed. These techniques have to be used in bulk CMOS only, because SOS CMOS is not affected by latchup. This immunity influences the layout in significant ways: for instance, butting n- and p-type diffusion regions together is acceptable practice in SOS — as a result, a diode is formed — while it is very dangerous in bulk.

Structured layout methodologies are briefly presented. These methodologies aim to increase productivity, which can be defined as the number of laid out devices per man/day, and decrease the chance of mistakes.

Plates referred to in this appendix are found in the PLATES section located between pages 172 and 173.

Finally, the layout of power and ground lines is dealt with. This topic requires a dedicated treatment because of its increasing importance. Modern chips have hundreds of thousands of transistors and several I/O signals. Even if a low static power dissipation technology — such as CMOS — is used, the frequency of operation can be so high (e.g., some state-of-the art chips are being aimed at operation frequencies of 40MHz), that power dissipation can reach 7W or more.

The high power dissipation is due mainly to the use of an increased number of I/O connections — 200+ pin chips will be common products in the near future. The input or output drivers contribute heavily to the total power dissipation (sometimes up to 50%). Distributing power and ground to provide the necessary current in all the areas of the chip, and maintaining the noise margin and chip reliability at acceptable levels can become a real challenge.

A.1. General Considerations on Layout

Until more powerful software systems replace the layout engineer, human ingenuity will still be indispensable for state-of-the-art, full-custom chip layout. We should, therefore, distinguish between *computer-assisted layout* and *computer-generated layout*. We concentrate our attention on the former; the latter — gate-array and standard-cell in particular — is aimed more at the fast turn-around, low- and medium-end of the digital market. This does not imply any technical judgement on the above methodologies; in fact, they are capturing a significant share of the market, and this share is drastically increasing.

Modern layout techniques should have — at least partially — the following features:

1. Minimization of the number of "drawn" devices.

2. "Regularity," which increases the productivity and decreases the chance

of mistakes.

3. Length minimization of conductors in critical paths.

Minimization of drawn devices is accomplished in two steps. First, the *logic design* must include as many identical cells as possible. This is a logic design issue, not a layout issue. Nonetheless, this is the necessary condition to achieve a minimization of drawn devices. The second step consists of *planning* the layout in such a way that all the logically identical cells share the same layout characteristics — that is, aspect ratio, transistor sizes, and location of signal and power/ground terminals. This is a very complex task and is not always possible, especially if area minimization is a critical requirement. To make an example, the same full-adder used as a building-block for an ALU in one part of the circuit might have less than optimal aspect ratio and/or terminal locations in another part of the circuit, where it is used for a different purpose (e.g., in the last row of a multiplier). Careful *floor-planning* can help to solve this problem.

Floor-planning is an essential step in any serious design. Industries have databases of building-blocks (e.g., ROM's, RAM's, multipliers, adders, latches), and before the actual design takes place it is possible to produce a floor-plan which is very close to that of the final product. A feasibility study can then assess whether the circuitry will fit in a single chip, which paths are the longest, etc.. This is not usually true in a university environment, where such databases are unlikely to be available. This leads to a typical chicken-and-egg problem: to come up with a good floor-plan, one needs to know the approximate size and aspect ratio of the basic cells. To know this, one should design them first. After designing them, the floor-plan might indicate that some of the cells have the wrong aspect ratio and/or terminal locations and have to be designed and/or laid out again. Doing the floor-plan with just a vague idea about the geometrical characteristics of the cells can lead to geometrical requirements on some cells that, when one attempts to lay them out, turn out to be impossible to meet. There is no easy solution to this problem, unless the designer has already had previous experience with the design of the particular building-blocks and, therefore, knows approximately the most likely size and aspect ratio (given the number of inputs, for instance).

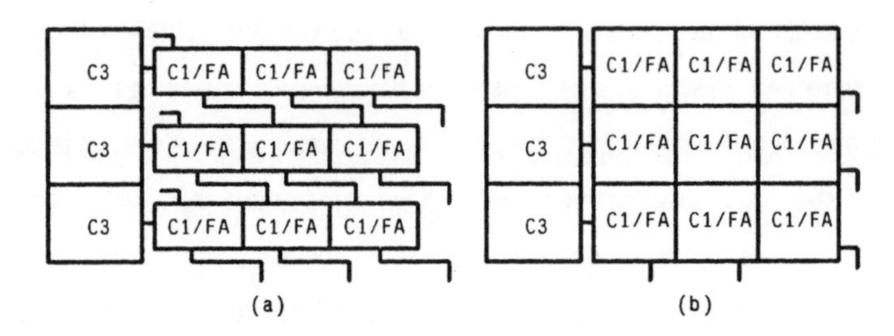

Figure A-1: A modified Booth multiplier with (a) optimized layout of the carry-save adder and (b) without. In both cases, the area of the cell multiplier does not change.

An important concept in correct layout techniques is pitch-matching. Let us consider the Booth multiplier presented in Section 6.6, Fig. 6-44. Laying out the carry-save adder and decode logic (C1/FA) first would be a mistake. Before laying out this circuitry, the C3 logic should be laid out; its height is likely to be higher than that of the carry-save adder and decode logic inside the array. It is a worthless effort to minimize the area of the C1/FA logic if the height of the C3 logic does not match. The multipliers in Fig. A-1(a) and (b) have the same area; Fig. A-1(a) shows the case where the carry-save adder and decode logic layout has been highly optimized. All the time spent on this optimization has been wasted.

The example above dealt with *vertical pitch-matching. Horizontal pitch-matching* is shown in Fig. A-2(a). The layout shown in Fig. A-2(b) occupies more area and requires cumbersome routing, without providing us with any real performance improvement. However, like most of the layout techniques, pitch-matching is not

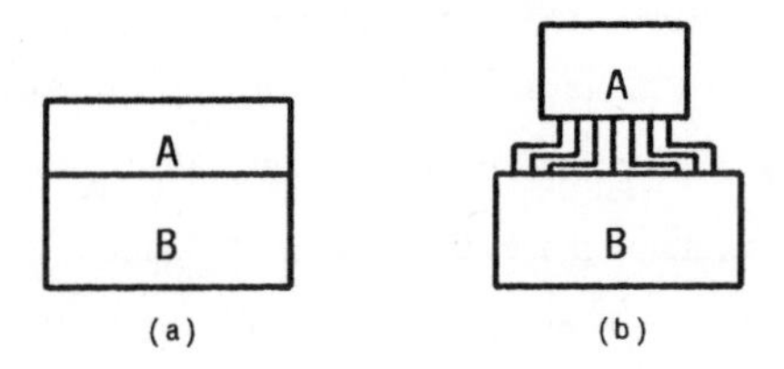

Figure A-2: "Regular" layout techniques call for (a), not (b).

always a valid, universal rule. To present a counter-example, let us now consider a floating-point adder. Without going into the details of floating-point arithmetic, it suffices to say that the first steps in adding two floating-point numbers together are "exponent comparison" and "mantissa normalization". For the sake of simplicity, let us assume that both exponents are positive and the two floating-point numbers are normalized. The smaller exponent is subtracted from the larger one, producing a result P. The mantissa of the smaller exponent is then right-shifted P places and its exponent incremented by P. We say that the two mantissas are now "aligned" and can be added together by a parallel adder.

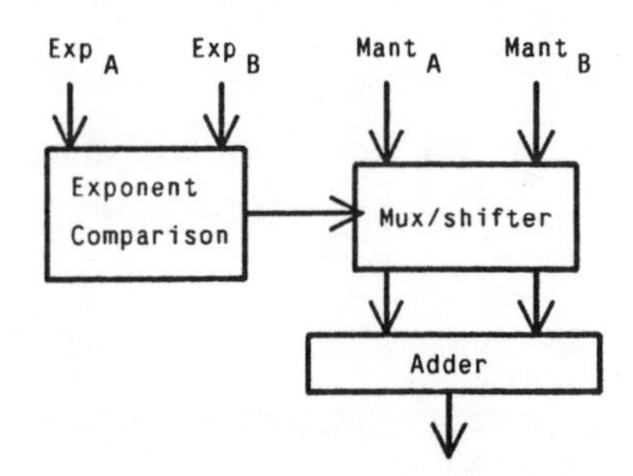

Figure A-3: Block-scheme of the input stage of a floating-point adder.

Fig. A-3 shows the logic blocks that are necessary to compare the exponents and to align and add the mantissas. The mux/shifter includes an input multiplexer, which inputs one mantissa to the shifter, while the other mantissa goes unchanged to the input of the parallel adder. The interconnection between the exponent comparison and the mux/shifter has six lines (32-bit floating point is assumed): one line indicates which mantissa has to be passed through (either $Mant_A$ or $Mant_B$), while the other five lines indicate how many places the other mantissa (either $Mant_B$ or $Mant_A$) has to be right-shifted. Since each mantissa has 23 bits, the largest possible shift is 22 — more than 22 shifts would zero the mantissa — that is, 5 bits are necessary. How should these blocks be laid out? The parallel adder at the bottom is much wider than the mux/shifter. If we follow the approach recommended before and shown in Fig. A-2(a), we have to widen the mux/shifter so that its width matches the width of the parallel-adder.

Although this approach reduces layout time, it might not give us the best performance, and the approach shown in Fig. A-2(b) is preferred in certain cases. If the mux/shifter had to be designed without the constraint of matching the width of the parallel adder, its width could well be about five times smaller than that of the parallel adder. Widening the mux/shifter makes some of the six lines coming from the exponent comparison block five times longer. Each line would be $1600\mu m$ instead of $300\mu m$ long ($2\mu m$ technology is assumed). This longer interconnection length requires much wider drivers in the exponent comparison to achieve the same speed. This increases the area, the power dissipation of the chip, and so on. However, this problem is not so serious in pipelined floating-point adders, where one can afford longer delay.

Another example is presented to illustrate the importance of floor-plan. The design of a high-performance, 16x16, 2-bit wide cross-bar chip. Each output can select one out of 16 possible inputs. There are 32 input data pads, 32 output data pads, and 16x4=64 input control pads. The cross-bar consists of a 16x16 grid of identical cells. Each column of 16 cells differs from the other columns only in a different decode logic. The layout of all 256 cells is identical, and the different decode logics are simply obtained by placing a few contacts in different places. A first-cut logic design

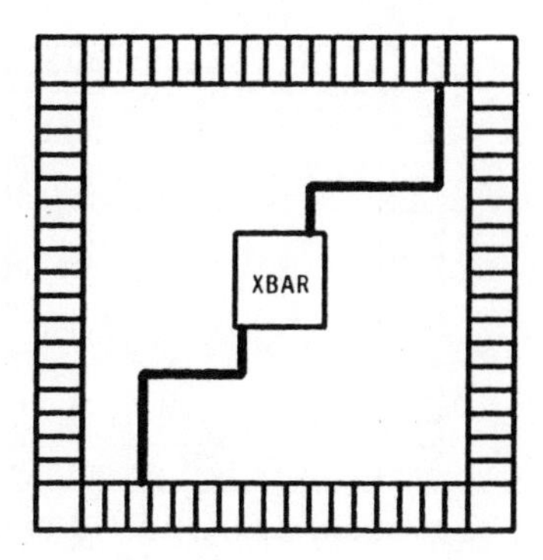

Figure A-4: 16x16, 2-bit wide cross-bar (XBAR) chip with a 138 pad frame.

and layout shows that a 16x16 grid can fit into a 1.5x1.5mm^2 (2μm technology is assumed). However, we have 32 output pads and 96 input pads (32 data and 64 control lines). Given 10 pads for V_{dd} and V_{ss}, the total number of pads is 138 and the pad frame is approximately 7.5x7.5mm^2. The packaging and bonding technology does not allow us different placement of the pads (e.g., staggered). The situation at this point is as depicted in Fig. A-4, where XBAR is the 16x16, 2-bit wide cross-bar logic. The signal has to be carried from the input pad to the cross-bar (bold top line) and from the cross-bar to the output pad (bold bottom line). Each line is approximately 3mm long. In order to drive such a long line, the output buffer of *each* cell in the 16x16 switch has to be drastically increased. Then, the logic inside the cell which precedes the driver must be increased as well. The final result is shown in Fig. A-5. The 16x16 cross-bar grid now occupies the entire available area inside the pad frame. The preliminary floor-planning of the cell saved time, because it prevented us from sizing and laying out a totally wrong circuit. Note that, given the extremely simple structure of the chip, the floor-plan took just a few hours, but saved many days of worthless design and layout.

Buses and other long interconnections must be laid out in metal. Under no

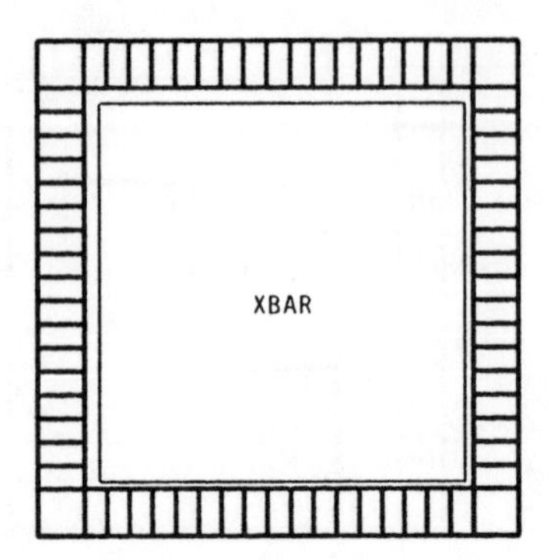

Figure A-5: The final floor-plan of the cross-bar chip.

circumstances should polysilicon be used for these lines. When the layout of a cell is carried out, metal and polysilicon lines are laid out perpendicularly, because this makes the layout easier, and — when possible — polysilicon lines go through the shortest dimension of the cell. Minimizing the length of polysilicon lines is a critical requirement in any layout. When two metal layers are available, the second (top) metal should be used for power and ground. The minimum width of the second-metal is always wider than that of the first-metal in order to minimize mechanical stresses on the material. Moreover, a second-metal to polysilicon contact is seldom available, and a connection between second-metal and polysilicon has to go through an intermediate second-metal to first-metal contact. Usually this contact and the first-metal to polysilicon contact cannot be stacked and must be kept at appropriate distance. Common practice is to use first-metal and polysilicon for signals and second-metal for V_{dd} and V_{ss}. This also has the benefit of reducing electromigration effects, because the second metal is closer to the chip surface than the first one, and, therefore, heat dissipation can be more effective.

The size, geometry, and placement of wells affects the area of bulk CMOS chips significantly, and it is important to optimize the use of wells. Because each well should be tied to the proper potential with substrate contacts, laying out small and sparse wells can waste a significant amount of real-estate. A few, large wells should be laid out. If two wells of the same type are contiguous but not connected together, they should be merged. Therefore, one should avoid alternating substrates of different types. A simple example is shown in Plate X, where the NAND gates are mirrored to reduce the total number of separate substrates.

As we saw in Section 7.3, wide devices need careful layout because of gate electrode RC delay. Note that even a 100μm-wide device can suffer from this phenomenon. Moreover, the diffusion resistance can slow down the device if few metal contacts are used to connect, say, the source to V_{dd}, and the device is very wide. Different layout techniques can be used depending on the transistor width. Wide transistors can be laid out as shown in Plate XVIII (a). Both source and drain diffusion resistances are decreased drastically by using metal and several diffusion-metal contacts. The source resistance of n-channel (or p-channel) transistors should be kept as low as possible [1], otherwise significant voltage drop takes place. The voltage drop is proportional to the current that the device can sink (source), and increases as the channel width increases. An insufficient number of source contacts increases the source to V_{ss} (V_{dd}) resistance, which worsens the noise margin and creates body effects that slow down the device, as we saw in Section 6.3. Moreover, this also increases the possibility of latchup.

Note that extensive use of contacts increases the diffusion area and, therefore, the junction capacitance. The increase in capacitance of the terminal connected to either V_{dd} or V_{ss} does not usually represent a problem. On the other hand, the increase in the capacitance associated with the output terminal of the gate (e.g., the two drain capacitances in an inverter) is a problem. It is usually preferred to increase the capacitance of the drain junction slightly while keeping its resistance as small as possible. In fact, this resistance contributes to the RC output constant, and the capacitance contribution can be very large if the gate has a large fan-out. As shown in Plate XVIII (b), the actual device width should be computed by considering the linear segments of the gate, that is, the corners do not contribute to total device width. A less conservative, yet common, approach consists of computing the actual gate width by averaging the outer width and the inner width. Finally, very wide devices are laid out as suggested in Section 7.3, that is, as many narrower devices in parallel, as shown in Plate XIX.

A.2. Layout Methodologies for Latchup Avoidance

Latchup hardening can be accomplished through fabrication processes, as we saw in Section 3.4. However, proper layout techniques should always be adopted to further enhance the chip's robustness against latchup. Floating substrates are more likely to favor latchup occurrence. Consequently, it is important that *every well* be tied to the respective potential by substrate contacts, unless the circuit explicitly requires floating wells, as in some sense amplifiers schemes presented in Section 6.4.3. P-well is connected to V_{ss}, while n-well is connected to V_{dd}. A p-well (n-well) substrate contact is a p-type (n-type) diffusion connected through metal to V_{ss} (V_{dd}). To keep every substrate area at a well-defined potential, many substrate contacts should be used in the primary substrate or in large wells. The distance between substrate contacts of the same type should never exceed 100μm. Latchup is favored by high well resistance; a way to decrease it is to fill the unused portion of the well with diffusion of the same type. The diffusion is covered by metal and connected to it. The metal layer is then tied to the proper potential.

Another layout protection against latchup consists of using *dummy collectors*. These structures act as "vacuum-cleaners," collecting carriers which, therefore, do not reach the areas of the circuit where logic gates are laid out. An n-type dummy collector is an n+ diffusion in n-substrate connected to V_{dd} and surrounding a p-type substrate. A p-type dummy collector is a p+ diffusion in p-substrate connected to V_{ss} surrounding an n-type substrate. Separation between n-channel and p-channel devices helps reduce latchup occurrence, although at the expense of decreased circuit density.

As far as the use of guard rings is concerned, some interesting results are reported in [5], where different layout structures for both bulk and epitaxial substrates have been studied. The fabrication process was n-well. The most robust structure is shown in Fig. A-6 and is now briefly discussed.

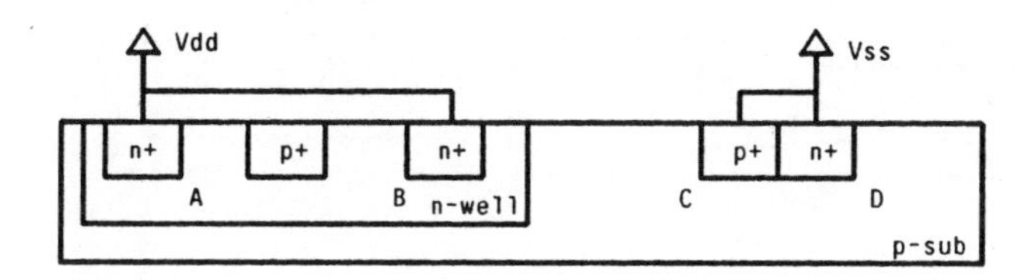

Figure A-6: Cross-section of a structure which features significant robustness against latchup.

Diffusion B, which together with diffusion A closes the guard ring around p+, has proved to be effective only in conjunction with substrate bias. However, diffusion B can help reduce the spreading resistance inside the well in actual layouts. Therefore, some benefit might be expected. However, the most important discovery is the relationship between diffusions C and D. Greater latchup hardening has been found by placing C and D as shown in Fig. A-6, with *the p+ diffusion closer than the n+ diffusion to the well edge*. Moreover, it is also essential to merge the two diffusions together. The rule that can be derived from this result is that *no n+ diffusion should be laid out closer to the well edge than the p+ diffusion.*

The above results give us some guidelines to lay out cells that share power lines, as it happens in actual layout. To simplify the discussion, let us consider two inverters. We have two possibilities: one is to make them share one power line, the other is to make them have separate power lines. From the point of view of circuit density, the first approach is preferable, because it leads to more compact layouts. The two approaches are shown, in cross-section, in Fig. A-7(a) and (b).

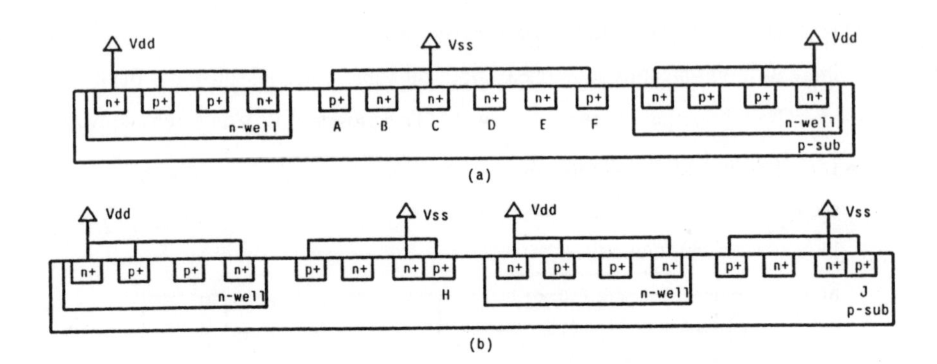

Figure A-7: Two inverters sharing one power line (a) and with separate power lines (b).

As Fig. A-7(a) shows, it is impossible to butt A-B and E-F together, because B and E are the drain junctions of the n-channel transistors. However, A and F should be as close as possible to B and E, respectively. Note, also, that there is no reason to insert a p+ diffusion between C and D. Fig. A-7(b) shows the second layout example. In this case, extra p+ diffusions (H and J) must be placed because of the possibly dangerous interaction between the pull-down structure of the leftmost inverter and the pull-up structure of the rightmost inverter. The layout takes more area, and we have to protect each pull-up or pull-down section from latchup, which can come now from both sides. Therefore, laying out gates as shown in Fig. A-7(a) not only occupies less area, but decreases the number of structures that, interacting together, can cause latchup. Finally, substrate bias is claimed to significantly decrease the likelihood of latchup [5]; epitaxial wafers also show much higher robustness against latchup.

Noise plays an important role in triggering latchup. In order to minimize it, some precautions should be taken; mostly, they involve the design and layout of I-O pads. First, pads should be separated from internal circuitry whenever possible. Then, power lines should be dedicated to the pad section with their own V_{dd} and V_{ss} pads. This allows us to achieve two goals:

- It makes it possible to evaluate the power dissipation of the pad section separately from that of the internal circuitry.

- It *decouples* the power supply of the internal circuitry from that of the pad section. Fluctuations or voltage drops in the power supply, which are likely to happen in circuits with several fast input and output pads, do not affect the operation of the internal circuitry. Note that a voltage drop of several hundred millivolts creates the right conditions to trigger latchup in gates that are not protected as carefully as the ones in the input-output section.

Finally, separation between pad power lines and internal buses decreases the likelihood of noise induction (due to capacitive coupling and inductive coupling) on the buses, which can trigger latchup when a bus signal reaches a gate. Avoiding bootstrapped gates is also highly recommended, unless SOI processes are used.

A.3. Layout with Structured Methodologies

Layout is a very time-consuming process, and the number of "drawn" transistors significantly affects the duration of a project. Large chips with more than 300,000 transistors do not usually require more than, say, 15,000. These figures refer to microprocessors, intelligent peripheral controllers, memory management units, and so on. Memory chips have completely different constraints: although the regularity can be much higher than that of the above circuits, the design time is not necessarily shorter, because each single block requires even more careful layout, and the interaction with the fabrication process parameters becomes stronger.

If we concentrate our attention to non-memory devices, the issues of regular design *and* regular layout are tightly coupled. The time necessary for the development of a new fabrication process plays a less significant role in this case, because new fabrication processes are used in more than a few different chips. A *regular* design aims to use the same logic blocks repeatedly. Same logic blocks should feature the same layout, that is, the same geometrical relationship between inputs, outputs, and power lines. Otherwise, some rearrangement is necessary — and this can be expensive.

A typical example is a parallel multiplier. In its simplest form — an array multiplier — very few different cells are necessary: a carry save adder, the bottom row full-adder (which is a modification of the carry-save adder, if a ripple-carry adder is used), and input/output buffers. Therefore, the number of drawn transistors is very small, and more attention can be paid to the design and layout of the carry-save adder, which will eventually determine the overall speed of the circuit.

However, when speed is not a critical requirement and other factors play a more crucial role — design time, costs, and so on — regular methodologies can be fruitfully applied throughout the entire design process, that is, from logic design to layout. One of the best examples of this fact is presented in [2], where a layout methodology for CMOS is presented. This methodology allows regular layout style with metal lines running in one direction and diffusion lines running perpendicularly. It is consistent with the necessary rearrangements of boolean equations, in order to achieve simple and effective pitch-matching. Transistors are not usually sized — allowing transistor sizing for delay optimization would make the layout irregular and contradict the whole purpose of increasing the productivity. This approach can boost productivity as much as five times over the more conventional "unstructured" layout methodologies. The only drawback — poor speed performance — is not a concern in those chips where short development time, ease of debugging, and yield maximization (i.e., area minimization) are major constraints.

A.4. Power and Ground Routing

Power distribution is becoming an increasingly critical issue in digital circuits, even when "low-power" technologies such as CMOS are used. Although static power dissipation is limited to leakage effects, dynamic power dissipation, which is proportional to the operating frequency, becomes significant in high-performance chips. A model for the V_{dd} and V_{ss} lines is shown in Fig. A-8 [3].

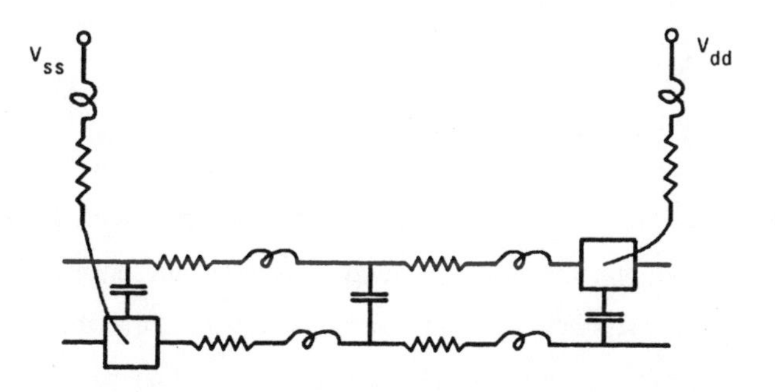

Figure A-8: Model for V_{ss} and V_{dd} lines which includes bonding wires.

The model includes the bonding wires and metal lines. The bonding wires consist of an inductance and resistance. The metal lines are also modeled with inductance and resistance, and a coupling capacitance connects them together. Besides electromigration effects — current density has to be *carefully* evaluated — the inductance represents a major danger to proper circuit operation. As we saw in Section 6.3, voltage drops on power lines (both V_{dd} and V_{ss}) not only slow down the circuit because of body effects, but also increase the likelihood of latchup.

A simple layout of V_{dd} and V_{ss} lines is shown in Fig. A-9. This simple topology of the top layer can be used when *multiple* metal layers, that is, three or more, are available [4]. This layer should be as thick and wide as possible to minimize voltage drops and current density. Multiple layers will be a must in the future.

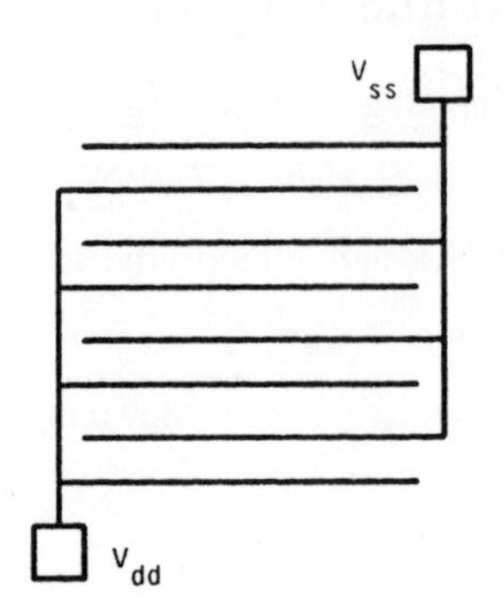

Figure A-9: Top layer layout.

When only two layers are available, power distribution becomes a very critical issue. First, current requirements are not evenly distributed throughout the chip. While internal circuitry may occupy up to 80% of total chip area, it might only consume 60% of total current. On the other hand, the input-output section, which represents 20% of the circuit area, may require the other 40%. It is recommended, therefore, to dedicate V_{dd} and V_{ss} pads to input-output buffers, especially if very fast drivers are used.

Fig. A-10(top) shows a layout pattern for a simple chip with input-output buffers on only two sides of the chip. Four pads are used. Four critical points, marked W1, W2, W3, and W4, are shown in the figure. The length of the metal wire from pad A to W4, from pad B to W1, etc., should be computed and the total resistance evaluated. To assure correct operation, it is necessary that the voltage drop on W1, W2, W3, and W4 be limited. By using the formulae presented in Chapter 5 it is possible to determine the *width* of the metal line that can minimize the voltage drop on the above four points. Note also that the requirements on these lines may not be symmetrical; TTL compatible input pads require to sink a larger amount of current than they need to source. Consequently, in this case the width of V_{ss} lines is more critical than the width of V_{dd} lines.

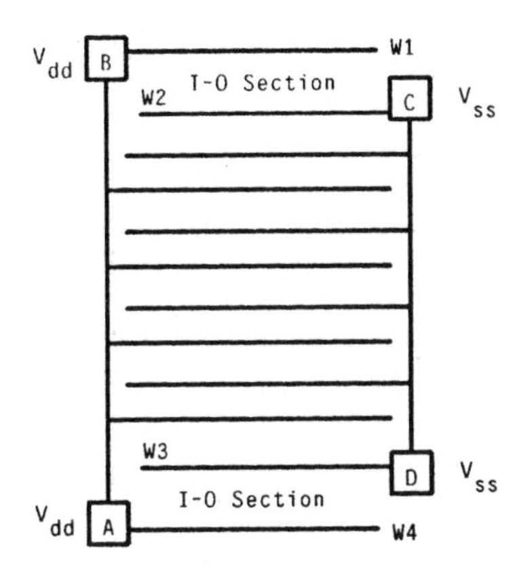

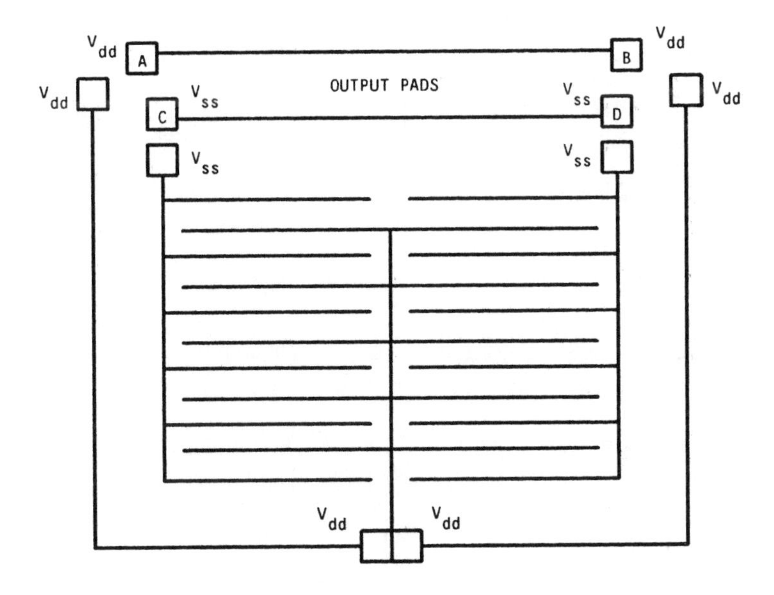

Figure A-10: Layout patterns for V_{dd} and V_{ss} lines.

If the chip has a particularly fast output section, and the output pads are grouped together, a power and ground distribution as shown in Fig. A-10(bottom) can be used. Pads A, B, C, and D are dedicated to the output pads; voltage fluctuations on these lines do not affect internal circuitry, because the power lines are decoupled. It is easier to size these lines, because the current requirements of the output section are separated from those of the internal circuitry (and input pads). Other pads can be added, depending on current requirements; future state-of-the-art CMOS chips are expected to use up to 20 V_{dd} and V_{ss} pads.

References

[1] Fang, R.C.Y., K.Y. Su and J.J. Hsu.
A Two-dimensional Analysis of Sheet and Contact Resistance Effects in Basic Cells of Gate-array Circuits.
IEEE Journal of Solid-State Circuits 20(2):481-488, April, 1985.

[2] Piguet, C. *et al.*.
A Metal-Oriented Layout Structure for CMOS Logic.
IEEE Journal of Solid-State Circuits SC-19(3):425-436, June, 1984.

[3] Saigo, T. *et al.*.
A 20K-Gate CMOS Gate Array.
IEEE Journal of Solid-State Circuits SC-18(5):578-584, October, 1983.

[4] Song, W.S. and L.A. Glasser.
Power Distribution Techniques for VLSI Circuits.
In *Proc. of 1984 Conference on Advanced Research in VLSI*, pages 45-52. M.I.T., January, 1984.

[5] Troutman, R.R. and H.P. Zappe.
Layout and Bias Considerations for Preventing Transiently Triggered Latchup in CMOS.
IEEE Trans. on Electron Devices ED-31(3):315-321, March, 1984.

Appendix B

Interconnect Capacitance Computation

This appendix shows an application of the results presented in Section 5.1.2. We compute the capacitance associated with a conductor coupled with an identical conductor. Both conductors run in parallel. Let the width W of both metal lines be equal to 2.5μm, their thickness T equal to 1500Å, the space S between them 2.5μm and the height of the oxide 6000Å: we will compute the capacitance by using both approaches presented in Section 5.1.2, using Eqs. (5-2), (5-6) and Eqs. (5-7), (5-8).

From Eq. (5-2) we have:

$$C_{intrinsic} = 2.366\text{pF/cm}\ ,$$

while from Eq. (5-6) we have:

$$C_{coupling} = 14.34\text{fF/cm}\ .$$

Let us now use the more accurate approach presented in the same section. We will consider two cases:

1. The dielectric *above* the two conductors is air and its dielectric constant equal to 1; the two conductors form a coupled microstrip structure.

2. The dielectric *above* the two conductors is SiO_2; the two conductors form a coupled stripline structure.

The number on the right indicates the corresponding equation in Section 5.1.2.

B.1. Case 1: Coupled Microstrip Structure

$C_p = 1.439\text{pF/cm},$ (5-13)

$Z = 4.473,$ (5-11)

$\varepsilon_{eff} = 333.9\text{x}10^{-15},$ (5-12)

$C_M = 2.35\text{pF/cm},$ (5-10)

$C_f = 455.5\text{fF/cm},$ (5-14)

$C_{f*} = 373.6\text{fF/cm},$ (5-15)

$C_{g1} = 279.7\text{fF/cm},$ (5-16)

$C_{g2} = 118.9\text{fF/cm},$ (5-17)

$C_{gt} = 10.63\text{fF/cm},$ (5-19)

$C_{ev} = 2.268\text{pF/cm},$ (5-7)

$C_{od} = 2.304\text{pF/cm}.$ (5-8)

Finally, we have:

$C_{intrinsic} = 2.286\text{pF/cm},$

$C_{coupling} = 18.0\text{fF/cm},$

and the error introduced by the first approach is less than 10%.

B.2. Case 2: Coupled Stripline Structure

$$C_p = 1.439\text{pF/cm}, \qquad (5\text{-}13)$$

$$Z = 4.473, \qquad (5\text{-}11)$$

$$\varepsilon_{eff} = 520.7\text{x}10^{-15}, \qquad (5\text{-}12)$$

$$C_M = 3.665\text{pF/cm}, \qquad (5\text{-}10)$$

$$C_f = 1.113\text{pF/cm}, \qquad (5\text{-}14)$$

$$C_{f*} = 730.8\text{fF/cm} \qquad (5\text{-}15)$$

$$C_{g1} = 683.1\text{fF/cm}, \qquad (5\text{-}16)$$

$$C_{g2} = 463.7\text{fF/cm}, \qquad (5\text{-}17)$$

$$C_{gt} = 41.44\text{fF/cm}, \qquad (5\text{-}19)$$

$$C_{ev} = 3.283\text{pF/cm}, \qquad (5\text{-}7)$$

$$C_{od} = 3.74\text{pF/cm}. \qquad (5\text{-}8)$$

We have:

$$C_{intrinsic} = 3.512\text{pF/cm},$$

$$C_{coupling} = 0.228\text{pF/cm}.$$

Appendix C

Figures from Section 5.4.2

Fig. C-1 shows the M/m load/drive ratio vs. the relative delay; this figure can be used to determine the best number of inverter stages. If we suppose, for instance, that M/m = 20, Fig. C-1 shows that the best number of inverters is 3, but it also shows that the difference in relative delay between a three-inverter chain and a four-inverter chain is not significant. If M/m = 8, the minimum Δ_r is achieved with a three-inverter chain. However, if non-inverting logic is required, we have the choice between a two-inverter ($\Delta_r \approx 9$) and a four-inverter chain ($\Delta_r \approx 7$); at this point other considerations, such as area and power dissipation, can make us decide which solution we want to choose.

Once the number of inverter stages in the chain has been decided, Figs. C-2 to C-6 determine the size of each inverter. If the number of inverters is, say, four and the load/drive ratio is 100, the second inverter will be 3 times larger than m, the third inverter 10 times, and the last inverter 32 times larger.

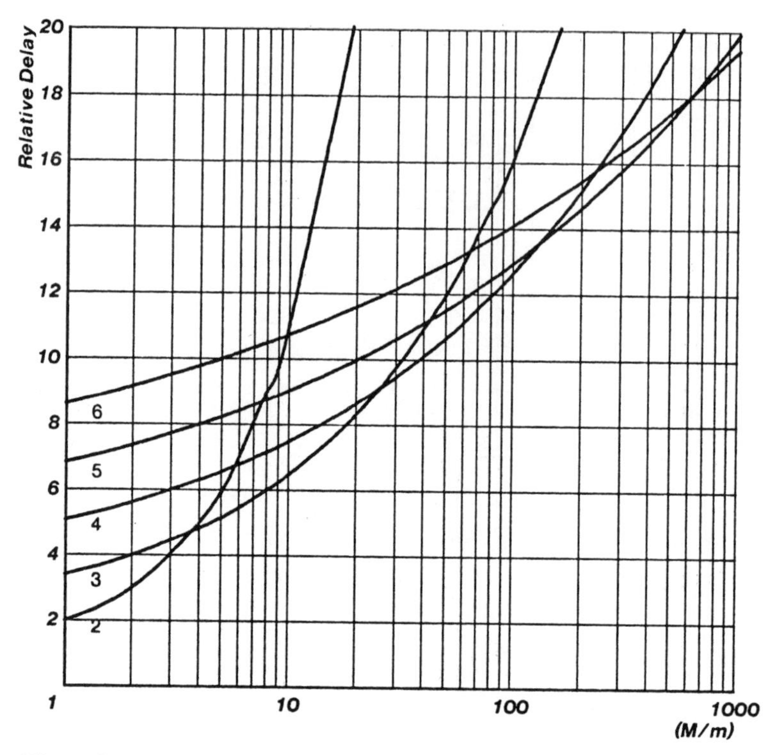

Figure C-1: Relative delay versus number of stages in the inverter chain.

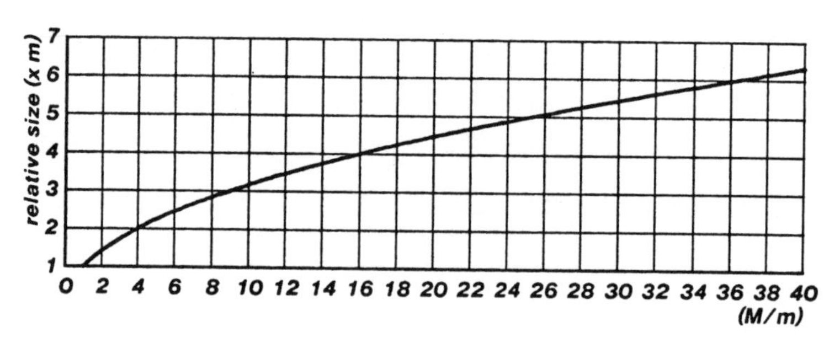

Figure C-2: Two-inverter chain: second inverter size versus load/drive ratio.

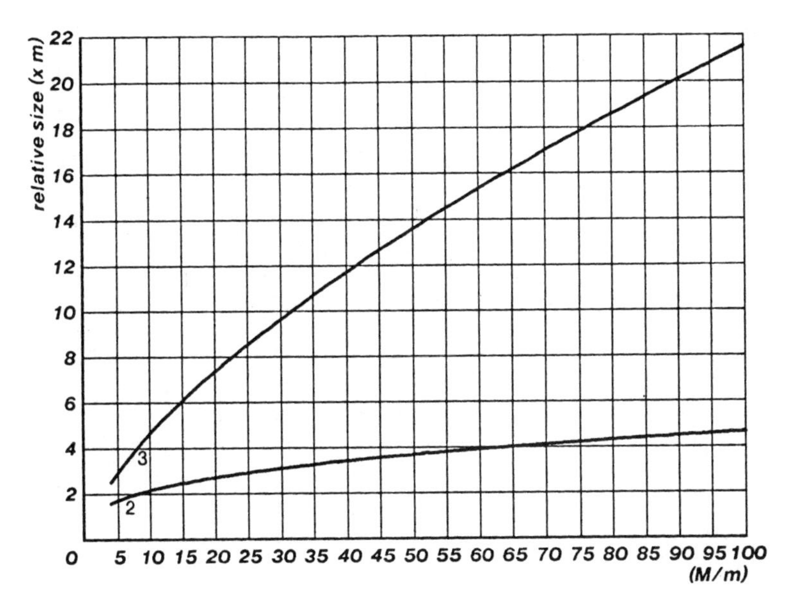

Figure C-3: Three-inverter chain: inverter sizes versus load/drive ratio.

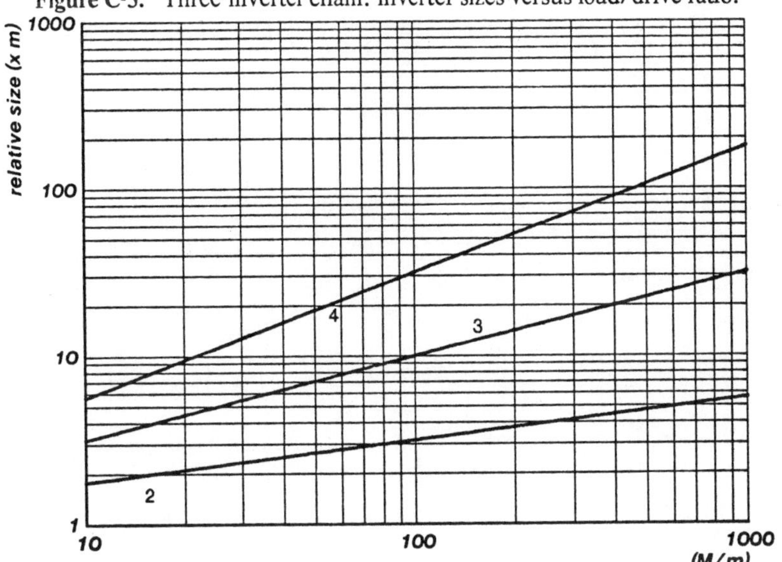

Figure C-4: Four-inverter chain: inverter sizes versus load/drive ratio.

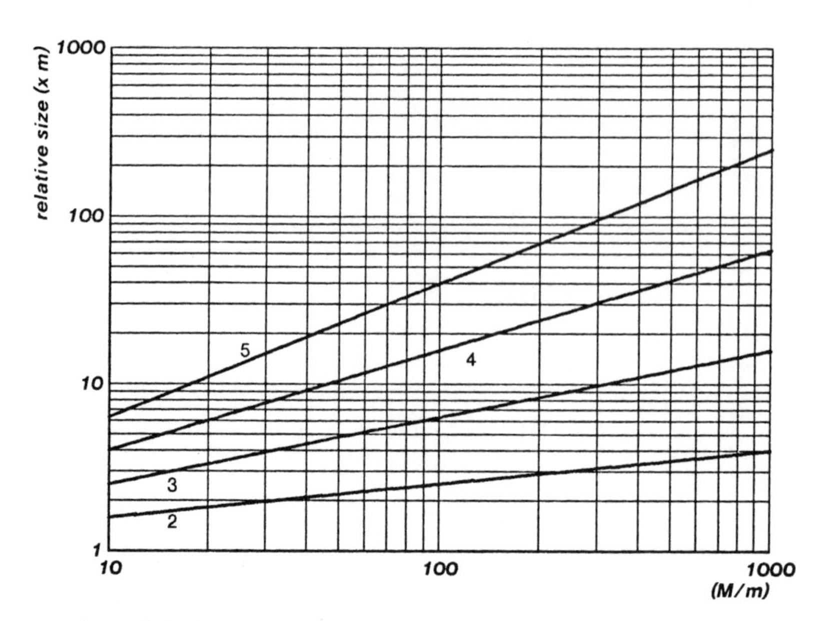

Figure C-5: Five-inverter chain: inverter sizes versus load/drive ratio.

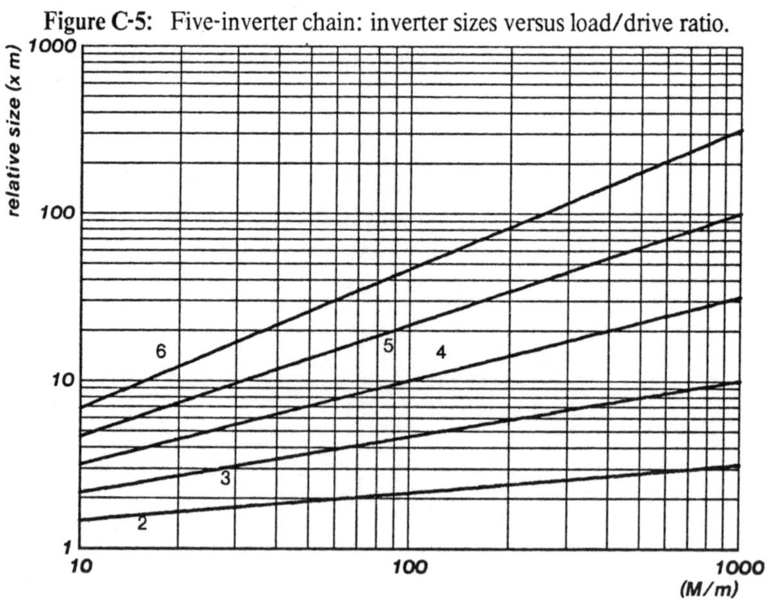

Figure C-6: Six-inverter chain: inverter sizes versus load/drive ratio.

Appendix D

Delay Minimization Based on Eq. (7-3)

The load $C_{L(i)}$ seen by the i-th inverter of the chain can be expressed as:

$$C_{L(i)} = DW_{n(i)}C_{jn} + DW_{p(i)}C_{jp} + C_{ox}(L_nW_{n(i+1)} + L_pW_{p(i+1)}) .$$

The total delay of the chain is the average of the delays corresponding to the two input transitions; by adding all the contributions together and considering the delay as expressed by Eq. (7-3) with the load shown above, the delay of an n-stage inverter chain driving a load C_L is:

$$\Delta = \frac{K_n}{2} \sum_{1}^{n-1}{}_i \frac{D(O_nW_{n(i)}+O_pW_{p(i)})+(L_nW_{n(i+1)}+L_pW_{p(i+1)})}{(\mu_n/2)(W_{n(i)}/L_n)}$$
$$+\frac{K_p}{2} \sum_{1}^{n-1}{}_i \frac{D(O_nW_{n(i)}+O_pW_{p(i)})+(L_nW_{n(i+1)}+L_pW_{p(i+1)})}{(\mu_p/2)(W_{p(i)}/L_p)}$$
$$+\frac{K_n^l}{2} \frac{D(O_nW_{n(n)} + O_pW_{p(n)}) + t}{(\mu_n/2)(W_{n(n)}/L_n)}$$
$$+\frac{K_p^l}{2} \frac{D(O_nW_{n(n)} + O_pW_{p(n)}) + t}{(\mu_p/2)(W_{p(n)}/L_p)} , \qquad \text{(D-1)}$$

where:

$$K_n = \frac{V_{Tn}}{(V_{dd} - V_{Tn})^2} + \frac{1}{2(V_{dd} - V_{Tn})} \ln \frac{1.5V_{dd} - 2V_{Tn}}{0.5V_{dd}} ;$$

$$K_p = \frac{V_{Tp}}{(V_{dd} - V_{Tp})^2} + \frac{1}{2(V_{dd} - V_{Tp})} \ln \frac{1.5V_{dd} - 2V_{Tp}}{0.5V_{dd}} \; ;$$

$$K_n^1 = \frac{V_{Tn}}{(V_{dd} - V_{Tn})^2} + \frac{1}{2(V_{dd} - V_{Tn})} \ln \frac{(2-k_n)V_{dd} - 2V_{Tn}}{k_n V_{dd}} \; ; \qquad \text{(D-2)}$$

$$K_p^1 = \frac{V_{Tp}}{(V_{dd} - V_{Tp})^2} + \frac{1}{2(V_{dd} - V_{Tp})} \ln \frac{(1+k_p)V_{dd} - 2V_{Tp}}{(1-k_p)V_{dd}} \; . \qquad \text{(D-3)}$$

K_n^1 and K_p^1 are relative to the last stage, and k_n and k_p in Eqs. (D-2) and (D-3) must be chosen according to the voltage level necessary to make the first external gate switch. For example, k_p may be such that $k_p V_{dd} = 2.4V$, and k_n such that $k_n V_{dd} = 0.4V$, if standard TTL logic is used outside the chip.

As a first approximation, note that K_n, K_p, K_n^1, and K_p^1 are fabrication process *constants*, that is, they do not depend on the actual size of the transistors. O_n and O_p are:

$$O_n = C_{jn}/C_{ox}\,, \qquad \text{and} \qquad O_p = C_{jp}/C_{ox}\,,$$

respectively. Note that the channel length of both devices is independent of the width, but the two devices can have different channel lengths. Finally, the load C_L driven by the last stage ("output load") has been set equal to tC_{ox}, where t is dimensionally an area.

The problem to solve is therefore a problem of unconstrained minimization, provided that all the widths are greater than zero, where the variables are, for a chain of n inverters:

$W_{n(i)}$ $\quad i = 2,\ldots n$;
$W_{p(i)}$ $\quad i = 2,\ldots n$.

$W_{n(1)}$ and $W_{p(1)}$ are assigned values. The function that we want to minimize is $\Delta(\mathbf{X})$, where $\mathbf{X}$[1] is a vector that contains all the widths of the inverter chain, that is:

$$\mathbf{X} = [W_{n(2)}, ..., W_{n(n)}, W_{p(2)}, ..., W_{p(n)}]^T .$$

We want to find an $\mathbf{X}_m$ such that:

$$\Delta(\mathbf{X}_m) = \min \Delta(\mathbf{X}) .$$

The function $\Delta(\mathbf{X})$ has its minimum when

$$\frac{\partial \Delta}{\partial W_{m(i)}} = 0 , \qquad \text{m=p or n , and} \qquad \text{i=2,...,n} .$$

Moreover, to guarantee that a minimum does exist, the Hessian of Δ (that is, the matrix of the second derivatives of Δ) must be positive definite. A common method used to solve the problem aboveinvolves a two-dimensional Newton-Raphson iterative procedure. We have:

$$\mathbf{X}_{i+1} = \mathbf{X}_i - \mathbf{H}_i^{-1} \nabla\Delta_i ,$$

where $\mathbf{H}_i$ is the Hessian of Δ computed in $\mathbf{X}_i$ and $\nabla\Delta_i$ is the gradient of Δ computed in $\mathbf{X}_i$.

From Eq. (D-1) we have:

$$\frac{\partial \Delta}{\partial W_{n(i)}} = -\frac{1}{\mu_n} K_n L_n^2 \frac{W_{n(i+1)}}{W_{n(i)}^2} - \frac{L_n L_p}{\mu_n} K_n \frac{W_{p(i+1)}}{W_{n(i)}^2} +$$

$$-\frac{L_n O_p D}{\mu_n} K_n \frac{W_{p(i)}}{W_{n(i)}^2} + \frac{1}{\mu_n W_{n(i-1)}} K_n L_n^2 +$$

---

[1]From now on vectors and matrices are shown in bold face.

$$K_p \frac{L_p O_n D}{\mu_p} \frac{1}{W_{p(i)}} + K_p \frac{L_n L_p}{\mu_p} \frac{1}{W_{p(i-1)}} \qquad i = 2,\ldots,(n-1)$$

$$\frac{\partial \Delta}{\partial W_{p(i)}} = -\frac{1}{\mu_p} K_p L_p^2 \frac{W_{p(i+1)}}{W_{p(i)}^2} - \frac{L_p L_n}{\mu_p} K_p \frac{W_{n(i+1)}}{W_{p(i)}^2}$$

$$-\frac{L_p O_n D}{\mu_p} K_p \frac{W_{n(i)}}{W_{p(i)}^2} + \frac{1}{\mu_p W_{p(i-1)}} K_p L_p^2$$

$$+ K_n \frac{L_n O_p D}{\mu_n} \frac{1}{W_{n(i)}} + K_n \frac{L_p L_n}{\mu_n} \frac{1}{W_{n(i-1)}} \qquad i=2,\ldots,(n-1)$$

$$\frac{\partial \Delta}{\partial W_{p(n)}} = \frac{1}{W_{p(n-1)}\mu_p} K_p L_p^2 + \frac{L_n L_p}{\mu_n W_{n(n-1)}} K_n$$

$$+ \frac{O_p L_n D}{\mu_n W_{n(n)}} K_n^1 - \frac{O_n L_p D}{\mu_p} K_p^1 \frac{W_{n(n)}}{W_{p(n)}^2}$$

$$- \frac{L_p t}{\mu_p} K_p^1 \frac{1}{W_{p(n)}^2}$$

$$\frac{\partial \Delta}{\partial W_{n(n)}} = \frac{1}{W_{n(n-1)}\mu_n} K_n L_n^2 + \frac{L_p L_n}{\mu_p W_{p(n-1)}} K_p$$

$$+ \frac{O_n L_p D}{\mu_p W_{p(n)}} K_p^1 - \frac{O_p L_n D}{\mu_n} K_n^1 \frac{W_{p(n)}}{W_{n(n)}^2}$$

$$- \frac{L_n t}{\mu_n} K_n^1 \frac{1}{W_{n(n)}^2}$$

We can now separate physical, process-dependent constants from the actual variables. We have:

$$A_n = \frac{1}{\mu_n} L_n^2 K_n ;$$

$$A_p = \frac{1}{\mu_p} L_p^2 K_p ;$$

$$B_n = \frac{L_n L_p}{\mu_n} K_n ;$$

$$B_p = \frac{L_n L_p}{\mu_p} K_p ;$$

$$C_n = \frac{L_n O_p D}{\mu_n} K_n ;$$

$$C_p = \frac{O_n L_p D}{\mu_p} K_p ;$$

$$D_n = \frac{L_n O_p D}{\mu_n} K_n^1 ;$$

$$D_p = \frac{O_n L_p D}{\mu_p} K_p^1 ;$$

$$E_n = \frac{L_n t}{\mu_n} K_n^1 ;$$

$$E_p = \frac{L_p t}{\mu_p} K_p^1 .$$

The gradient of Δ is therefore:

$$\frac{\partial \Delta}{\partial W_{n(i)}} = \frac{A_n}{W_{n(i-1)}} - C_n \frac{W_{p(i)}}{W_{n(i)}^2} - B_n \frac{W_{p(i+1)}}{W_{n(i)}^2} - A_n \frac{W_{n(i+1)}}{W_{n(i)}^2} + \frac{C_p}{W_{p(i)}} + \frac{B_p}{W_{p(i-1)}}$$

$$i = 2, \ldots, (n-1)$$

$$\frac{\partial \Delta}{\partial W_{p(i)}} = \frac{A_p}{W_{p(i-1)}} - C_p \frac{W_{n(i)}}{W_{p(i)}^2} - B_p \frac{W_{n(i+1)}}{W_{p(i)}^2} -$$

$$A_p \frac{W_{p(i+1)}}{W_{p(i)}^2} + \frac{C_n}{W_{n(i)}} + \frac{B_n}{W_{n(i-1)}}$$

$$i = 2, \ldots, (n-1)$$

$$\frac{\partial \Delta}{\partial W_{n(n)}} = \frac{A_n}{W_{n(n-1)}} - D_n \frac{W_{p(n)}}{W_{n(n)}^2} - \frac{E_n}{W_{n(n)}^2} + \frac{B_p}{W_{p(n-1)}} + \frac{D_p}{W_{p(n)}}$$

$$\frac{\partial \Delta}{\partial W_{p(n)}} = \frac{A_p}{W_{p(n-1)}} - D_p \frac{W_{n(n)}}{W_{p(n)}^2}$$

$$- \frac{E_p}{W_{p(n)}^2} + \frac{B_n}{W_{n(n-1)}} + \frac{D_n}{W_{n(n)}}$$

We can now write the second derivatives to compute the Hessian of Δ. We have:

$$\frac{\partial^2 \Delta}{\partial W_{n(i)} \partial W_{n(j)}} = 0 \qquad j \neq i, (i+1), (i-1)$$

$$\frac{\partial^2 \Delta}{\partial W_{n(i)} \partial W_{n(j)}} = - \frac{A_n}{W_{n(i-1)}^2} \qquad j = i-1$$

$$\frac{\partial^2 \Delta}{\partial W_{n(i)} \partial W_{n(j)}} = \left[A_n W_{n(i+1)} + B_n W_{p(i+1)} + C_n W_{p(i)} \right] \frac{2}{W_{n(i)}^3} \qquad j=i$$

$$\frac{\partial^2 \Delta}{\partial W_{n(i)} \partial W_{n(j)}} = \frac{A_n}{W_{n(i)}^2} \qquad j=i+1$$

$$\frac{\partial^2 \Delta}{\partial W_{n(i)}\, \partial W_{p(j)}} = 0 \qquad j \neq (i-1), i, (i+1)$$

$$\frac{\partial^2 \Delta}{\partial W_{n(i)}\, \partial W_{p(j)}} = -\frac{B_p}{W^2_{p(i-)}} \qquad j = i-1$$

$$\frac{\partial^2 \Delta}{\partial W_{n(i)}\, \partial W_{p(j)}} = -\frac{B_n}{W^2_{n(i)}} \qquad j = i+1$$

$$\frac{\partial^2 \Delta}{\partial W_{n(i)}\, \partial W_{p(j)}} = -\frac{C_n}{W^2_{n(i)}} - \frac{C_p}{W^2_{p(i)}} \qquad j=i$$

$$\frac{\partial^2 \Delta}{\partial W_{p(i)}\, \partial W_{p(j)}} = 0 \qquad j \neq i, (i+1), (i-1)$$

$$\frac{\partial^2 \Delta}{\partial W_{p(i)}\, \partial W_{p(j)}} = -\frac{A_p}{W^2_{p(i-1)}} \qquad j = i-1$$

$$\frac{\partial^2 \Delta}{\partial W_{p(i)}\, \partial W_{p(j)}} = \left[A_p W_{p(i+1)} + B_p W_{n(i+1)} + C_p W_{n(i)}\right] \frac{2}{W^3_{p(i)}} \qquad j=i$$

$$\frac{\partial^2 \Delta}{\partial W_{p(i)}\, \partial W_{p(j)}} = \frac{A_p}{W^2_{p(i)}} \qquad j = i+1$$

$$\frac{\partial^2 \Delta}{\partial W_{p(i)}\, \partial W_{n(j)}} = 0 \qquad j \neq (i-1), i, (i+1)$$

$$\frac{\partial^2 \Delta}{\partial W_{p(i)}\, \partial W_{n(j)}} = -\frac{B_n}{W^2_{n(i-1)}} \qquad j = i-1$$

$$\frac{\partial^2 \Delta}{\partial W_{p(i)}\, \partial W_{n(j)}} = -\frac{B_p}{W^2_{p(i)}} \qquad j = i+1$$

$$\frac{\partial^2 \Delta}{\partial W_{p(i)} \partial W_{n(j)}} = -\frac{C_p}{W^2_{p(i)}} - \frac{C_n}{W^2_{n(i)}} \qquad j = i$$

$$\frac{\partial^2 \Delta}{\partial W_{n(n)} \partial W_{n(j)}} = 0 \qquad j \neq n-1, n$$

$$\frac{\partial^2 \Delta}{\partial W_{n(n)} \partial W_{n(j)}} = -\frac{A_n}{W^2_{n(n-1)}} \qquad j = n-1$$

$$\frac{\partial^2 \Delta}{\partial W_{n(n)} \partial W_{n(n)}} = D_n \frac{2W_{p(n)}}{W^3_{n(n)}} + \frac{2E_n}{W^3_{n(n)}} \qquad j = n$$

$$\frac{\partial^2 \Delta}{\partial W_{n(n)} \partial W_{p(j)}} = 0 \qquad j \neq n,(n-1)$$

$$\frac{\partial^2 \Delta}{\partial W_{n(n)} \partial W_{p(j)}} = -\frac{D_n}{W^2_{n(n)}} - \frac{D_p}{W^2_{p(n)}} \qquad j = n$$

$$\frac{\partial^2 \Delta}{\partial W_{n(n)} \partial W_{p(j)}} = -\frac{B_p}{W_{p(n-1)}} \qquad j = n-1$$

$$\frac{\partial^2 \Delta}{\partial W_{p(n)} \partial W_{p(j)}} = 0 \qquad j \neq n-1$$

$$\frac{\partial^2 \Delta}{\partial W_{p(n)} \partial W_{p(j)}} = -\frac{A_p}{W^2_{p(n-1)}} \qquad j = n-1$$

$$\frac{\partial^2 \Delta}{\partial W_{p(n)} \partial W_{p(n)}} = D_p \frac{2W_{n(n)}}{W^3_{p(n)}} + \frac{2E_p}{W^3_{p(n)}} \qquad j = n$$

$$\frac{\partial^2 \Delta}{\partial W_{p(n)} \partial W_{n(j)}} = 0 \qquad j \neq n, (n-1)$$

$$\frac{\partial^2 \Delta}{\partial W_{p(n)} \partial W_{n(j)}} = -\frac{D_p}{W^2_{p(n)}} - \frac{D_n}{W^2_{n(n)}} \qquad j = n$$

$$\frac{\partial^2 \Delta}{\partial W_{p(n)} \, \partial W_{n(j)}} = -\frac{B_n}{W_{n(n-1)}} \qquad j = n-1$$

The computation of the inverse of the Hessian is not particularly expensive in this case because of the limited number of variables involved: a 5-stage inverter chain has eight variables and requires the inversion of an 8x8 Hessian. Several methods exist to compute the inverse of a matrix and can be found in any introductory book on linear algebra. For instance, Gaussian elimination with back-substitution can be used, but any other method, either direct — that is, requiring an explicit computation of the determinant — or indirect, will work as well.

Appendix E

Equations Related to Fig. 7-10

The following list of equations refers to Fig. 7-10 and is used when computing the fall-time (rise-time) of an inverter with load C_L and input signal.

$$v_{in}(t) = \frac{V_{dd}}{t_u} t = v_{in}$$

$$C_L \int_{V_{dd}}^{v_{in} - V^e_{Tn}} dv_{out} = -\frac{\beta_n}{2} \int_0^{t_s} (v_{in} - V^e_{Tn})^2 \, dt \tag{E-1}$$

$$C_L \int_{\frac{V_{dd}}{t_u} t_s - V^e_{Tn}}^{V_Y} dv_{out} = -\beta_n \int_{t_s}^{t_u} \left[(v_{in} - V^e_{Tn}) v_{out} - \frac{v^2_{out}}{2} \right] dt \tag{E-2}$$

$$C_L \int_{\frac{V_{dd}}{t_u} t_s - V^e_{Tn}}^{V_o} dv_{out} = -\beta_n \int_{t_s}^{\tau_n} \left[(v_{in} - V^e_{Tn}) v_{out} - \frac{v^2_{out}}{2} \right] dt \tag{E-3}$$

$$C_L \int_{V_Y}^{V_o} dv_{out} = -\beta_n \int_{t_u}^{\tau_n} \left[(V_{dd} - V_{Tn}^e) v_{out} - \frac{v_{out}^2}{2} \right] dt \tag{E-4}$$

$$C_L \int_{V_{dd}}^{V_X} dv_{out} = -\frac{\beta_n}{2} \int_0^{t_u} (v_{in} - V_{Tn}^e)^2 \, dt \tag{E-5}$$

$$C_L \int_{V_{dd}}^{V_o} dv_{out} = -\frac{\beta_n}{2} \int_0^{t_s} (v_{in} - V_{Tn}^e)^2 \, dt \tag{E-6}$$

$$C_L \int_{V_X}^{V_o} dv_{out} = -\frac{\beta_n}{2} \int_{t_u}^{t_s} (V_{dd} - V_{Tn}^e)^2 \, dt \tag{E-7}$$

$$C_L \int_{V_X}^{V_{dd} - V_{Tn}^e} dv_{out} = -\frac{\beta_n}{2} \int_{t_u}^{t_s} (V_{dd} - V_{Tn}^e)^2 \, dt \tag{E-8}$$

$$C_L \int_{V_{dd} - V_{Tn}^e}^{V_o} dv_{out} = -\beta_n \int_{t_s}^{\tau_n} \left[(V_{dd} - V_{Tn}^e) v_{out} - \frac{v_{out}^2}{2} \right] dt \tag{E-9}$$

Appendix F

Symbols and Physical Constants

α_n	n-channel transistor current gain
α_p	p-channel transistor current gain
β_n	n-channel transistor voltage gain
β_p	p-channel transistor voltage gain
ε_o	Permittivity in vacuum = 8.854×10^{-14} F/cm
ε_{ox}	SiO_2 dielectric constant = 3.9
ε_{Si}	Si dieletric constant = 11.9
ϕ_B	Bulk potential
ϕ_F	Fermi potential (V)
ϕ_{MS}	Workfunction difference (V)
ϕ_S	Schottky barrier height (V)
μ_n	n-channel majority carrier mobility (cm^2/Vs)
μ_o	Permeability in vacuum = 1.25663×10^{-8} H/cm
μ_p	p-channel majority carrier mobility (cm^2/Vs)
C_n	n+ junction capacitance per unit area (pF/cm^2)
C_o	Gate oxide capacitance (pF)
C_{ox}	Gate oxide capacitance per unit area (pF/cm^2)
C_p	p+ junction capacitance per unit area (pF/cm^2)
h	Planck's constant = 6.62617×10^{-34} (J-s)
k	Boltzmann's constant = 1.38×10^{-23} (J/K)
L_n	n-channel length (drawn) (μm)
L_p	p-channel length (drawn) (μm)

$L_{n(eff)}$	n-channel effective length (μm)
$L_{p(eff)}$	p-channel effective length (μm)
m_n	Electron effective mass $\approx 0.9m_o$ (kg)
m_o	Electron rest mass $= 0.91095 \times 10^{-30}$ (kg)
m_p	Hole effective mass $\approx 0.2m_o$ (kg)
m^*	Effective mass of charge carrier (either m_n or m_p) (kg)
n_i	Si intrinsic carrier concentration 1.45×10^{10} (cm^{-3})
N_a	Acceptor concentration (cm^{-3})
N_B	Substrate doping concentration
N_d	Donor concentration (cm^{-3})
q	Elementary charge $= 1.602 \times 10^{-19}$ C
t_{ox}	Thin-oxide thickness (Å)
T	Absolute temperature (K)
V_{Tn}	n-channel threshold voltage (V)
V_{Tp}	p-channel threshold voltage (V)
W_n	n-channel width (drawn) (μm)
W_p	p-channel width (drawn) (μm)
$W_{n(eff)}$	n-channel effective width (μm)
$W_{p(eff)}$	p-channel effective width (μm)

Physical constants are for T = 300K.

Index